Water and Liberalisation

European Water Scenarios

Water and Liberalisation

European Water Scenarios

Edited by

M. Finger, J. Allouche &
P. Luís-Manso

IWA
Publishing

Published by IWA Publishing, Alliance House, 12 Caxton Street, London SW1H 0QS, UK

Telephone: +44 (0) 20 7654 5500; Fax: +44 (0) 20 7654 5555; Email: publications@iwap.co.uk
Web: www.iwapublishing.com

First published 2007

British Library Cataloguing in Publication Data
A CIP catalogue record for this book is available from the British Library

Library of Congress Cataloging- in-Publication Data
A catalog record for this book is available from the Library of Congress

ISBN: 1843391139
ISBN13: 9781843391135

Contents

Abbreviations

AMD	Asset Ownership, Management and Development
ART	Autorité de Régulation des Télécommunications
CAPEX	Capital Expenditure
CEC	Commission of European Communities
CRE	Commission de Régulation de l'Energie
DBFO	Design, Build, Finance and Operate
DGI	Dimensions of General Interest
DPM	Direct Public Management
EC	European Communities
EEB	European Environmental Bureau
EU	European Union
FCR:	Full-cost recovery
GIS	Geographic Information System
ICT	Information and Communication Technology
IFEN-SCEES:	Institut Français de l'Environnement, Service Central des Enquêtes et Etudes Statistiques
ILO:	International Labour Organization
IMF:	International Monetary Fund
LRMC	Long Run Marginal Costs
OECD:	Organisation for Economic Cooperation and Development
OFGEM	Office of Gas and Electricity Markets
OFTEL	Office of Telecommunications
OFWAT	Office of Water Services
O&M	Operation and Maintenance
OPEX	Operational Expenditure
PPP	Public-Private Partnership
PSO	Public Service Obligations
PSP	Private Sector Participation
PSIRU	Public Services International Research Unit
PTS	Post & Telestyrelsen
ROI	Return on Investment
RoR	Rate of Return
RPI-X	Retail Price Index – Expected efficiency savings (formula for price-cap regulation)
SGEI	Services of General Economic Interest
SGI	Services of General Interest
SME	Small and Medium-sized Enterprise
TNC	Trans-national Corporation
TPA	Third Party Access
UN	United Nations
UK	United Kingdom
USO	Universal Service Obligations
WFD	Water Framework Directive
WSS	Water Supply and Sanitation
WTO	World Trade Organisation

Introduction

Liberalisation, privatisation and network industries: a similar path for water?

Jeremy Allouche, Patrícia Luís-Manso & Matthias Finger

Liberalisation and private sector participation (PSP) in water supply and sanitation (WSS) systems, and in network industries in general have been the subject of major debates in Europe. This book is the result of a large European research project entitled *"Water liberalisation scenarios - an empirical analysis of the evolution of the European water supply and sanitation sectors (EUROMARKET)"*. This project was funded by the Directorate-General Research under the thematic programme Energy, Environment and Sustainable Development of the Fifth RTD Framework Programme and by the Swiss State Secretariat for Education and Research. It focuses on European Union (EU) countries before the 2005 enlargement (EU-15) and Switzerland.

The topic of water liberalisation is usually portrayed as an ownership issue. However, the project was designed to go beyond the ideological divide between public and private, and proposes to analyse the issue along an institutional and organisational perspective. This kind of analysis is, we think, particularly pertinent in the water sector, as this sector is increasingly characterised by a multiplication of actors and organisational forms, such as networks, associations, organisations and private corporations. Indeed, it is not private sector participation that should be questioned, but rather the coherence of the overall institutional arrangements of the water sector, whether in Europe or elsewhere. How can decentralisation, user-participation, inter-regional co-operation, river basin management, and other approaches enhancing local responsibilities be reconciled with growing financial needs, technological expertise, and the growing power of trans-national corporations in the sector? Applied to Europe, this certainly constitutes one of the intellectual and practical challenges that will be addressed in this book.

Privatisation and liberalisation are often confused, although they refer to different processes. Privatisation refers to the transfer of assets ownership (including infrastructure) and management from the public to the private sector. Private sector participation is a more general term, encompassing privatisation but also the delegation of the management of an activity to a private firm (without transfer of asset ownership).

As for the term liberalisation, it refers to a process by which competition is introduced in situations or sectors hitherto characterised by exclusive or special rights or monopoly granted to historical operators. Some authors use the term deregulation in the same sense. Liberalisation therefore may involve public or private operators.

Since the end of the 20th century, globalisation and world capitalism are seen as an unstoppable trend that has no viable alternatives. The collapse of Marxism as a paradigm and of communism as an alternative to capitalism left no substitutes for liberalism. In this regard, liberalisation is the result of this major driving force. Liberalisation, which concerns the broader issue of the transformation of the public sector, should thus be understood within the evolution of the changing role of the State in economic affairs (see Swann, 1988). Different approaches explain this critical evolution in modern economies.

The first one is embodied in neo-liberal ideology. In this regard, the corresponding reduction of the role of the state is one principal feature of neo-liberal economic theories, or, as some would argue, ideologies. Opposed to the

Keynesian theory, which had inspired most countries from the fifties onwards, the liberal and especially the neo-liberal school of thought of the so-called "Chicago boys", openly criticise state intervention in economic activities as being too costly and, in any case, counter-productive for the global competitiveness of a country (Friedman, 1953). Public Choice theories moreover argue that regulation, or any form of state intervention for that matter, is ultimately driven by lobbies, and thus in the private, rather than in the public interest (Lane, 1985). Their roots are grounded on a clear mistrust in public authorities and a consequent push for introducing market forces into every area of life. The basic vision of this school of thought is market **"deregulation"**.

The second school of thought considers that although market forces dominate, there is a need to introduce some form of regulation, taxation or welfare systems to ensure social cohesion. For example, Karl Polanyi, one of the earliest and most trenchant critics of the free market economics of von Mises and Hayek (von Mises, 1981; Hayek, 1960), considers that the free market is a utopian conception and that all markets are regulated in the sense that their parameters are established by state institutions (Polanyi, 1957). As he argues, *contrary to expectations, the introduction of free markets, far from doing away with the need for control, regulation and intervention, enormously increase their range* (Polanyi, 1957). Polanyi believes that free market creates social catastrophes that entail in the longer run the imposition of limits upon it. There is in fact an implicit catastrophic learning-regulation model in Polanyi or, as Allan Scott has suggested, a forgetting-deregulation-relearning-re-regulation cycle to it (Scott, 1997: 14). Snyder is another author challenging the assumption that neo-liberalism results in a relatively free and unregulated market. His research (e.g., on the coffee industry) reveals an environment in which governments, far from receding from economic management, have found new ways to re-regulate markets in the wake of neo-liberal reforms. Snyder, following Polanyi, finds almost the opposite: by destroying old policy domains, neo-liberal reforms open space for "**re-regulation**", and the creation of new public-private institutional designs.

In Europe, the creation of the European internal market has triggered this polarised debate. Some authors believed that Europe will become a "stateless market" (Kapteyn, 1992; see also Strange, 1996). Other authors consider that regulatory policies continue to condition market relations within Europe (see Egan, 2001; Scharf, 1999).

These two contrasting visions offer a very interesting parallelism to what is currently occurring in the network industries. The purpose of the book is to understand the general evolution of the WSS sectors in Europe. In particular, it aims at analysing the major challenges pertaining to private sector participation and liberalisation, and to compare the WSS sectors with other network industries. This question is crucial at a time when the Directorates-General of the European Commission are discussing the possible extension of internal market rules to the water sector, following the same model as other network industries.

This introductory chapter is divided into three main parts. The first one highlights how private sector participation has become an important trend in network industries, including in WSS, and the need to go beyond the ideological debate. The second part focuses on liberalisation as the ultimate step in the reform of these sectors, and on understanding the evolution of this process. It investigates whether this is a uniform process across network industries, including for the WSS sectors. Finally, the main structure of the book is presented in the third part.

PRIVATE SECTOR PARTICIPATION IN NETWORK INDUSTRIES: AN IMPORTANT TREND

Network industries (or public utilities) have been radically transformed during these last twenty years. One of these changes comes from the involvement of the private sector in the management of these entities. This participation is in our view part of a greater paradigm change in the state's role in economic activities, which cannot just be simply stopped, as the following figures and argument show.

Facts and figures

Since the late 1980s, private sector participation has been an undeniable trend in all infrastructure sectors. Between 1990 and 1998 the World Bank has tracked 1700 projects newly owned or managed by private companies in the water, transport, and telecommunications sectors in developing countries. The total amounts invested increased from US$16 billion in 1990 to US$120 billion in 1997 (World Bank, 2000: 167-68). However, between 1997 and 2004 the growing rate of PSP was in sharp decline. This can be explained by several factors: firstly, these markets were in the early nineties overrated by large private operators; secondly, the lack of experience from local public authorities and governments in involving private actors in the managing of network industries has often led to confusions and disputes; and thirdly the resistance against PSP by non-governmental organisations, association and political parties was perhaps underestimated in the different countries concerned. The momentum in terms of investment in infrastructures is now gaining back. In 2004, investment flows to infrastructure projects with private participation in developing countries grew by 12 percent to US$ 64 billion (Izaguirre, 2005).

Overall, one can see that, since the early nineties, private sector participation in infrastructure projects has become a significant trend with an important impact on the institutional and organisational setting of these sectors.

Looking specifically at the water sector, one can observe the same development. Though, it is important to recall that WSS lagged far behind other infrastructure sectors in terms of private activity. Indeed, water projects represented just five percent of the investment commitments of the private sector in infrastructure projects between 1990 and 2004 (Izaguirre & Hunt , 2005).

Across the world, it was estimated that private water corporations were supplying 545 million people in 2005 while this figure was only of 51 million in 1990 (Masons Water Yearbook, 2004/5). In 2005, the three biggest multinationals active in this sector were Suez, Veolia Environment and RWE (with 117.4, 108.2 and 69.5 million people served worldwide, respectively). According to some water industry analyst's forecast, private sector participation should reach 1.16 billion people by 2015 (Masons Water Yearbook, 2002/3)

In EU-15, it is roughly estimated that there are more than 30'000 water operators and about 35% of the population is supplied by private operators and 10% by mixed ones (EUROMARKET, 2004a). France, England and Wales account for the majority of private provision of water supply services. Private sector participation has increased in other countries since the early nineties, such as Spain for water supply and Belgium for sanitation (EUREAU, 1993 and 1997, EUROMARKET, 2004b). Such facts and figures are proof that the private sector has already become a lasting presence also in the European water sector.

Having looked at the current dynamic of private sector participation, let us now see how these issues have been dealt and what the value added of this book is within the literature.

The need to go beyond the ideological debate

In the current literature, most of the debates portray in some sense public and private participation as two opposite directions, with one being the privatisation of the resources and the assets while the other a direct public management model. Interestingly, when one analyses the different models of management in the WSS in Europe, one can see that the privatisation and the direct public management models are not the dominant ones (EUROMARKET, 2004b). The privatisation model has only occurred in England and Wales (and, outside Europe, in Chile). The direct management model is more widespread but many public operators are becoming increasingly autonomous and do not correspond anymore to this model. One should also note that the frontier between public and private involvement is very much blurred. It would be surprising that even in the most extreme model of public management, there would be no private sector participation (e.g., selling of industrial products for construction of the network; outsourcing contracts, for example for the measurement of pressure and flow in the water distribution network).

Besides the fact that the line between public and private models of management is difficult to draw, the literature, which has focused on the differences between the performance of so-called private and public models, is not very conclusive. Those in favour of private sector participation usually put forward the efficiency of private systems compared to public ones (e.g., Martin & Parker, 1997; Vickers & Yarrow, 1988). On the other hand, the opponents to private sector participation, and in particular to the participation of trans-national corporation (TNCs), emphasise the increase of water prices, little or no evidence of service improvements, under-investments and corruption problems (see for e.g. Lobina and Hall, 2003). However, these results are rather ambiguous and it is very difficult to show differences just according to a question of ownership as the regulatory, economic, cultural and social conditions may in fact play a greater role in the (dis)functioning of either private or public WSS systems (see EUROMARKET, 2005).

Rather than evaluating a model according to its ownership features, the approach followed in the book is to analyse the weaknesses and advantages of each model of management. Each model of management is defined according to its organisational and institutional framework rather than just to ownership. Thus, the analysis of the reforming features of network industries, and liberalisation in particular, is of utmost importance in understanding different modes of management.

LIBERALISATION: A UNIFORM PROCESS ACROSS NETWORK INDUSTRIES?

Liberalisation is seriously questioning the status and role of the State in the economy (Yergin & Stanislow, 2002). Besides the theoretical arguments for a reduction of the State in a national economy, the need to increasingly compete on an international level and the often high public debt, have exerted strong financial pressures onto numerous States. They now have to reduce their debt, diminish public expenditures, or otherwise "adjust" to the new global economy. Such financial pressure is indeed in our view one of the main, if not the main force that has pushed the state to undergo substantial transformation since the late 1980s.

One can distinguish different approaches when analysing the current liberalisation process of network industries. Nonetheless, they all tend to assume that network industries are similar in nature and that, consequently, the liberalisation process will lead to a certain convergence in the institutional and organisational settings. This section questions this assumption. It is divided into three main parts. The first part focuses on the major theoretical underpinnings of the liberalisation process in network industries. In this regard, we look at the two main approaches to liberalisation in network industries and see how they perceive the evolution and final outcome of this process. The second part analyses the liberalisation process in network industries in Europe. The third part investigates whether the WSS fits one of these approaches and puts forward the specificities of water compared to other network industries.

Models for network industries' liberalisation

Over the last twenty years, network industries have been radically transformed in the way they are managed. Two different models describe and forecast this evolution based on the degree of liberalisation and the need for regulatory control.

The first one is grounded on the assumption that network industries are not different from any other industries, and consequently, the final outcome of their liberalisation is that they will be ruled under general competition rules (i.e. not sector regulation). In terms of this so-called "deregulation" process, the telecommunications sector is usually described as the most advanced one in the network industries. Some specialists have therefore used this sector as a 'role model' for the other sectors (Finger & Varone, 2006). In this regard, it was suggested that the opening up of network industries to competition has paved the road for a possible replacement of the traditional regulatory paradigm by competition policy. For example in New Zealand, regulatory agencies in most network industries have been abrogated altogether. And many telecommunications experts predict that the regulatory environment as we know it is bound to disappear in the wake of generalised competition. The ultimate goal is therefore competition. Regulation is a temporary function, used to manage the transition from a monopoly to competition. In this context, the aim of regulation is to protect new entrants in an environment with large and often dominant incumbents.

The most famous and cited study that follows this approach is the one carried out by the Centre for Economic Policy Research and the Centre for Business and Policy Studies under the leadership of Bergman. It suggests a stylised evolution of the network industries in three phases. The first phase is when the liberalisation process has not started and corresponds to the old structure of legal monopolies where the public interest is normally deemed to be safeguarded by public ownership. In a second step, monopolies are broken up, and regulatory activity increases. In this phase an independent regulatory authority is usually established. Regulatory authorities have to promote entry and prevent foreclosure. This implies that they resolve problems of interconnection and access to the existing network infrastructures for new entrants. In this step, mechanisms are also put into place to preserve services of general interest and protect consumers. At some point during the second phase, the gradual introduction of competition should make it possible to reduce the intensity of regulatory control. In a third step, when competition is more effective, the need for sector-specific rules diminishes, eventually to the point where only competition rules are necessary (Bergman et al, 1998).

Some authors defend the need to introduce more competition in the water supply and sanitation sector, reducing regulation both in scale and scope (e.g., Robinson, 1997: 167). The British government, for example, in its proposal to increase competition in this sector, argued the virtues of competition and the need to limit regulation. *Wherever possible, it wants to introduce effective market competition as the best guarantee of the customer's interest in the short and long term* (Department of the Environment and Welsh Office, 1996).

The second model, the "re-regulation" approach challenges this three-step methodology to liberalise. It is based on the assumption that the gradual introduction of competition will not make it possible to reduce the intensity of regulatory control. In this view, regulation is not a transitional period but an end state in itself. One may say that the evolution in the European Union has been approaching this perspective. *Contrary to what some of its promoters may initially have thought, the deregulation undertaken in the context of the internal market nevertheless failed to transform network industries into activities likely to be governed solely by market laws and competition rules* (Stoffaës, 2003: 60). This re-regulation model is actually been put forward by Majone in his various books where he argues that the regulatory state is needed for the correction of market failures, as well as to increase the welfare of consumers (Majone, 1990; Majone, 1996).

This argument has also been taken up in a document emanating from the EU directorate-general for economic and financial affairs, which stated that the view that network industries regulation is to be phased out is not generally accepted. The legacy of a century or more of legal monopoly is not likely to be overcome quickly and, in most of the network industries, natural monopoly elements will persist for the foreseeable future (European Commission, 1999: 36). The differences between these two approaches, and especially the distinction between a competition and a regulatory policy, are summarised in table 2.

Table 2: Comparison of the two approaches

	Approach A: Deregulation	*Approach B: Re-Regulation*
Scope	Multi-sector	Sector specific
Objectives	Competition	Services of general interest
Intervention	Sporadic	Continuous
Control	Anti-competitive behaviours	Price, quality of services

None of these approaches under-considers the specificities of different network industries but both of them believe that the liberalisation process will lead to a common outcome. In this regard, water supply and sanitation have usually been depicted as one of the last remaining networks to be completely transformed and, as such, they are seen as being in a preliminary stage of this evolution. In the light of the findings of this book, we can indeed question whether the water supply and sanitation sector is likely to evolve towards one of these models.

Liberalisation in European network industries

The European Union is responsible to ensure that competition in the European internal market is not distorted. For this purpose, Article 86(3) of the EU Treaty entrusts the Commission with a specific surveillance duty *in the case of public undertakings and undertakings to which Member States grant special or exclusive rights*. The Commission must *where necessary, address appropriate directives or decisions to Member States* which enact or *maintain in force any measure contrary to the rules contained in the Treaty, in particular to those rules provided for in Article 12 and Articles 81 to 89*. Underlying this is the motivation of the EU to create a European internal market in terms of good, services and people as defined in the Treaty of Rome and the subsequent treaties.

This process has had a strong impact on the management of network industries. Until far into the 1980s, network industries have been largely ignored from the European economic integration process. In fact, it was implicitly considered that services of general economic interest were not part of the European political agenda and, as such, they remained under national competences. In the late eighties, this exception was slowly reconsidered at the European level following new economic arguments as well as some liberalisation experiences in certain industries at national level. In fact, the liberalisation of one network industry at the EU level led to a closer scrutiny of the scope of pro-competitive options in other network sectors. The pro-competitive options in network industries that were laid by the Commission were much influenced by new developments in economic theory. These theories argued for unbundling of formerly integrated activities, thus separating production, from transport, and distribution (e.g., Demsetz, 1989), as well as for tendering and competitive bidding (e.g., Baumol, 1969). These ideas slowly imposed themselves and the European Commission clearly decided to pursue this agenda across the different network industries. This strategy was based on the idea that by separating the network infrastructure (transmission lines for electricity, cable telecoms for telecommunication or water pipes for water supply) from the services provided over this infrastructure, one could introduce new forms of competition.

Looking backwards, it is interesting to see that the European Commission essentially pursued, at least in the early nineties, a very sector specific approach to network industries. The first proposals for a common gas and electricity market were laid down in 1988, and then followed by the liberalisation of the telecom sector in 1990. Other sectors soon followed. The following year, the European Commission looked at the railway sector. In 1994, the European Council declared that the postal sector would be subjected to a 'gradual and controlled' liberalisation process. As emphasised by Pelkman, the liberalisation of network industries very much looked like *an ad-hoc approach, on a sector-by-sector basis, and – in the absence of overall political guidance, driven by the Competition Directorate-General* (Pelkman, 2001: 435-6).

However, it slowly tends to become a general framework policy for all network industries (Pelkmans, 2001). The major turning point was the 1996 paper on services of general interest, which showed the first signs of a general reflection on the nature and so-called specificities of public utilities. Whether this was implicit or explicit, the European Commission regarded the example of end user service competition in the telecommunications sector, as the best case for opening up network industries to service competition. With this policy extended to all network industries, the same ideas have been slowly reproduced for each sector, and WSS are now looked under the same scrutiny.

The logic of the Commission is that European rules should apply to all undertakings with an effect on trade between Member States. The European Court of Justice has consistently held that the concept of undertaking covers any entity which is engaged in an economic activity. These include any activities consisting of offering goods and services on a given market. According to this logic, the Commission has considered that the provision of network industries' services should be regulated by EU competition rules as they correspond to economic activities. The implementation of a general framework policy to network industries clearly raises the question whether these industries can be considered under the same light.

This logic is currently been thought and tentatively applied to all network industries. The water sector, which is usually depicted as the last network to liberalise, is currently been approached in the same light. The European Commission does not favour, at least to our knowledge, one particular type of model as described in section II.1. Its logic is to pursue a pro-competitive policy, which would involve third party access. This logic, in combination with the different models described for network industries, would in our view result in two outcomes. The question therefore remains whether these European network industry models are applicable for the water supply sector.

The likely evolution of water supply in Europe: towards one of these models?

In the following paragraphs, we explore to what extent it is feasible to extrapolate these models to the water supply sector. In the light of the research made throughout this book, it appears that the water supply will not likely evolve towards one of these models.

The so-called "deregulation model" if applied to the water supply sector would imply the introduction of competition through the so-called third party access, namely the granting of rights to use the infrastructure of the incumbent. Regulation would be insured not through a sector specific independent authority but rather through a multi-sector competition authority (see table 2). Furthermore, one could imagine that these competition rules were set at the European level in order to better guarantee the competitiveness of the so-called European internal market.

On the other hand, the "re-regulation model" would also be characterised by third party access competition but its main difference would be the creation of independent regulatory bodies. They would be responsible not only for competition matters but also for guaranteeing a certain level of quality in terms of security of supply, environmental and consumer protection.

In this book, we have identified six possible scenarios for the water supply and sanitation sectors (see chapter five for more details):

- Scenario 1: Delegation Contracts & Strong Regulation
- Scenario 2: Delegation Contracts & Extreme Competition
- Scenario 3: Outsourcing
- Scenario 4: Regulated Monopoly
- Scenario 5: Direct Public Management
- Scenario 6: Community Management

Box 1: Summary description of European scenarios

Scenario 1: Delegation contracts with strong regulation
The first scenario is a tendered market with strong regulation. Suppliers compete for geographically delimited markets and competition is strictly regulated at national or even EU levels. The European market is dominated by large private operators. Independent regulatory authorities exist in all EU countries and have a mandate to control water prices and the quality of services (drinking water and wastewater treatment quality standards, leakages in networks, water cuts, and water shortage). The main particularity of this scenario is that it associates *ex ante* and *ex post* regulation. *Ex ante* regulation is guaranteed through competitive bidding. Bids are retained mainly along quality and public service criteria. *Ex post* regulation is ensured through the independent regulatory authorities and aims at forging incentives for best performance, i.e. the optimal level of effort to reduce cost. Additionally, social issues are considered as an important objective. In this regard, disconnection is forbidden and social funds are put into place across Europe.

Scenario 2: Delegation contracts and extreme competition
The second scenario is the tendered market scenario. By tendered market, we mean a competitive customer markets, whereby suppliers of products, services and financing compete for geographically delimited (i.e., often local) markets. This scenario is therefore characterised by competition for the market. Large private operators dominate the European market. This can largely be explained because of their large experience in this particular type of contracts (i.e. concessions and lease), as well as on their technological expertise, access to financial resources and management know-how. Small and medium-sized companies also remain in the European WSS market but these are slowly being taken over by the large ones. The market is not unbundled; drinking water markets are integrated (extraction, treatment, transport, distribution, and customer service), as are sanitation markets. Most contracts take the form of delegated management contracts, mostly lease, but also concession contracts. Lease contracts are the most typical form of contract since there is no need to expand the network. All major rules for bidding have been defined under a compulsory EU Directive, which pushes for strong competition by limiting the duration of delegation contracts to a minimum. Furthermore, environmental standards are more and more stringent at the EU level as a result of new toxic substances that are affecting water quality. This pushes for higher technological expertise and favours the growth of large operators.

Scenario 3: Outsourcing
The third scenario is characterised by a high diversity of operations due to the 'hollowing out' of the operators who outsource their main activities (from metering, via maintenance of the infrastructures, to the handling of the customers' complaints). Outsourcing is seen as the major answer to the growing pressure of achieving greater efficiency and innovation. The general success of the use of outsourcing contracts in other governmental and former utility sectors is an important driver for the implementation of these practices in the water sector. One can find a large variety of management models across Europe ranging from direct public management to full divestiture and including delegated management. The difference with other scenarios lies in the fact that there is certain dissatisfaction with regulation (regulatory failure, cost of regulation) and that there is growing financial pressure to seek more efficient operations thanks to specialisation and technological developments. Outsourcing is furthermore triggered by numerous suppliers who emerge and start offering corresponding services. In addition, and unlike delegated management, the revenue risks are not transferred to the winning bidder. As a consequence of this outsourcing phenomenon, the EU decided to lower the threshold values beyond which contracts have to be awarded through public tendering.

Scenario 4: Regulated monopoly
The fourth scenario is a regulated monopoly scenario. This scenario is characterised by a large diversity of operators and by numerous superimposed regulations and rules. If there is competition, it operates mainly through compulsory benchmarking, even though this is made very difficult because of the variety of rules and practices. Nonetheless, the European Union favours the development of these benchmarking practices. There may also be competition for large users. Common carriage has only been introduced in a limited number of Member States. The type of operators ranges from private companies to highly autonomous municipal undertakings that work at supra-municipal level. These entities are usually regulated under private law and capital may either be public, private or mix. Most of these operators rely on a multi-utility diversification strategy and therefore are involved in other network industries.

Scenario 5: Direct Public Management
The fifth scenario, direct public management, is characterised by no competition apart from traditional procurement arrangements. Contracting out is generally restricted to large turnkey (design and build) infrastructure provision and to the high technology domain. Competition for finance may also be obtained through bond markets. The local operator is awarded the responsibility to provide the integrated water services to the community, i.e.: to operate as the only provider. Water supply and sanitation remains entirely in public hands, most likely at the municipal or at the regional level. It is not strictly a return to the old public management model as it incorporates various innovations in technology and management structures (e.g., more flexible accounting standards, flexibility in employment contracts). Tariffs are structured according to volumetric charges, and disconnections are banned. There are no direct subsidies, except for infrastructure development in isolated and less developed regions. In compliance with the Water Framework Directive, full cost recovery is implemented. Finally, public water operators modernised their governance structures in order to increase public participation in water management.

Scenario 6: Community management
The sixth scenario is a community management scenario. Basically, communities of interest – e.g., local communities, agglomerations, users groups, associations of industry and private users – own and manage the water supply and sanitation infrastructure. These communities decide the service level and the corresponding investments. There is also outsourcing of construction and even other services, yet the ownership of the infrastructure remains with the community. The local community does not forcibly operate the whole integrated services, i.e. water supply and sanitation services. Normally, sanitation services are managed by individuals (i.e. septic tanks) but new technologies enable some communities to set up a decentralised system for wastewater. Regarding the operation and maintenance of WSS, there are two alternatives: the community may be involved in the day-to-day operation and maintenance or it can delegate this task or some other aspects to a professional. The institutional arrangements for this particular scenario may vary from voluntary organisations (e.g. user cooperatives) to water management associations formed by landowners, private enterprises or public corporations. Customers may also own water assets or can contribute to the WSS management through representation in water company boards. Public participation is surely the main distinctive element of this scenario.

In the light of the six scenarios identified in this book, one can see that only two correspond to the models identified for network industries. The first one, delegated contracts & strong regulation, recognises the need for regulation despite competition. It clearly corresponds to the second model presented above. The second one, delegation contracts with extreme competition, follows the path of increased competition. Both of these scenarios are influenced by a key driving force, i.e.: the desire to introduce more competition. In fact, this reveals the main assumption in network industry literature – competition, at least in the long-term, is always possible for every sector and only technical barriers slow down this process. This assumption may be proved to be wrong, at least for water supply and sanitation. Indeed, even these two first scenarios do not consider the idea of third party access as it has clear limits in the water supply sector. None of them actually corresponds to the deregulation or re-regulation models as described above.

The four remaining scenarios follow different paths. Their main features are not forcibly competition but a combination (or not) of the perceived specificities of water services provision compared to other network industries and/or the modernisation of the management methods. By modernisation of management methods, we mean the professionalisation and autonomisation of different services run by water and sanitation operators, as well as the need to increase accountability and define clear responsibilities in the organisational setup. The third scenario, outsourcing, is probably the best illustration of this process, at least in terms of professionalisation and definition of responsibilities. All the services are clearly defined and separated in order to benefit from specialised firms and/or individuals in the various services. In this scenario, one can see a strong combination of private and public actors according to their specific expertise. In the fourth scenario, regulated monopoly, one of its main features is the autonomisation of operators, which is directly related to the modernisation of the management methods. In the fifth scenario, direct public management, new measures such as the separation of the accounting system also contribute to this so-called modernisation of the management methods. Finally, in the last scenario, the community management scenario, this modernisation of management methods takes place with the introduction of decentralised technologies that leads to a higher degree of responsibilisation of the different users.

According to the scenarios identified in the book, many aspects seem to indicate that WSS may not follow the same pathway as other network industries, which may be explained along three reasons.

The first one concerns the limits of introducing competition in this sector. Compared to other network industries, the main form of competition is competition for the market. One can notice a tendency to opt for shorter term contracts in order to introduce more competition but the limits of such form of competition are clear-cut. Another possible form of competition in water supply and sanitation is comparative competition. Germany, Italy, Netherlands, and the UK have already introduced this form of competition. The logic behind is that competition by comparison would prove useful in stimulating backward companies to emulate those which are performing better. However, some specialists emphasise that such a system is not very successful as it is very difficult to standardise sufficiently to make cross-company comparison useful (e.g., Robinson, 1997: 167). Furthermore, many specialists are sceptical about the real possibilities of introducing competition in the market (e.g., Gee, 2004: 38-9). England and Wales have introduced compulsory third-party access to the water networks for the supply of industrial customers following the adoption of the Water Act 2003. With the Water Act, several tools are prescribed in order to increase competition in water supply and sanitation management, namely inset appointments, cross-borders supplies, connections agreements, and common carriage (Robinson, 1997: 171-5). However, first results seem to indicate that the possibilities and dynamics of this type of competition are restricted, at least for the moment. Overall, the scope of competition in the WSS is very limited compared to other sectors.

The second reason explaining why WSS may not follow the same route as other network industries is the absence of major technological innovations altering the structure of the industry, as it was the case in other sectors. This is linked to the scope of competition. With major innovations, the current situation in the water sector could radically change. For example, as suggested by Van Dijk (2003: 6), if filters would allow people to make good quality drinking water from lower quality drinking water coming from the grid at a low cost, it would then be economically viable to build a national grid, shared by a number of drinking water companies who could sell the filters and provide other services for customers for competitive prices.

Finally, the main theories of network industries also considerably underestimate the specific perception people have of water compared to the other network services. In the minds of many peoples water is (still) something very special, and as such reflects an ancestral view of our civilizations' relationships to one of the most vital resources: water, it is said, is an essential good, something that humans and cultures rely upon for their existence and development, not to mention the fact that water has religious meanings. In short, water, in overall perception, remains public and special, something that cannot be commodified, commercialised, or traded. In fact, the few attempts that were made around the world were rapidly abandoned due the strong political and popular resistance. For what concerns the management of water supply and sanitation systems, this perception remains the same. Private sector participation in the management of these systems is also seen with suspicious eyes. PSP in other network industries (i.e.: electricity, telecommunication) have certainly been a matter for political/ideological debates but they have rarely been followed by strong popular protest. In the water supply and sanitation sectors, this protest occurred for the most part in developing countries (e.g.: Jakarta - Indonesia, Cali - Columbia, Cochamba - Bolivia, Johannesburg - South Africa or Bahia Blanca – Argentina) (Vlachos & Braga, 2001: 53) but new legislation in the Netherlands (Hall, 2004) shows that this phenomena is clearly international.

STRUCTURE OF THE BOOK

In this book, the water supply and sanitation sectors and their possible evolution in Europe are examined. Throughout, there is an emphasis on institutional and organisation issues as opposed to an ideological debate based on ownership. The book seeks to address the trade-offs between different modes of management. After having identified the main strengths and weaknesses of each model, the purpose is to see how they can in some way be improved.

The book comprises two parts: the first one describes the WSS sectors today in Europe; and the second one focuses on possible scenarios for the evolution of the water sector, and their main implications in social, economic, and environmental terms. The book concludes with the main theoretical contribution of our project in the field of water management.

Part 1: The WSS sectors in Europe

Part one provides a clear picture of the main dynamics of the WSS sectors in Europe, and it is divided into four chapters. Chapter I presents a framework designed for analysing the WSS market structure and its dynamics. Then this framework is applied to single out distinctive markets within this framework. Four closely interrelated markets are identified and their relevant driving forces to change analysed – the market for drinking water and sewerage services, the market for suppliers, the market for abstraction and discharge, and the market for delegated contracts. The conclusion of this chapter provides an assessment of the main characteristics of all markets and an overview of their limits and scope.

Chapter II features a comparative policy analysis of the legislation at national level. It aims at presenting common trends and differences in the current administrative regulatory frameworks that govern the WSS sector at national level in eight EU member-states, namely Belgium, Germany, Spain, England & Wales, France, Italy, the Netherlands, Portugal, and Sweden. These countries were selected because of they are representative of the main water policies conducted in Europe. Thus, by taking into account the most different institutional arrangements, water situation and political systems, the comparative study seeks to isolate the relative influence of the different dimensions on the evolution of WSS policies.

The following two chapters look more precisely at the two most important actors in the water sector in Europe, namely the different types of operators and the European Commission. Chapter III explores whether the current EU policy in water standards, resources management and liberalisation of network industries could constitute a general framework for analysing the possible evolution of the WSS sector with regards to liberalisation. Chapter IV describes the different types of European WSS operators, and analyses their corresponding strategies. WSS operators are very much diverse, ranging from small non-autonomous public entities to large multi-utility TNCs, and encompassing regional public, mixed and private organisations. The strategies of water operators change from case to case and in fact, different types of operators react differently and are exposed to varying realities and dynamics.

Part 2: Scenarios for the evolution of the sector

Part two focuses on the future evolution of the WSS sectors in Europe, and comprises five chapters. Chapter V integrates the main dynamics of the WSS markets identified in Part One in the formulation of likely European water scenarios. This set of contrasted scenarios represents what is plausible, not necessarily what is either desirable (i.e., normative) or probable (i.e., trend based). The scenarios are: Delegation Contracts with Strong Regulation; Delegation Contracts and Extreme Competition; Outsourcing; Regulated Monopoly; Direct Public Management; and Community Management.

Chapter VI addresses the economic implications of these scenarios. It aims at going beyond the analysis of the traditional indicators for financial-economic performance, by addressing the degree of effectiveness and efficiency with which the WSS systems, under the several scenarios, are able to respond and react to various, diverging, societal and environmental demands.

Chapter VII analyses the environmental implications of each of the six scenarios on the basis of a set of sustainable water management criteria. One of the key aspects concerns the incentives that the scenarios' institutional and management framework sets in terms of sustainable water management.

Chapter VIII examines the social implications of the scenarios. The term "social" includes two different and complementary dimensions: firstly, it concerns the social corpus of the sector, meaning the workforce that contributes to its existence and its internal dynamics; secondly, it also concerns the relationships of the WSS sector with society, between users and providers.

Chapter IX addresses the main cross cutting issues per scenarios. It provides a discussion of the main regulatory issues characterising PSP in the WSS sector, and develops an assessment model aiming at evaluating the outcomes of alternative scenarios, considering environmental, economic and social aspects. In the WSS sector, many factors are promoting competition as well as a greater involvement of the private sector. The main issue to tackle is not whether to increase competition or involve the private sector but rather finding the most effective and efficient solutions to do that given that WSS remains a service implying significant market failures.

REFERENCES

Baumol, William Jack (1969), Welfare economics and the theory of the State, London: The London School of Economics and Political Science University of London.

Bergman, Lars et al, (1998), "Europe's network industries: conflicting priorities – telecommunications", Monitoring European Deregulation, No. 1, Centre for Economic Policy Research (CEPR), & Centre for Business and Policy Studies (SNS).

Demsetz, Harold (1989), Efficiency, Competition, and Policy. The Organisation of Economic Activity, II, Oxford: Basil Blackwell.

Department of the Environment and Welsh Office (1996), Increasing Customer Choice: Competition in the Water and Sewerage Industries: The Government's Proposals, Department of the Environment and Welsh Office, London: HMSO, April 1996.

Egan, Michelle (2001), Constructing a European Market : Standards, Regulation, and Governance, Oxford: Oxford University Press.

EUROMARKET (2004a), "Analysis of the Strategies of the Water Supply and Sanitation Operators in Europe", Deliverable 3.

EUROMARKET (2004b), "Analysis of the European Water Supply and Sanitation Markets and its Possible Evolution", Deliverable 2.

EUROMARKET (2005), "Analysis of the Economic Implications of the Scenarios", Deliverable 6.

European Commission (1999), Liberalisation of network industries: Economic implications and main policy issues, Directorate-General for Economic and Financial Affairs, No. 4.

Finger, Matthias & Frédéric Varone (2006), "Bringing technical systems back in: towards a new European model of regulating the network industries", Competition and Regulation in Network Industries, Volume 1, No. 1, pp. 89-108.

Friedman, Milton (1953), Essays in positive economics, Chicago, London: University of Chicago Press.

Gee, Alexander (2004), "Competition and the water sector", Competition Policy Newsletter, Number 2 (Summer), p. 38-40.

Hall David (1999), Trends in the water industry in the EU, 1999, February, PSIRU Reports Report number: 9902-W-Eur.

Hall, David, Emmanuel Lobina & Robin de la Motte (2004), Making water privatisation illegal: - new laws in Netherlands and Uruguay, London: PSIRU.

von Hayek, Friedrich August (1960), The Constitution of Liberty, London: Routledge.

Izaguirre, Ada Karina (2005), "Private Infrastructure", Public Policy for the Private Sector, The World Bank Group, Note Number 299, October.

Izaguirre, Ada Karina & Catherine Hunt (2005), "Private Water Projects", Public Policy for the Private Sector, The World Bank Group, Note Number 297, July.

Kapteyn, Paul (1993), The Stateless Market: The European Dilemma of Integration and Civilization, New York: Routledge.

Lane, Jan-Erik (1985), State and the market: the politics of the public and the private, London, Beverly Hills: Sage Publications.

Lobina, Emmanuel & David Hall, (2003), Problems with private water concessions: a review of experience. On www.psiru.org

Majone, Giandomenico (ed.) (1996), Regulating Europe, London: Routledge.

Majone, Giandomenico (1990), Deregulation or Re-Regulation?, New York: St. Martin's Press.

Martin, Stephen & David Parker (1997), "Privatization: the conceptual framework", In: Martin, S & D. Parker (eds.), The impact of privatisation: ownership and corporate performance in the UK, London: Routledge, pp. 1-30.

Owen / Pinsent Masons Solicitors (2004/5), Masons Water Yearbook, Pinsent Masons Solicitors.

Owen / Pinsent Masons Solicitors (2002/3), Masons Water Yearbook, Pinsent Masons Solicitors.

von Mises, Ludwig (1981), Socialism, First published 1922, Translated from German by J. Kahane, Indianapolis: Liberty Fund.

Pelkmans, Jacques (2001), "Making EU Network Markets Competitive", Oxford Review of Economic Policy, Vol. 17, No. 3, pp. 432-456.

Polanyi, Karl (1957), The Great Transformation, Boston: Beacon Press.

Robinson, Colin (1997), "Introducing Competition into water", In: Beesley, M.E. (ed.), Regulating Utilities, broadening the debate, London: Business School and Institute for Economic Affairs, pp. 153-87.

Scott, Alan (1997), The Limits of Globalization: Cases and Arguments, London & New York: Routledge.

Stoffaës, Christian (2003), Towards European Regulation of Network Industries, Report by the discussion group chaired by Christian Stoffaës, July 2003.

Scharf, Fritz (1999), Governing in Europe, Oxford: Oxford University Press.

Strange, Susan (1996), The Retreat of the State: The Diffusion of Power in the World Economy, Cambridge: Cambridge University Press.

Swann, Dennis (1988), The Retreat of the State: Deregulation and Privatisation in the UK and US, Hemel Hempstead: Harvester Weatsheaf.

Van Dijk, Meine Pieter (2003), Liberalisation of drinking water in Europe and developing countries, Inaugural Address, Delft: UNESCO-IHE.

Vickers, John & George Yarrow (1988), The water industry privatization: an economic analysis, Massachusetts: MIT Press.

Vlachos, Evan & Braga, Benedito (2001), "Les Défis de la gestion de l'eau urbaine" In : Maksimovic, Cedo, Tejada-Guibert, José Alberto & Roche, Pierre-Alain, Les Nouvelles frontières de la gestion urbain de l'eau : Impasse ou espoir ?, Paris: Presses de l'école nationale des Ponts et chaussées, Chapitre 1.

World Bank (2000), Partnerships for Development, Washington: The World Bank.

Yergin, Daniel & Joseph, Stanislow (2002), The Commanding Heights: the Battle between Government and the Marketplace that is Remaking the Modern World, New York: Simon & Schuster.

Chapter I

The European Water Supply and Sanitation Markets

Authors[*] Marco Schouten and Meine Pieter van Dijk

INTRODUCTION

Providing drinking water and wastewater services to a consumer seems a relatively straightforward practice. Water is abstracted from a river, an aquifer or in some cases even the sea. This water is treated and pumped into underground pipes, ending up at the premises of consumers where it flows out of their taps. The wastewater that comes from the shower pit, the latrine and sometimes from the drains in the street flows into another underground piping system, ending up at a wastewater treatment plant. There, the materials that really harm the environment are removed before the wastewater is discharged into the environment. Water providers, all over the world, have managed this cycle for more than 100 years, and the fundamentals of the processes remain largely untouched (Thomas and Ford, 2005). The simplicity and clarity of the delivery process seems to be in stark contrast with the complexity and diversity in which the sector is organised.

About 30,000 water operators in the European Member States (excluding the ten Member States that entered recently) are managing the described delivery processes, generating a total production value of about 37 billion Euro annually (Eurostat, 2004). The size of, and diversity within, this group of operators makes it exceptionally hard to make general statements. This heterogeneity is due to the local public authorities' individual path and preferences in organising the service provision. Since the start of the 19th century, each local authority made continuous explicit and implicit policy choices shaping the way water services are to be provided, basing itself on changing local circumstances, as for example demographical, social and political dynamics. A dimension of water services, adding to the heterogeneous character of the sector, is the possibility to partially unbundle the different activities related to water services provision. Water services are not an amorphous whole but can be divided into numerous autonomously managed activities, as bulk water supply - water treatment - local water distribution - sewage collection - sewage treatment. All of these activities can be provided by one organisation – for example very recently the three separate water management of Amsterdam announced to become as of 2006 the first integrated water service organisation in the Netherlands, claiming 5% cost savings due to the prospected merger (VNG, UvW and DWR, 2005). Nevertheless, it is

---

* This chapter is based on Deliverable 2 of the EUROMARKET project. This chapter would not have been made possible without the valuable help of the University of Birmingham (Andrew Nickson and David Musco) and Bocconi University - IEFE (Antonio Massarutto, Vania Paccagnan and Massimo di Domenico).

not necessary so, and even quite uncommon, that one single service provider is responsible for the whole range of activities in one spatial area. The majority of European member states divide the water supply activities from the sanitation activities. Moreover, in Member States like France and Spain, each and every activity of the water cycle is split over separate service providers (see Table 1).

Table 1: Heterogeneity of service suppliers

	Split Between Different Activities - Yes-No		
	Split Between Water & Wastewater	Water Supply; Split between Bulk supply & Distribution	Sanitation; Split between Collection & Treatment
Italy	No	Yes (limited)	No
Ireland	No	No	No
UK	No (limited)	No	No
Finland	No	No (limited)	No
Sweden	No	No (limited)	No
Greece	No	Yes	No
Denmark	Yes	No	Yes
Netherlands	Yes	No	Yes
Portugal	No	Yes	Yes
Austria	Yes	Yes	Yes
Germany	Yes	Yes	Yes
Belgium	Yes	Yes	Yes (Flanders)
Luxembourg	Yes	Yes	Yes
Spain	Yes	Yes	Yes
France	Yes	Yes	Yes

Especially the Southern Member States tend to split water supply into bulk water transfers and local water distribution. In these Member States water resources are not always available locally and regional/national coordination is required. Transporting water over long distances is costly. On one hand because water is a relatively low value product, with the selling price normally ranging between € 1 and € 2 per cubic metre, and on the other hand because water is very dense - meaning that it has a high mass per unit volume. In fact, water is one of the few compounds that is actually denser in the liquid form than in solid form (this is the reason why ice floats on water). Examples of distinct service providers dedicated to the bulk transfer of water can be found in at least seven, mostly Southern, Member States. In Portugal Aguas de Portugal provides bulk water supplies (via 20 water systems) to municipalities that in turn serve over 7 million people. In Spain the 11 water basin authorities and 2 island water boards provide (i.e. sell) bulk water supplies to the local service providers. In Greece Eydap Assets is responsible for bulk water supply to Athens (supplying over 4 million people). In Germany there are the various "Verbande" which provide bulk water supplies to many local water utilities (one supplies over 3.5 million people in South West Germany). In Belgium there are various regional bulk supply organisations. In Brussels, for example, there are currently two separate water supply entities – CIBE is responsible for water production and IBDE is responsible for its distribution. In Austria there are around 150 local authority associations who supply the local water companies with bulk water. In Finland there are at least 3 examples of regional water production companies.

Also in some Member States a split can be found between sewage collection and treatment (e.g. Denmark, Benelux, Spain, Portugal, Germany and the Netherlands). Here, the municipality typically manages the sewage collection system and the treatment works is managed by a regional or multi-municipal entity. For example, in Germany local authorities manage the sewage collection, while the Länder administrations are responsible for identifying the public bodies for wastewater treatment. The Länder administrations may then delegate the wastewater treatment responsibilities to private parties. If the service provision is currently organised via a multi service provider, the ongoing liberalisation of energy sector may have impact of public-private partnerships (PPP) in water sector, as can be noticed in Sweden and the Netherlands.

The markets in which each water provider operates are of a heterogeneous character. The ways in which markets are organised are highly dependent on local arrangements and circumstances. It is therefore not an easy task that we have set ourselves in this chapter to discuss the structure and dynamics of the markets. Especially of interest to us is to analyse how the private sector is currently involved, as liberalisation of the water sector is one of the major issues of the book. In this chapter, we will address the different forms of private sector involvement, the emerging market opportunities and the extent to which they are exploited by the private sector. The relevance of our analysis is that it might help policy makers and practitioners in the water sector to understand how markets are arranged elsewhere, and to see whether alternative market structures might also be adaptable to their situation.

Of course, to identify the structure and dynamics of the European water and sanitation markets at the pan-European and EU Member State level cannot be done without generalising some of the characteristics of the sector. To do this we constructed a framework to analyse market structures and market dynamics. Our analysis of the European water supply and sanitation markets takes as point of departure the 30,000 service providers that are operating in various markets. The objective of our analysis is not to assess in depth the characteristics of these markets in each individual Member State but to be able to provide a broad overview of the European water supply and sanitation markets.

The chapter is structured as follows. First, we will introduce the framework we designed for analysing the market structure and dynamics. Then we will apply this framework to single out distinctive markets within this framework. This description of each of the four markets is followed by an analysis of the driving forces to change relevant to these markets. The conclusion of this chapter will provide an overview of the assessment of the main characteristics of all markets, and an overview of the market forces.

THE INTEGRATED MARKET FRAMEWORK

Constructing a framework for analysing the water supply and sanitation market, with the aim to provide a better understanding of liberalising tendencies in the water sector, requires a relatively unconventional approach. Usually markets are in the water sector discussed in a fragmented and one-dimensional manner. Mostly the focus is solely on the users of water services; market analysis has in this instance a more demographical character, aiming to identify (growth patterns of) different user groups. Especially for the purpose of water demand management such studies are valuable. Other studies aiming to shed light on sustainability issues of the water resources, have as primary focus the amount of water that is abstracted (and discharged) into the natural environment by water service organisations. These studies are only placed in the context of markets in the (few) cases where the rights to abstract and discharge are transacted between market parties. A third approach towards analysing water supply and sanitation market is from the point of view of international operators that work on contract basis. Market opportunities for such companies, encompasses the potentially interested governments that may be willing to contract them. Indicative in this respect was the confusion on the term ‘customer’ in an interview with two representatives of an international operator. For the interviewees the ‘customer’ could be either the mayor or the water user (Interview SUEZ, 2005). The last and probably the most fragmented dimension towards analysing markets is from the point of view of the companies that sell their products and services to water service providers. If you are a producer of 9-inch pipes you will judge the market potential by analysing the demand for 9-inch pipes of water service providers. Each and every product that water service providers are or could be, procuring provides a potential for a market.

For the purpose of our analysis we find it important to make an effort to construct a multi-dimensional framework of the water supply and sanitation markets. As the central objective of our research is to analyse the introduction of competition in the water sector (liberalisation), we want to point out that there is already quite some competition in the water sector; it only depends to which market one is referring to. Moreover, by constructing a multi-dimensional framework, we will also be able to indicate the interdependencies between the various markets. This is important, as we will show that modification in the structure of one market can alter the structure and level of competition in other markets. Another reason for us to take an integrated approach is to enable us to present also the different degrees and types of competition that are possible in the various markets. We will show that in some water supply and sanitation markets the barriers to competition are very high; while in others many types of competition are possible. Other considerations that led us to construct an integrated framework was the desire to indicate the different types of dynamics in each of the markets and the different types of regulation to which each market is subordinated.

Engaging into an analysis of the European water supply and sanitation markets requires a proper definition of what constitutes a market, as there are many possible definitions available. Traditionally a market was the place where buyers and sellers meet to exchange their goods, such as a village square. Households use the term to the companies from which they can buy a well-defined commodity, for example to buy groceries at the super-market. From the point of view of the industry it is the other way round; the buyers are arranging their market as the buyers are the ones to sell their product to (Lipsey et al, 1987). We prefer the terminology as is commonly used by economists, being a market is a collection of buyers and sellers that transact a particular product or product class. Hence, a market is created when exchange takes place. Exchange is the act of obtaining a desired product from someone by offering something in return. For exchange potential to exist, five conditions must be satisfied (Kotler, 1998):

- There are at least two parties.
- Each party has something that might be of value to the other party.
- Each party is capable of communication and delivery.

- Each party is free to accept or reject the exchange offer.
- Each party believes it is appropriate or desirable to deal with the other party.

Whether exchange actually takes place depends upon whether the two parties can agree on terms of exchange that will leave them either better off (or at least not worse off) than they were before the exchange. Exchange is frequently described as a value-creating process because it normally leaves both parties better off. In the water supply and sanitation sector numerous types of exchange take place.

Probably the first type of exchange that comes to mind is the delivery of water services, as water from the tap or wastewater handling, to customers in return for payment of a tariff. The regulated transaction that makes up this market is the provision of drinking water or sewerage services to customers by the service provider. To refer to this specific market we address it as the '**market for drinking water and sewerage services'**. Each of the services provided composes a separate sub-market. For instance, there is a sub-market for industrial water, for domestic water and for wastewater. Even there is a sub-market for making the water connections of customers.

Another type of exchange is when the water company acts as the buyer, purchasing inputs vital to its' business process from various suppliers. The regulated transaction that makes up the market for suppliers is the delivery of all kinds of inputs by external suppliers to the service provider. Again, each of these inputs composes a separate sub-market. For example, one refers to the labour market when the companies 'purchases' human resources, to the capital market when the company seeks finances, the energy market to purchase the necessary energy or even the market for 9-inch pipes when the company needs this type for their operations. For our analysis we combine all these sub-markets under the umbrella term of '**market for suppliers**, being the market in which the utility is operating to purchase inputs for its' business processes.

For our analysis we excluded one specific input from the 'market for suppliers' in view of its uniqueness, being the right to abstract raw water and discharge wastewater. This '**market for water abstraction and discharge'** relates to the need for a water utility to 'purchase' raw water from the natural environment and discharge wastewater to the natural environment. In return sometimes the water utility needs to pay a fee for abstraction or discharge to the 'owner' of the water resources. The owner of the water resources needs to accept that the service provider is implicating its natural water resources via these transactions. The market for abstraction and discharge involves the ownership and possible trading of rights to abstract water from, and discharge wastewater into the natural environment. In this sense the "buyer" is the service provider, the "supplier" is the entity that controls or owns the water resources and the "regulator" is typically a national or regional body.

A fourth and final type of a market relevant for the water sector is the so-called '**market for delegated contracts**'. This is a relatively young market in which the seller is the responsible entity for service provision. It 'sells' the right to act as service provider for a geographical area to an operator. In return the operator can pay a fee to the responsible entity or can get compensated by gaining a profit from the service provision. The regulated transaction in this market is the granting of the opportunity to serve a monopolised water region. The entity that is finally responsible that the population receives water might choose from different parties to grant the right to operate the service provision. The 'market for delegated contracts' is especially relevant in view of the discussions about liberalisation of the European water supply and sanitation market. Most often when liberalisation is discussed it refers only to the 'market for delegated contracts'.

Hence four different markets emerge that together form the overall water supply and sanitation markets. As the water sector is a heavily regulated industry, each of these markets is subject to stringent regulatory regimes. The transactions in each of the markets are monitored and controlled by regulators. The nature and severity of the regulation will depend on the characteristics of each of the markets. In the market for drinking water and sewerage services the regulatory regime will be aimed at protecting customer interests, as the water service provider is a monopolist. Consequently regulatory interferences in this market will focus on controlling tariff and service levels. In the market for suppliers the regulatory interferences are of a different nature. They will be aimed at trying to establish a perfectly competitive market by establishing rules for tendering and regulating mergers between suppliers. In the market for delegated contracts the regulatory interventions will be similar to the market for suppliers in the sense that these trying to simulate a perfect competitive market, but moreover the regulatory regime will protect the general interest once the service provider has been installed. The regulator will see to it that the originally contracted and agreed upon targets are monitored and lived after. In the market for abstraction and discharge the intentions of the regulator will be aimed at ensuring environmental sustainability, for example through licensing of water abstractions.

Based on the above segmentation in markets, an analytical framework emerges that supports the analysis of both the structure and dynamics of the market (see Figure 1). The framework is based on the concept of regulated transactions.

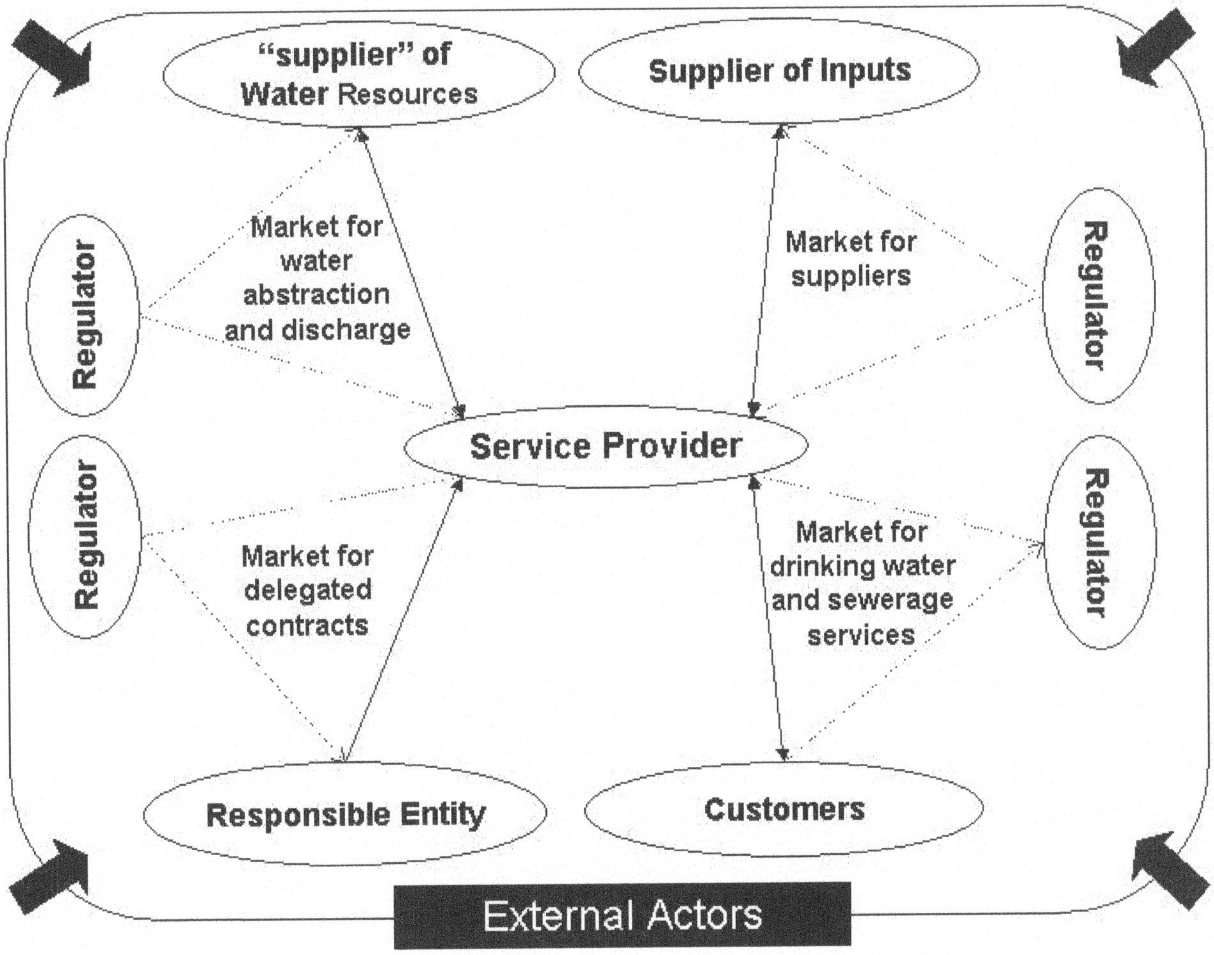

Figure 1: The butterfly-shaped integrated market framework

The, above presented, butterfly-shaped framework is based on the segmentation of the overall water supply and sanitation markets into four inter-related markets. Each of the four markets consists of a trilateral governance structure: a buyer, a seller and a regulator. By segmenting the overall market into four sub-markets we can explore the various interactions. Indeed some of the more complex linkages are not shown in the figure for presentational reasons. For example the environmental regulator might also interfere in the market for suppliers by imposing minimum requirements on supplier transactions as to avoid more indirect environmental implications. Another example of such more complex linkage is the interference of the regulator of the market for delegated contracts. The way the customers perceive the service delivery by the provider is an essential indicator for this regulator to possibly make regulatory impositions in the arrangement with the responsible entity. Often it is seen that one and the same regulatory body incorporates the regulatory regimes of several markets. For presentational reasons four separate regulators are identified, emphasising that each of the transactions that make of each of the four markets are subjected to different types of regulation.

Whilst the proposed analytical framework identifies four major markets, it is important to understand the interrelations between the markets. Attempts to introduce competition in one market will potentially alter the structure and dynamic of the other markets. Indeed the linkages between the responsible entity and the service provider in the market for delegated contracts, and the suppliers responsible for delivering the inputs and the service provider, can be seen to operate in tandem. The background behind this argument is the assumption that especially multinational private parties operating upon a delegated management contract will be inclined to make use of their own supply chain companies to realise corporate profit margins, instead of organising competitive open tenders for these services on the open market. For example when transnational companies (such as Veolia, RWE or Suez) act as service providers they bring with them their large network of their own in-house suppliers, enabling them to purchase the various inputs outside market for suppliers' competition. In franchise bidding in the market for delegated contracts, the various input prices for operation, equipment supply, and cost of capital are effectively subsumed into a single bidding process that typically produces a single all-in price. Private Service providers in the market for delegated contracts may also be able to manipulate the market for suppliers by awarding contracts to associated companies on favourable terms. For example, the economic regulator Ofwat (Office of Water Services) in England and Wales is closely monitoring the transactions between the private operators and associated suppliers of inputs. When competition in the market for delegated contracts is fierce, the competition in the market for suppliers tends to weaken and the other way round. This situation is what can currently be observed in Sweden (illustrated in Box 1).

Box 1: A Swedish illustration of the interrelations between competition in the market for delegated contracts and the market for suppliers

99% of the Swedish 289 municipalities are currently managing their water and wastewater provision either as a municipal administration (Direct Public Management) or as a municipally owned limited company (Delegated Public Management). Competition in the Swedish market for delegated contracts is consequently almost absent as only a handful of examples exist of municipalities organising competitive tendering for the right to serve a monopolised area (today there are only 6 management contracts in Sweden). On the other hand the market for suppliers in Sweden is very competitive. As much as 70% of all goods and services needed to operate municipal water and wastewater works are brought on the market for suppliers in open competition. Most of these contracts are short-term contracts with renewal once a year or every second year (Lannerstad, 2003). The argument made by pro-public representatives is that within the Swedish water sector the combined level of competition in the operators and the market for suppliers is higher compared to countries with higher levels of competition in the market for delegated contracts, as for example France (Lannerstad, 2002).

In contrast, increased competition in the market for delegated contracts itself will probably require greater competition in the market for abstraction and discharge as it will increase the need to free up existing abstraction licenses and 'consents to discharge' through some form of trading, time limitation and/or regulatory intervention. Neither of the identified water and sanitation markets is operating in a vacuum. Each of them is constantly subject to influences from external actors; as non-governmental organisations, labour unions, environmental activist groups, academics, etc. These external actors provide a constant source of turbulence to each of the defined water supply and sanitation markets, although the level of influence is dependent on various factors. One of the major determinants of the level of influence is the presence in the market of powerful interest groups aligned to change or maintain the status quo. In cases where there is a large public interest and press coverage on the (potential) liberalisation of the water sector, as for example in Sweden, the water sector and its' structure finds itself in the middle of a public discussion. Also the lobbying of labour organisations can be influential as can be witnessed in Northern Ireland, where employees are lobbying to maintain their existing position. In other countries, as Greece, the interest of external actors is far less and the water sector can operate relatively autonomously.

Of major influence to any structural changes of the markets are the experiences gained with reform measures, such as in the framework of liberalisation policies. These experiences can be gained either within the water sector or in other sectors. For example, a driver to change for other Swedish publicly organised water providers were the experiences in the municipalities of Karlskoga, Norrköping and Norrtälje that experimented with private sector involvement. Specifically the English experiences with privatising water are often used as a benchmark. For example in the capital city of Scotland, Edinburgh, a massive campaign of many levels of the Scottish society was organised in the beginning of the 1990s rejecting English-style outright privatisation (Juuti and Katko, 2005). In countries like Sweden and the Netherlands, the English experience is often used as an indication of the dangers of involving private parties in the water sector (see for example Gustafson, 2001). Also successes and failures of the liberalisation of the electricity, the gas and the railway sector can be influential. In the Netherlands the problems related to the liberalisation of the railway system have created a lot of distrust in liberalising also other sectors. The same can be seen in Northern Ireland, with the relative failure of liberalising the electricity sector. On the other hand, if the experiences in other sectors, as for example in Scotland, are perceived as positive, this can be an important trigger to change also in the water sector.

The following section is dedicated to analyse the four identified markets in the framework. The diversity and heterogeneity of each of the markets will be a reoccurring issue in the analysis. Not only because each of the markets is a complex set of sub-markets and actors, also the markets can differ a lot from one country to another. The emphasis in our analysis will be on two specific markets, being the market for delegated contracts and the market for drinking water and sewerage services, as these markets are the most discussed ones in the whole debate pertaining to the liberalisation of the water sector. We want to underline that the mere fact that we elaborate more on these two markets does not mean that we consider the other two markets of less importance in the discussion on liberalisation. Putting it even stronger, it is our clear conviction that all four markets and their interrelations should receive due attention in any decision making on reform measures. That is exactly the reason for us to construct the market framework, pointing out that it is of eminent importance to view reform measures in the context of all of the four markets.

The market for abstraction and discharge

In Europe the "supplier" of water resources and the "regulator" in the market for abstraction and discharge are in most cases the same body, often in the form of an environmental agency or a water basin agency. The state retains the right to allocate the resource between uses and fix environmental standards. Mostly the environmental standards are based on European Directives (see in particular the Water Framework Directive and chapter three). This environmental regulation affects the service provider responsibilities by introducing constraints over some activities such as wastewater treatment. Due to the regulatory impositions there is a need for more expertise, know how and infrastructure investments. The quantity of water resources available to distribute varies greatly across Europe (see Chapter two for a more detailed analysis).

The market for abstraction and discharge is not typically a true market because water rights to abstract water and discharge wastewater across the EU are typically issued and controlled by the nation state. Currently all Member states have some form of abstraction permits, even if implementation deficits are noticed in Belgium, France and Portugal for abstraction permits. Also it can be observed that abstraction charges are quite ineffective everywhere, except in Sweden. Implementation of these charges is lacking in Portugal and the amount withheld in Italy remains symbolic, as the charges are negligible. Concerning sewerage, the prohibition of direct discharges in surface water is implemented in most countries, except Spain and Portugal where the measures exist but implementation is partial (see also Chapter two).

In the market for abstraction and discharge there are only limited possibilities to introduce competition. The existing water rights in the form of abstraction licenses and 'consents to discharge' could be subjected to some form of trading, time limitation and/or regulatory intervention. Introducing such type of competition in this market has proved to be very difficult. There are only a handful of examples in parts of California, Australia, Chile and Spain in which water rights are traded. The reasons are manifold but can be traced back to traditional reasons for market failure. Incumbent licence holders, especially the private service providers, have little incentive to trade the right to abstract water and discharge wastewater, since these are fundamental to their operations. Also informational bottlenecks proved difficulties for organising such trades in a transparent way and setting acceptable maximum volumes for the trading system. The shifting of abstraction rights or discharge rights within a catchment area as a result of trades resulted in environmental externalities. In general, it could be argued that the risk of market failure is severe enough to restrict any wide scale adoption of liberalising the market for abstraction and discharge.

The market for suppliers

In providing water and sewerage services the provider must undertake and manage a number of specific activities. These can be defined as individual, possibly contestable, services or products that together represent the building blocks or inputs of the water and sewerage service provision. These inputs can be provided either by the service provider directly or purchased from external suppliers. They either relate to the day-to-day operational management of the service or to the infrastructure required to deliver the service. Three main actors are active in this market, namely the service provider, the suppliers of inputs and the competition authority. The European water providers purchase annually a total amount of 20 billion Euro (Eurostat, 2004), of which France is by large the biggest purchaser. The supplier of inputs can deliver individual products or services to the service provider or they may be partly or wholly bundled together through various contractual arrangements e.g. turnkey construction contracts or operational management contracts. Both infrastructure provision and operation services may also be bundled together in the form of design (D), build (B) and operate (O) contracts. The market for suppliers can be broadly divided in three main sub-markets:

1. <u>Labour market</u>

The sector in the EU15 member states employs in total almost 186,000 workers (Eurostat, 2004). Total employment in the water sector fell from 1995 to 1999 with an average of 1.2% per annum. Since then the employment index registered two years of 1.6% growth, followed by a fall of 0.3% in 2002. Larger member states as Germany and the UK did not exhibit similar patterns. In both of these countries the employment in the water sector was marked by a contraction, with an annual decline of minus 2.6% in Germany and minus 0.9% in the UK. A country like Spain showed an annual employment growth of 5.7% during the years between 1997 and 2002. Average personnel costs were Euro 38,100 per employee. The majority of the work force in the water supply and sanitation sector is male. The men make up almost 80% of the workforce. There is small portion of labour contracts that have a part-time character. Almost 94% of the contracts have a full time employment arrangement. An indication of the efficiency of the water providers is the indicator of the wage adjusted labour productivity, which should express how much value each employee is generating. The 15 Member States recorded a labour productivity of Euro 83,799 per person employed in 20001, which results in a labour productivity ration of 219.7%, indicating that the value added was more than double the level of personnel costs. Also this figure is varying in the diverse member states:

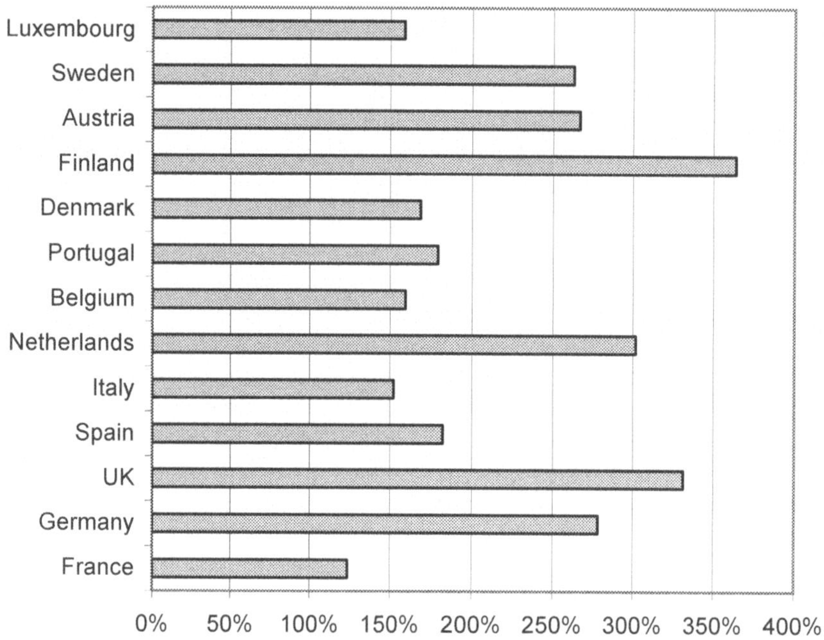

Figure 2: Wage adjusted labour productivity (Eurostat, 2004)

2. <u>Markets for products and services</u>
Numerous products and services are required and purchased separately by service operators. Service provider procures all kinds of materials. Not only pipes are procured, also chemicals and many other types of operational materials need be acquired. Moreover, in almost all cases, energy needs to be purchased. Typically these products are acquired in a competitive setting from private parties. Also numerous supporting services are procured to operate the infrastructure, such as leakage management, sludge disposal, and regular inspection of assets. Also support is hired to do the activities related to customer liaison and account management, such as billing, meter reading, debt recovery, customer complaint and enquiry handling. More general supporting activities that are required are activities as regulatory affairs, human resource management, accounting and finance, research and development, information technology, planning, property management, public and employee relations, and vehicle management. Also supporting scientific activities such as sampling, measurement, monitoring, data analysis and presentation are sometimes procured. Across the EU, the above-mentioned inputs are more service-oriented and are generally provided by the service provider itself. However, the private sector is playing an increasing role, either directly through being contracted for a fixed time period to provide one or more of the above services, or indirectly through being used as benchmarks for public utilities.

3. <u>Capital market</u>
The service provider requires access to capital finance such as bonds, loans, grants, retained earnings and possibly share equity, to fund both infrastructure provision and to a lesser extent working capital. Water providers are continuously in need for financing of their investments. Annually investments in water infrastructure encompass an amount of almost 9 billion Euros (Eurostat, 2004), distributed in the following manner over the member states:

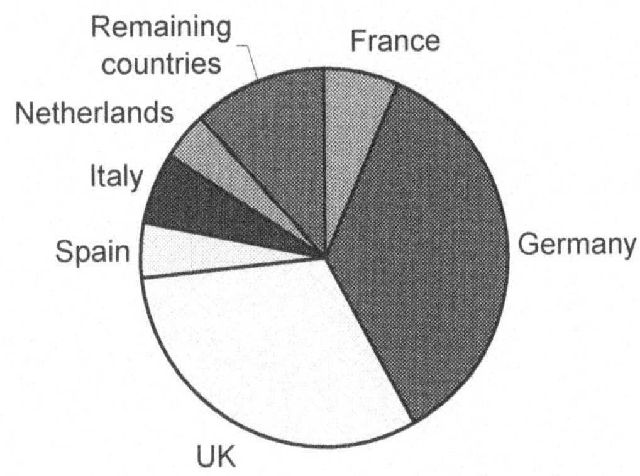

Figure 3: Gross investments in tangible goods in the European water sector in millions of Euro (Eurostat, 2004)

It is often argued that the discipline of the financial markets in the management and allocation of different risks is a vital ingredient in the potential success of private sector participation. Mainly banks are offering their services as suppliers of capital. In most European Member States public service providers have the opportunity to attract capital finance from the government or public banks at a lower rate than the market rate, outside the competition.

Competition in the markets for suppliers can be introduced as it involves various operational inputs, including contestable service contracts and turnkey infrastructure provision. Competition in the markets for suppliers is sometimes seen (in Member States such as Sweden and Germany) as a viable surrogate to competition in/for the operators and market for drinking water and sewerage services. There is also a possibility for comparative competition within the market for suppliers, e.g. market testing to assess transfer prices as undertaken by the responsible entity or the competition authorities (i.e. incentive based regulation of conduct). The market for suppliers is typically competitive, while the degree of competition varies between different product-service market segments. A high portion of the service provider's activities can be supplied directly by private parties. Member States with highly competitive market for suppliers generally produce internationally competitive suppliers. Germany and to a lesser extent Sweden, Finland and Denmark are the home of internationally renowned suppliers of water treatment technologies, pumps, process control and automation systems, and water quality monitoring equipment. An analysis of intra and extra European trade in 1995 for liquid pumps, water treatment equipment and water instruments highlighted that Germany was by far the leading exporter.

The market for delegated contracts

Typically the responsible entity for service provision is the municipal council. There are some notable exceptions to the municipal council being the sole responsible entity - especially where water and sewerage services are provided at a regional scale by service providers that operate under private law. Whilst the structure of local government varies across Europe it is evident that local government, by its very nature, is fragmented (see Figure 4). In 1998 Europe had about 73,460 municipalities, of which the vast majority had less than 5,000 inhabitants. This large number, along with cultural preferences, may help to explain the heterogeneity adopted as local authorities attempt to gain economies of scale. Local government has been reorganised in most Member States. The general move has been towards smaller municipalities either disappearing in larger authorities or working together in supra-municipal bodies. This has been mainly driven by the recognition that it is very difficult to provide the full range of municipal services in the smallest local authorities. The latest examples of such reorganisations are Greece, Portugal and Italy. Specifically for Italy the aggregation was also created to enable the establishment of an integrated system for water management. The larger the service area, the higher the attractiveness to the private sector to become involved, although as a counter force one could also assume that the larger the service area, the better incumbents are able to avert interference from international operators, as for example in Sweden and the Netherlands. Exceptions to this trend of consolidation of municipalities are France and Spain that have retained their smaller municipalities for a variety of cultural and political reasons. They have obtained the required economies of scale in service provision through franchising with large private sector companies and the extensive use of multi-municipal service provision. Hence the adoption of these institutional arrangements in the water supply and sanitation sector may partly be driven by the political decision to maintain the existing fragmented local authority structure.

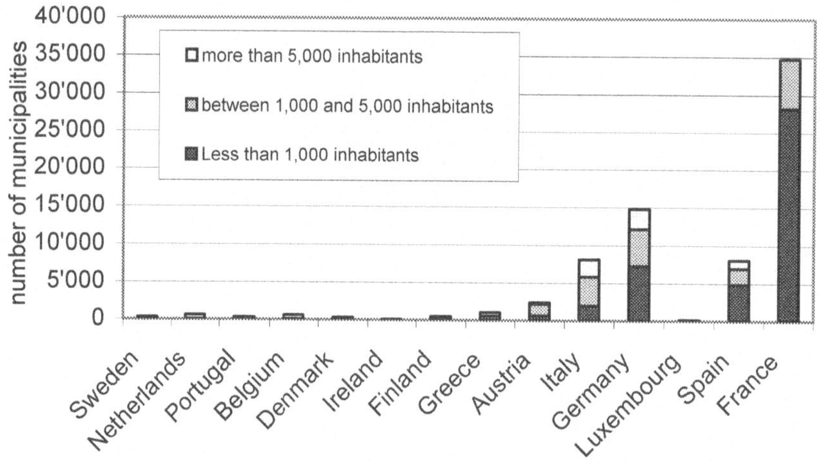

Figure 4: Number of municipalities distributed according to size (Council of Europe, 1998)

A large number of small municipalities acting as responsible entities and service providers may encourage associations, which may then be attractive to PPP. As indicated also by Camdessus (2003) there is a minimum size of a contract needed for private parties to become interested in a potential partnership. Added to this is that in cases of small-scale responsible entities often the expertise is missing to act as a full worthy partner in a PPP. Nevertheless, the need for private sector skills and finances may be felt harder if the municipality is smaller. Small municipalities find it difficult to maintain expertise so they are tempted to buy it in from private sector. Large municipalities, acting as responsible entities and service providers may be able to finance and manage systems very well, like in Sweden.

The responsible entities have several choices to make related to how they structure their relation to the service provider. It has something to sell, being the (temporary) right to serve inhabitants of a monopoly region with water services. It is this temporary right, which is the regulated transaction that makes up the market for delegated contracts. A system of regulation – informal or formal- is required to ensure that the public interest is served properly in transacting this right. The regulatory interventions in the market for delegated contracts are quite similar to those in the market for suppliers that also focus to create a level-playing field for competitors. Its' intention is to facilitate that the terms for competition are in place in transacting the right to serve a monopoly region.

The responsible entities need to structure their relationship with the service provider. In two consecutive studies across the European water sector, Eureau made this relationship their object of research (Eureau, 1992 and 1996). Although there are a number of differences between the Eureau studies and Euromarket, such as the exclusion of the wastewater sector in the Eureau studies, the concepts supporting the Eureau analysis are useful. Eureau managed to identify a number of communalities at pan-European level for the institutional relationship between the responsible entities and the service provider. These communalities provided a classification of management systems. Eureau's focus on a limited number of features reduced the large diversity of management systems in this manner. Although it needs to be acknowledged that systems in the same class might show significant differences. If we analyse the classification Eureau ends up with, two strong variables are paramount in shaping an institutional arrangement for the water supply and sanitation sector, namely:

1. Direct management or delegated management. The character of water as a good of public interest implies that the public sector (as the responsible entity) will assume the ultimate responsibility for service provision but the way that it arranges the management of the service provision is for the responsible entity to decide upon. It may choose to execute the management of the service provision itself, with a very limited degree of separation with a service provider. If the responsible entity decides to go for full divestiture as in England and Wales, also a situation of direct management is created since a new (private) entity takes over the service provision and assumes both the responsible entity's and the service provider's role. Another option is to rely on a service provider to execute the management of the services at arms' length, by providing the service provider with a certain degree of autonomy.

2. Public management or private management. Another dimension that reflects the institutional arrangement is whether the service provision is managed by public or by private actors. A responsible entity might choose to

involve the private sector to carry out the management of service provision, thereby bringing with it commercial attitudes, external financial resources and know how.

According to the EUREAU studies the identified groups may be placed on a continuum, where the link between the responsible entity and the service provider clearly decreases in strength from total dependence to total independence.

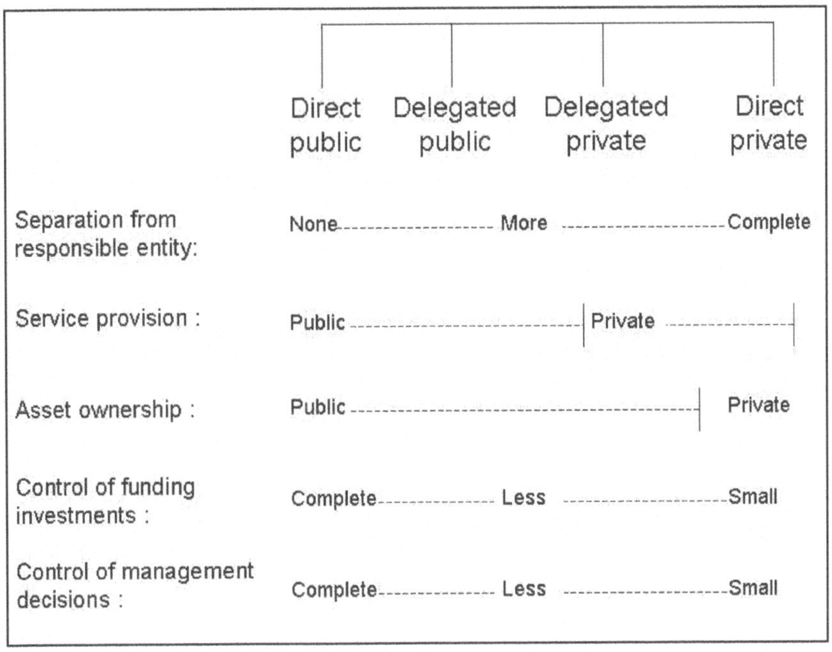

Figure 5: Comparison of the 4 defined institutional arrangements (IWA, 2003 modified by authors)

In some Member States responsible entities choose to have a service provider very close by, so they are able to directly influence its operations. If such an arrangement is chosen, the responsible entity needs to realise that there is a trade off. On the one hand, it might be easier for it to influence the service provider's behaviour, but it negatively affects the autonomy of the service provider. On the other hand, there is also a trade-off if the responsible entity delegates the service provision. Delegation might increase the degree of autonomy of the service provider, but the responsible entity needs to put in place a regulatory system to ensure that the operator does not abuse his autonomy status. Since it is common in the water supply and sanitation sector to separate ownership from managerial control, responsible entities find difficulty to ensure that the service providers work on their behalf, instead of letting their self-interest prevail. Therefore responsible entities set up regulatory impositions within which the service provider operates. In general, as the degree of separation between the service provider and the responsible entity increases, the more there is a need to put regulatory guarantees in place that the service provider keeps serving the public interest. In direct public management, regulation is accomplished largely through community governance (political oversight) rather than explicit regulation, while in delegated private management, the regulation is largely embedded in the terms and conditions of the service contract, which is negotiated with and endorsed by the responsible entities. For the most distanced arrangement, direct private management, regulation is typically accomplished through formal and independent regulatory authorities. The following paragraphs will discuss each of the four identified institutional arrangements in more detail.

Direct Public Management

In Direct Public Management, the responsible entity chooses to execute the management of the service provision itself. In this respect there is no real separation between the responsible entity and the service provider. As such direct public management can be characterised by the following distinctive features:

- Usually a department of the municipality undertakes the management of the water services, on behalf of the municipal council (being the responsible entity).
- Usually the service provider is not autonomous in selecting or hiring its' management.
- Since the service provider is a department of the responsible entity, the service provider is only in exceptions capable of making decisions on hiring and firing of staff.

- Almost always tariff setting is conducted by the responsible entity
- Usually the service provider is not autonomous in setting its' budgets, since they are part of the larger municipal administration and budgeting process.
- Usually the service provider is not autonomous in attracting external funds since its' access to funds is via the municipal administration.
- The service provider is not autonomous in deciding upon the quantity and quality requirements of investments, in view of the tight relation with municipal budgeting and investment planning.
- The responsible entity owns the infrastructure.
- Since the service provider has very limited autonomy, regulation and control is very direct from the responsible entity to the service provider.

In the past, Direct Public Management was by far the most widely adopted institutional arrangement that the responsible entities chose in order to organise the management of water services. The following table indicates which services are dominantly organised in this manner for each European country. As in all the clusters, variations between the Member States within this cluster are large.

Table 2: Countries dominantly organised as Direct Public Management

	The full range of water supply and sanitation services	***Only sanitation services***	***Only sewerage collection services***
Countries	Denmark, Luxembourg, Switzerland, Sweden, Austria, Northern Ireland, the Republic of Ireland, Finland	Germany	The Netherlands, Belgium

In many countries, for example Northern Ireland, this model is in transition towards alternative structures, but especially in countries as Luxembourg, Switzerland and Denmark the constraints for change are too strong for structural changes in the relationship between the responsible entity and the service provider to occur. In some countries still this urge to change can be found. For example, in Sweden smaller rural municipalities are rapidly merging their water and sanitation activities and entering Delegated Public Management arrangements to attain efficiency improvements and find a solution for the lack of funding for capital investment. A factor in some countries is the desire to create large players that can participate in the emerging market for abstraction and discharge in Europe, especially South East Europe. The main constraint to change in countries that are currently dominated by a Direct Public Management arrangement is the strong record of the incumbent service providers. Therefore the urge to change is absent. If change is planned, opposition mobilises in the form of labour unions, incumbent service providers and customer organisation, supported by the public. For example in Switzerland, with its' tradition of the public directly influencing political decisions through public polls, the opposition of the public towards liberalisation is large. In 2002 the public voted in a poll against liberalisation of the energy market, and after this result also liberalisation of the market for abstraction and discharge has not appeared on the political agenda.

Delegated Public Management

In Delegated Public Management, the responsible entity appoints a service provider to execute the management of water services on its behalf at arms' length. Often several neighbouring responsible entities (e.g. municipalities) combine the execution of water services for a region in one service provider.

- The service provider takes the form of a separate public company operating at arms-length from the responsible entity. The service provider's management is controlled as such by its' public shareholders. Although the service provider remains public, it is able to operate in a relatively autonomous fashion.
- Normally the nomination of the managers of the service provider is arranged for by the responsible entity.
- The service provider is autonomous in hiring and firing of staff.
- Tariff setting is usually delegated to the service provider
- The service provider is autonomous in setting its' budgets, although it is common that responsible entities have the final say on it since it affects tariff setting.
- As semi-autonomous entities, service providers have the ability to access funds beyond the responsible entity, although subsidies from the responsible entity are common for major investments.
- The service provider is autonomous in deciding upon the quantity and quality requirements of investments.

- The responsible entity normally owns the infrastructure. It needs to be noted that it is possible for the responsible entity to sell a minority of the shares to private parties. In that case still the institutional arrangement will be categorised as delegated public management. If the majority of the shares is sold to private parties, then a situation of devolving is occurring.
- Responsible entities are usually shareholders in the service provider, in order to be able to control the management. In some cases the responsible entity allows a minority shareholding from private investors.

The following table indicates which services are dominantly organised in this manner for each European country. As in all the clusters, variations between the Member States within this cluster are large.

Table 3: Countries dominantly organised as Delegated Public Management

	The full range of water supply and sanitation services	***Only water and sewerage treatment services***	***Only water services***
Countries	Italy, Portugal, Scotland, Greece	The Netherlands, Belgium	Germany

In these countries service providers operate in a more autonomous fashion. They remain within the public sector, although a possibility of minority private shareholding exists. Variations within this group of countries are considerable. In some countries experiments with delegated private management and majority selling of shares are undertaken but these are still on a small scale. For example, in Greece the two major water companies of Athens and Thessaloniki have allowed minority shareholding by the private sector and in Belgium the private sector is greatly involved in the wastewater treatment sector.

The main driver for change towards changing the relationship between the responsible entity and the service provider is the increased financial pressure due to investments required for rehabilitation and upgrading. For example, in Greece the need to improve the operating efficient of existing public sector bodies combined with the perceived failure of direct public management where it failed to cope adequately with a long drought from 1989 to 1993, has resulted in an increase in the separation between the responsible entity and the service provider. In the Netherlands, water utilities were historically set up under Direct Public Management, but were transformed in the last decades of the 20th century into delegated public management bodies. Today, initiatives are being considered in the wastewater sector for delegated private management but this is still in its infancy.

Delegated Private Management

In the Delegated Private institutional arrangement, the responsible entity appoints a private company for the management of tasks to be conducted on the basis of a time-bound contract (e.g. a lease or concession contract). In this respect the private entity assumes the role of the service provider.

- The private entity is independent from the responsible entity and acts as a contractor.
- The service provider is autonomous in nominating / appointing / selecting its' managers.
- The service provider is autonomous in hiring and firing of staff.
- The setting of the tariff structure is mostly arranged in the contract.
- The service provider is autonomous in setting its' budgets.
- Funding is arranged according to the type of contract. If it is a lease contract, the responsible entity will need to provide the funding for investment. If it is a concession contract, responsibility for funding is delegated to the private entity.
- The quantity and quality requirements of investments are regulated in the contract.
- The responsible entity retains ownership of the infrastructure. Although again the possibility for minority shareholding of the private sector is possible.
- The regulation is mostly decided in the contract, otherwise mediators and courts will be asked to solve conflicts

Delegated private management is becoming increasingly important in Europe, but only two countries can be dominantly characterised by it: France and Spain. The so-called French model attracts a lot of attention. The franchise bidding regulation is usually referred to as competition for the market; enabling a responsible entity to choose amongst several operators. The idea of competition for the market was first proposed by Chadwick (1859), although it was Demsetz who 're-discovered' Chadwick's work in 1968. In practical terms, competition for the market is established to grant an exclusive right to provide services for a given period of time. Competition for the market (gestion déléguée) is particularly common in France, where municipalities sub-contract their duties to a private company. Capital assets remain public property. If the winner is a private company, a Public-Private-Partnership is established, i.e. partnership between the public sector and private sector for the purpose of delivering a project or a service traditionally provided by the public sector (European Commission, 2003). The degree of private sector involvement varies, from the operation of assets to financing of new investments. If there is any competition it only occurs at the bidding moment, when several companies compete to win, while during the contracting period natural monopoly exists. The right is mostly assigned by the responsible entity on the basis of minimum cost for the provision of the services required. As a result, theoretically, delegation is made to a more efficient service provider and prices are not monopolistic. Prices tend to average costs if the number of competitors is several.

Major drivers found in Spain and France are pushing towards further extending the degree of separation between the responsible entity and the service provider. The major drivers are targeted primarily at increasing the degree of delegation, not so much towards entering direct private management. There seems to be few drivers and many constraints to also consider direct private management. Most of the constraints have to do with the resistance to place goods of 'public nature', such as water services, completely in the hands of the private sector. Delegation to the private sector softens the financial pressure that the public sector is coping with and the belief is that it enhances efficiency. Successful local and international PPP experiences also trigger others to initiate more PPP arrangements.

Direct private management

The Direct Private institutional arrangement is characterised by the fact that the public authorities take a major step back as responsible entity and limit themselves only to control and regulation. All tasks, responsibilities and ownership are placed in the hands of private parties. The direct private institutional arrangement has the following features:

- The private party assumes full responsibility and is independent from the responsible entity, apart from regulatory issues.
- The service provider is autonomous in nominating / appointing / selecting its' managers.
- The service provider is autonomous in hiring and firing of staff.
- The service provider carries out tariff setting although this is mostly subject to a regulatory regime. In England & Wales the role of the Office of Water Services (Ofwat), being the economic regulator for water and sewerage services, is very strong with regard to the tariff setting
- The service provider is autonomous in setting its' budgets.
- As a private entity, the service provider can access funds as it chooses.
- The service provider is autonomous in deciding upon the quantity and quality requirements of investments.
- The service provider owns the infrastructure.

Direct private management is still only applied in England and Wales, where the private party has acquired the complete ownership and management of the water and wastewater services.

The market for drinking water and sewerage services

Customers can be conveniently split into three groups: households, non- households and neighbouring utilities. The majority of households are connected to both the public water supply network and the public sewerage system. Non-households encompass a broad range of different customer groups ranging from major manufacturing sites, to commercial sites, to government owned sites. The markets for drinking water and sewerage services really consist of a myriad of specific sub markets, as drainage, foul sewage, trade effluent, potable water supply, non-potable water supply, water for fire-fighting, interruptible or standby water supply, peak or off peak water supply, and connection to network. The service is therefore not that homogenous and this has important implications for any liberalisation policy. These services are supplied to a range of different customers, such as households and non-households, including the manufacturing industry, commercial sites and public buildings. Competition in the market for drinking water and sewerage services occurs when the position of the incumbent service provider is contestable by new entrants or neighbouring service providers. In principle there are four options for introducing competition in the market for drinking water and sewerage services:

1. *Competing networks*. The high costs associated with installing competitive networks prevent to implement this option. For the water industry, duplicability of assets is not feasible. This form of competition in the market may be allowed in the form of inset competition. In this case, competition may occur in the form of competitive bidding to supply new groups of customers. Once these are connected, however, a monopolistic situation is established.

2. *Private supply*. Private supply occurs when one consumer (self-supply) or a group of consumers (co-operative supply) supply themselves rather than rely on an incumbent service provider. In this case the local monopoly is broken by the presence of these self-suppliers. A condition to make this option feasible is that water resources are

accessible to customers, which are implicit in the monopolistic structure. The regular form to dispose wastewater is through the sewerage network although on-site trade effluent treatment facilities are also common across the EU. In some regions manufacturing industries club together to construct and operate joint industrial wastewater collection and treatment systems. For some industrial wastewaters it may be appropriate to tanker it to a neighbouring merchant wastewater treatment facility. This will depend on the nature of the industrial wastewater, the volume produced and the distance between the source of the wastewater and the merchant facility. The ability of manufacturing industry to adopt self-supply is effectively determined by the environmental regulatory agencies that control the right both to abstract water from and discharge wastewater to the water environment. In some Member States it appears that the right to self-supply may potentially be curtailed by the granting of exclusive rights to the service provider. It may also be curtailed by excessive standby/reservation charges levied by incumbent service providers. Hence we can say that self-supply is a potentially important source of competition within manufacturing industry but it is less so for households, commercial and public buildings.

3. *Retail competition.* Retail competition occurs when an entrant takes over the service provision in an area while continuing to purchase bulk water from the incumbent service provider. In this case the new entrant simply exploits a differential price between bulk and retail supply and does not invest in distribution facilities.

4. *Common carriage competition.* Common carriage competition occurs when several water utilities use a single network to supply customers, and customers can choose their water supplier, like in telecommunications. Competition can occur between vertically integrated suppliers sharing access to a single network or between many retailers that share access to a single network and purchase water from competitive bulk suppliers. In order to allow several operators to use the same network a basic issue is the network access problem. This way of introducing competition stated that provision of services over the network is not monopolistic and can be done in a competitive manner. Several suppliers compete for customers using a single network. In this system, the players are the customers; the retail suppliers, who purchase water from bulk supplier and sell it to customers; the network operator, who manages water network, compares retail demand and bulk supply through a water balance market; and the bulk suppliers, who sell water to retail suppliers and communicate supply to water balance market. They abstract and treat raw water. In common carriage, the market serves to equilibrate water demand and supply through an adjustment in tariffs. This system is competitive only if new entrants have the possibility to use incumbent networks. The implementation of this form of competition entails a significant institutional framework to develop an efficient bulk supply and network access regime. Apart from institutional design, problems may arise concerning water quality. Sharing a common network entails mixing different kind of water of different chemical composition, bacteria and other parameters, and consequently contamination risks arise. However, minimum standard obligation, in terms of maximum levels of harmful substances, colour and turbidity, could reduce this risk less likely.

Obviously the voice and opinions of the customers play a role in how the markets for drinking water and sewerage services are shaped, as they are the ones that procure the services. Some consumers may lobby if they feel they will gain from restructuring. Households may lobby against it, if they feel they will loose from any proposed restructuring. In England such can be noticed with large industrial consumers lobbying for more competition while domestic household representative groups are lobbying for a more public character of the water industry. It is of particular interest to focus on the current levels of customer perceptions. Eurobarometer (2002) has undertaken a pan-EU market research exercise for the water supply service in 2000 and 2002. This highlights customers' perceptions of the service that they are receiving rather than an objective assessment of the service that they actually receive. The results from the survey indicate that in a number of Member States there is a high level of dissatisfaction about water pricing levels. Dissatisfaction levels reached up to 50% in France and Eastern Germany. On the other hand, with some notable exceptions, as Italy, Greece, Portugal and Eastern Germany, the perception of the quality of service appears to be relatively good. Generally, the perception of service levels appears to improve as you move northwards. This may reflect a variety of factors as climatic, historic investments, and institutional arrangements. In terms of overall satisfaction levels there are substantial differences between the top and bottom performers. Some of the poor performers as Italy, France and Greece, are currently in the process of changing or modifying their existing institutional arrangements. It is interesting to note that customer dissatisfaction does not always correlate with higher prices. Satisfaction with price levels generally reflects a perception of good 'value for money' and not simply low prices.

The complexity of tariffs can vary – reflecting the extent to which the market for drinking water and sewerage services is a local/regional one, rather than a national one. In the below presented Table 4 different tariff measures are identified as currently present all over Europe.

Table 4: Tariff structures across the EU Member States

	Block Tariffs	*Free block*	*No Fixed fee*	*Unmetered households*	*Regionally Averaged charges*	*Permission for disconnections*	*Social tariff*
Sweden						Yes	
Finland							
Denmark				Yes			
Austria			Yes			Yes	
Germany						Yes	
Switzerland						Yes	
Netherlands					Yes		
UK			Limited	Yes	Yes	Yes	Yes
France	Yes/No					Yes	
Luxembourg	Yes						Yes
Belgium	Yes	Yes			Yes (limited)	Yes	Yes
Spain	Yes				Yes (limited)		Yes
Portugal	Yes						Yes
Italy	Yes				Yes		
Greece	Yes						Yes
Ireland	*No Household Charges Levied in Ireland*					Yes	

The responsible entity always maintains some control over the tariff (re)setting process. Statistics from Eurostat (2004) show that the price to supply water services increased every year during the period of 1993 to 2003, averaging 4 percent per year. It is the nature of this control that needs further investigation. It can be controlled by: an organised competitive bid with formal contractual tariff resetting mechanisms (contract regulation), active regulatory control (rate of return or price regulation), passive regulatory control (investigation only if tariff increases proposed or complaints made by customers), or some form of self regulatory control (municipal rubber stamping of budgets or prices proposed by the service provider). The relationship is further defined by the extent to which the other aspects are controlled by the responsible entity. This will ultimately determine the degree of separation between the responsible and service provider. Indeed it is evident that as control of aspects as investment decisions, financing, senior management appointments and ownership are ceded to the service provider, the responsible entity generally takes a firmer grip on price control.

In England and Wales, for example, each individual market segment has approved charges associated with it. Other Member States may have more or less complicated charging arrangements. Surprisingly the water supply market is probably more varied across the EU than the sanitation market. In a number of EU Member States no distinction is made between household and non-household customers. All customers pay the same unit price for water supply. This can be contrasted with countries as Greece, Italy, Portugal and Spain with increasing block tariffs, or countries as England and Austria that have declining block tariff structures. The degree of complexity of the sanitation market is relatively uniform across the EU. Most Member States appear to make a distinction between foul sewage and drainage services – with the former being charged according to the volume of water consumed and the latter as some form of property tax possibly determined by site area. The majority of service providers also apply trade effluent tariffs where industrial customers are charged according to their pollutant load. The service providers also apply connection charges for sanitation and water supply services and charge for the reception of tankered waste from septic tanks and cesspools. Here the market is effectively characterised by the method of structuring the blocks, typically, but not always, based on the volume consumed. In addition, there may be individual contractual arrangements for large industrial customers, as can be observed in Germany, France, and United Kingdom.

The regulation of the prices can be exercised by many different types of regulatory bodies. In some cases the price control can be exercised by State Departments as in the case of Ofwat in England and Wales and the Ministry of Public Works in parts of Greece. In other countries the control of the prices is in the hands of Regional Government Bodies either directly, such as the Regional Governments of Spain, or appointed, as the Italian ATO authorities or the Dutch water boards. In price monitoring rather than setting, also other governmental departments appear to play an important, if somewhat passive role. For example, the Competition Authorities in Finland, Austria and Germany continuously monitor price-setting regimes in the water supply and sanitation sector, as does the Ministry of Economic Affairs in Belgium and Luxembourg. New central regulatory bodies, with a more supervisory role, have also recently been established in Portugal and Italy. However, as yet, they do not appear to have any price control responsibilities.

There are targeted social tariffs in a few Member States. In England and Wales households with specified medical conditions or who are on specified benefits with three or more children can opt for a capped tariff. In some municipalities in Greece there are reduced bills for households with three or more children. In the Spanish region Murcia for example pensioners can obtain a reduced bill. In the Belgium region of Flanders retired people, families on minimum income and disabled people are exempt from sanitation charges. The heterogeneity of the water supply and sanitation markets is also influenced by the extent of direct government subsidisation and the extent of cross subsidisation between customers. The types of subsidisation measures along with their presence across the EU are profiled in Table 5.

Table 5: Types of subsidisation measures

	Large (>30%) subsidies	*Reduced VAT*	*Reduced WWT*	*Income support*	*Targeted Assistance*
Sweden				Yes	
Finland				Yes	Yes
Denmark				Yes	
Austria				Yes	
Germany		Yes		Yes	
Switzerland	Yes	Yes		Yes	
Netherlands		Yes	Yes	Yes	
UK		Yes		Yes	Yes
France		Yes		Yes	Yes
Luxembourg				Yes	
Belgium		Yes	Yes	Yes	Yes
Spain	Yes	Yes		Yes	
Portugal	Yes	Yes		Yes	
Italy	Yes	Yes		Yes	
Greece	Yes			Yes	
Ireland	Yes			Yes	

More generally, three Member State groups can be identified according to the extent of tariff measures and subsidisation measures:

Table 6: The extent of tariff and subsidy measures by Member State grouping

Three groups of Member States	***The extent of*** *Tariff structure measures*	*Subsidisation measures*
Austria, Denmark, Finland, Germany, Sweden, Netherlands, Switzerland	Limited	Limited
Luxembourg, France, UK	Extensive	Limited
Belgium, Greece, Portugal, Italy, Ireland, Spain	Extensive	Extensive

Increased competitive activity will tend to unwind cross subsidies and possibly undermine direct subsidies. Competitive activity could result in substantial tariff rebalancing which could undermine public support for any proposed management system change. This may be an important market constraint in those Member States with extensive direct/cross subsidisation. It is also interesting to note that the nature of the institutional arrangement (or more specifically whether it is private or public in nature) does not necessarily correlate to the extent of subsidisation (especially cross subsidisation). For example, Member States with direct private management (e.g. England and Wales) and extensive delegated private management (e.g. France) appear to have important cross subsidies built into the tariff systems. This may be a reflection of the disparity in household incomes and the political desire to protect lower income families from what some experts like to call "water poverty". Another dimension of tarification that indicates of the heterogeneity of the market for drinking water and sewerage services is the element of time. The water supply and sanitation service provision could be considered as being time dependent. In other words the timing of the demand is just as important as the nature of the demand. For example, the service provided in summer is not the same as the water service provided in a non-critical period as the winter. Some Member States apply seasonal tariffs for water supply to reflect the importance of the timing at which customers make their demands for the service.

MARKET DYNAMICS

A myriad of different drivers and constraints (for and against liberalisation) exist across Europe at any given point in time affecting the structures of water supply and sanitation markets. While analysing these forces, a categorisation of driving and resistance forces can be identified according to three tiers (see Figure 6). The first tier of forces is related to the directly involved incumbent service providers and responsible entities. As these entities are either responsible for the service provision or managing the service provision, the influence they exercise is large on possible liberalisation. The second tier is the group of forces related to the more indirectly involved stakeholders, being the consumers, the press, the workers' unions and the private parties. More indirectly this group shapes the propensity and direction of liberalisation. The last tier that influences possible liberalisation of the water and sanitation markets has more the character of an exogenous factor, being the (perceived) experiences with liberalisation, either in other regions in the water sector or in other networking sectors. The influence of failure of success of flagship liberalisation project cannot be underestimated; they shape the perceptions of stakeholders and provide the context for policy making.

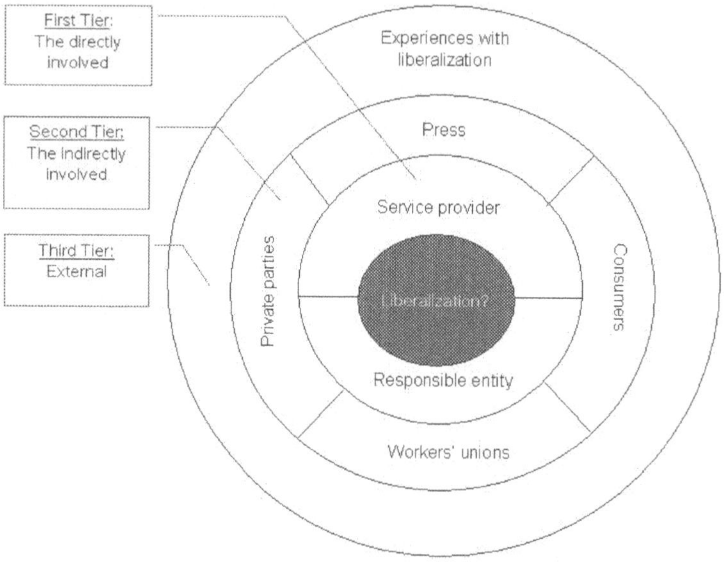

Figure 6: Overview driving and resistance forces towards liberalisation

In any given market, it is the balance between these driving and resistance forces that will determine the *propensity to change* the existing institutional arrangement. Historically, this balancing act has resulted in predominantly direct public management of the European water and sanitation sector. The nature of the driving and resistance forces will influence the *direction of change*. For example, the presence of powerful lobby groups for or against liberalisation may determine if the responsible government prefers to delegate the service provision to a public or a private entity.

The tier of entities that are directly involved

Two main actors shape the service provision, being the service provider that is executing the service provision, and the responsible entity that bears the final responsibility for the service provision to the public.

The service provider

Since it this entity that is currently in charge of executing the service provision, its' influence on any changes on the service provision is quite large. Several characteristics (see figure 8) of this incumbent service provider might trigger or discourage liberalisation. It is interesting to see that in different countries the same characteristics produces an opposite effect:

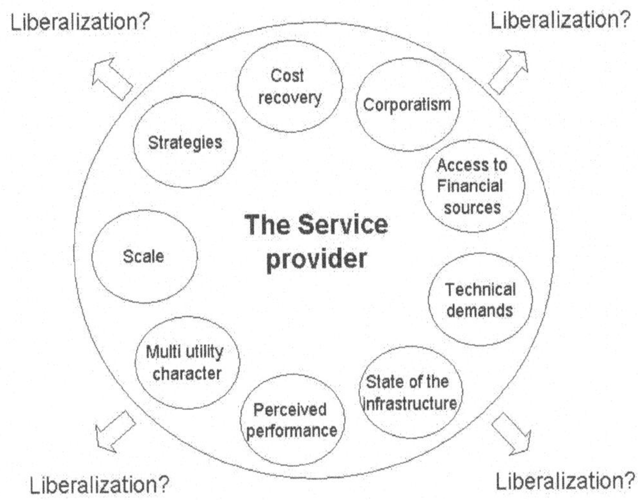

Figure 7: The service provider

- *The scale of the incumbent service provider.* For example in Sweden, Italy, Belgium, Switzerland and in the Netherlands it is believed that large incumbent service providers might be better able to maintain the required expertise themselves and avert interference from international operators. On the other hand, size also attracts private parties (Camdessus, 2003) since a minimum size is required to have a concession contract profitable in view of the transaction costs associated.
- *Corporatism of the service provider.* For example in Belgium, there is a situation where for a long time the service providers were used to work without political interference, except on retail price. The service providers as such do not specifically favour solutions that would increase competition, even if they remain passive to (partial) privatisation processes.
- *The strategies of incumbent service providers towards market expansion.* For example in Greece the ambition of the incumbents to exploit emerging market opportunities in the water sector, triggered partial privatisation of the two main water companies. An opposite effect can be witnessed for example in Scotland were worries about possible market entry by UK water companies that are searching for expansion, triggered a restrain on liberalising its market for abstraction and discharge. Also this can be seen in Northern Ireland where employees of the incumbent service providers are lobbying to maintain the existing distribution of economic rents.
- *The current level of cost recovery of the service provider.* If there are limited opportunities for full cost recovery, the incumbent is a less attractive target in the eyes of private parties, as for example in Northern Ireland.
- *The multi-utility character of the service provider.* An existing integration between energy, gas and water in one service provider might blur the distinctions between the sector characteristics. As such, a multi-utility character makes the influence of the ongoing liberalisation in the other networking sectors stronger (especially the implementation of the EC Directive 96/92 on energy liberalisation) since the perception will be more dominant that what goes for one in-house sector should also go for the other. An example of this driving force can be found in Germany and Switzerland.
- *The present state of the infrastructure of the service provider.* The need to upgrade existing infrastructure, leads to foreseen increased financial pressure. Liberalisation might pave the way for enlarged access to funds. This is at hand in many countries as Greece, Italy, the Republic of Ireland and the Dutch sanitation sector. If there is no real need to upgrade existing infrastructure, such as in Scandinavia, there is also no need to search for other financial sources. Apart from the state of the infrastructure also the readiness of sound information on the infrastructure can be important. For example there is a lot of sound information on the assets of Scandinavian incumbent service providers, which is particularly interesting for private parties since it enables them to make a sound risk assessment.
- *The (perception of) performance of the service providers.* If incumbent service providers are (believed to be) performing well as in Switzerland and the Netherlands, the urge to change might be lacking. While if there is, possibly due to high profile failures as the drought in Greece, a strong desire to improve the operational efficiency of existing public service providers in combination with a perceived efficiency of the private sector, this might trigger liberalisation.

- *The present access to financial sources of the service provider.* If the public incumbents currently have the possibility to access cheap government loans, as in the Netherlands, this might hinder liberalisation since a private party would not have access to these cheap government loans as it has increased costs of finance.
- *Technical demands on the service provider.* If there is a situation of increasing complexity of water and wastewater technology as for example in Greece, Italy, France and Germany due to environmental requirements on ground water abstraction and surface water discharge, there might be a need to involve private sector expertise and finance. Also for example the implementation of the European Directives affected in many countries, for example in Belgium, Portugal and the Netherlands, the required technical complexity and created subsequently a demand for private sector expertise and involvement.

The responsible entity

The responsible entity that is currently in place is another major factor that needs be acknowledged. Several drivers and constraints (see Figure 8) are identified related to its' position, set up and abilities.

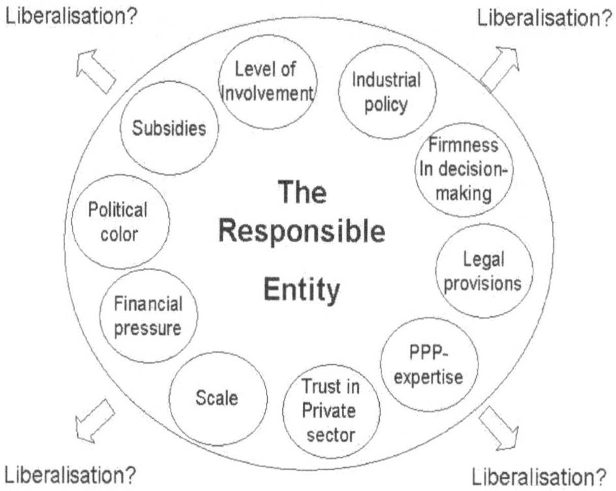

Figure 8: The responsible entity

- *The current level of influence of the responsible entity on the service provision.* If the incumbent service provider is highly dependent on government vagaries, this might seed the urge from the service provider to reduce the political vagaries in service provision by searching for alternative institutional arrangements. For example in the Scandinavian countries, the incumbent service providers are quite independent from government vagaries, and as such do not feel the urge to change. In other countries you see the opposite effect, such as in Northern Ireland and the Republic of Ireland, where the strong involvement of the central government, either as a direct service provider or a major source of finance, prevents private sector involvement.
- *The current level of subsidies the responsible entity provides to the service provider.* If subsidies are in place, as for example in Austria, the government might be more inclined to liberalise the sector trying to relieve the financial burden of subsidizing.
- *The current political colour of the responsible entity.* In general one could say that a reigning liberal or right wing national/local political majority tends more towards adopting a pro-Liberalisation approach, while a reigning social democratic or more left wing national/local political majority is more resistant towards liberalising public services. Although one could say that political parties are acting within the national political dogmas. For example in Sweden and Belgium liberalisation is neither by right wing nor left wing politicians advocated, while in Greece both political sides choose liberalisation.
- *The current financial pressure on the responsible entity.* For example, conforming and/or joining the Economic and Monetary Union and the stability pact criteria pressured the Greek, Spanish and Belgium governments to reduce public debt and triggered the partial privatisation of water companies to generate additional financial resources. Other examples are the Republic of Ireland and Portugal, in which a decline in EC Regional Funds led to a search for alternative sources of finance. Also in Swedish and German municipal governments, the desire to reduce the financial burden on the public budget triggered the involvement of private parties.

- *The current scale of responsible entity.* A large number of small municipalities acting as responsible entities may encourage associations, which may then be attractive to PPP. For example in France, Belgium, Portugal and the Republic of Ireland, PPP is used to create multi-municipal structures and hence gain economies of scale.
- *The current level of trust the responsible entity has in the private sector.* For example, the long history of interaction and hence trust between private and public sectors in France provides a good basis for partnerships, while for example in the Republic of Ireland there is only little confidence and trust in the private sector possibly due to a lack of historical relations.
- *The current level of PPP-expertise of the responsible entity.* If regulatory expertise or general expertise about PPP is lacking, such as for example in Sweden, this attributes to the hesitance of introducing private sector involvement, while if it is available or readily obtained from neighbours, such as in Scotland at least this element does not form a constraint to a future liberalisation.
- *The firmness of the responsible entity in decision-making.* If politicians are unable or unwilling to take the blame for an increase of tariffs they might be inclined to use the private sector as an instrument to undertake a politically difficult task. While if politicians are firm in their decision-making, such as in Scandinavia, Denmark and Luxembourg, where politicians are willing and able to introduce tariff increases and undertake considerable efforts to comply with the Water Framework Directive.
- *The current legal provisions in place that encourage PPP.* For example in the Republic of Ireland, Spain and some German Länder, private sector involvement in the water and sanitation sector is already currently arranged for, while in other countries, such as the Netherlands, Italy and Sweden, there still are major legal constraints to PPP. On the other hand the current "pipeline developments" at EU level indicate that the European Commission is looking for ways to heighten the exposure of the European water sector to competitive forces. If these developments would materialise in legislation it would definitely trigger liberalisation processes.
- *The industrial policy of the responsible entity.* For example in Belgium, the industrial policy of the government triggered an up scaling of regional companies to allow them to develop activities abroad. Politicians hope that the current investment in the wastewater sector will benefit to regional construction companies and will contribute to a regional capacity building in the WSS sector. Also the German Bundestag initiated a so-called "Sustainable water management policy" (BT-Drs. 14-7177 of 21 March 2002) that targets at reaching optimisation and the efficiency gains through modernisation of the present system. The constituent policy elements will affect the organisation of the sector and may lead to shifts in the present structure as well as to increased competition and private involvement. In contrast, Spanish politicians use the argument of public responsibility for sustainable aquatic ecosystems to enhance the need for public management responsibilities in the water sector.

The indirect involved stakeholders

The second tier of forces affecting liberalisation is the influence that stakeholders and pressure groups are able to exercise. The following main groups can be identified:

- *Consumers*: If the consumers pay a relatively low price for water and sanitation services, as in Switzerland or Northern Ireland, or if they feel they might loose from any proposed restructuring, as in the Netherlands and England and Wales, the consumers might be inclined to block changes through public consultations. A specific group of consumers that exercises influence are the large industrial consumers, they might be able to install a lobby if they feel they will gain from restructuring, such as in hand in the Netherlands and England and Wales with regard to expanding competition.
- *Workers' unions.* The fear of job losses due to private sector involvement in the water and sanitation sector might trigger them to lobby against liberalising the sector, such as happened in Northern Ireland.
- *Press.* If the press is extensively involved, such as in Sweden, this obstructs the liberalisation process since some politicians might be inclined to play the public sentiment on this controversial topic. If the media does not feature the liberalisation, such as in Greece, implementing private sector involvement might turn out easier.
- *The private sector.* Lobbying by private companies who want access to economic rents might be a driving force for liberalising the sector, such as at hand in France and Spain. Also for the Belgium situation, an unpredictable element in the transformation process of the water sector is the interest that private parties, as Suez and Veolia, might have in their regional operators.

Perceived liberalisation experiences

Apart from the direct and indirect incumbent parties within the water and sanitation sector, an import factor that influences possible liberalisation is the experience with liberalisation elsewhere, either in other parts of the country, other countries or other sectors. Even before liberalisation is on the agenda, one of the first things one will look for is to assess the success or failure elsewhere, and to use these learning experiences for their own situation. Local evidence that PPP in neighbouring utilities in the water sector is working well might convince other to also enter a PPP process, as in Sweden or the pioneering Dutch DBFO contract in Delfland. Also the experiences abroad of private parties successfully executing the water services provision might convince. For example in Scotland, the efficiency of English water companies is shown as a success and something to aim for. Interesting enough, the experiences in England and Wales with involving private parties are used, by for example the Swedish media, as a negative experience of private sector involvement in the water sector. Also in international benchmarking the Swedish water sector comes out positively. On the other side if international benchmarking surfaces a poor performance, such as for the Greek public sector, this might trigger private sector involvement. Also liberalisation experiences in other sectors might provide either a driving force or a resistance force towards liberalising the water and sanitation sector. Local evidence that PPP works well in others sectors, as in Scotland, definitely triggers PPP in the water and sanitation sector. On the other hand when liberalisation in other sectors turns out negatively, as for example the privatisation of electricity in Northern Ireland, of the railways in the Netherlands, opponents use these disappointments as arguments against liberalisation of the water sector. Another influence on the perception on possibly policy development with regard to liberalisation is shaped by examples of corruption related to private sector involvement experiences (for example Grenoble and Milan).

CONCLUSION

The European water supply and sanitation markets are both complex and multi-faceted. In practice it consists of four closely interrelated markets – the market for drinking water and sewerage services, the market for suppliers, the market for abstraction and discharge, and the market for delegated contracts. It can be concluded that some of the identified markets are more open to competition than others. Hence, it is important not simply view the water supply and sanitation market as one amorphous whole but as a set of services with associated transactions that require detailed consideration on a case by case basis. The potential for competition *in* the market appears to be restricted to a relatively small part of the infrastructure system, being those parts that do not exhibit the potential for major market failure. It appears that the majority of transportation/storage network is monopolistic in character. Competitors in the potentially competitive parts of the market will therefore require access to these assets via unpractical common carriage arrangements. The potential for common carriage is also hindered by the need for unbundling and the high regulatory costs associated with the need to minimise information asymmetries and the need to maintain coordination following possible deverticalisation. Adequate regulatory procedures also need to be introduced to deal with the management of the health and environmental externalities. Access pricing methodologies will also need to be developed to deal with the subsidies that are inherent in the water supply and sanitation market. All of the above regulatory activity to control potential market failures will be very costly to implement with unproven market efficiency rewards. Potential market failures are not insurmountable but remedies do involve potentially substantial transaction costs. The underlying question is whether these costs are outweighed by the potential economic benefits of this form of liberalisation.

Public management still predominates in the EU water and sanitation market, although there has been some change in the nature of this public management. For example, there has been a shift from direct public management to delegated public management in Belgium, Switzerland, the Netherlands and Germany. It is evident that there has generally been no large-scale move toward direct or delegated private management in most Northern Member States. This is in contrast to a number of Southern Member States, as Spain, Greece, Portugal and Italy, where private sector participation has gradually increased over the past decade. The increase in private sector participation in the Southern Member States is mirrored in the sanitation sector. Interestingly, and in contrast to *water supply*, in a number of more Northern Member States, as the Netherlands, Belgium, Austria, Ireland and Scotland, there also appears to be increasing interest in private sector participation within the *sanitation* sector. In the two Member States with the highest level of private sector participation, France and the UK, there have been limited moves to increase the competitive pressures on the incumbents.

In determining the extent and limitations of competition in the water supply and sanitation markets, the key issue appears to be the assessment of possible economic benefits of liberalisation when compared to the transaction costs associated with the need to control various possible market/regulatory failures. This assessment of costs and benefits can only be undertaken on a case-by-case basis, as it will be determined by the conditions that exist prior to any proposed change. Essentially the optimal liberalisation path (if there is one) will be different for each local or national context. It is evident from our analysis that liberalisation is either:

- Very difficult in the case of competition *in* large parts of the market as a result of the potential for severe market failure, and the associated large transaction costs; or
- Particularly intensive and possibly costly, from a regulatory perspective, in the case of competition *for* the market, or comparative competition to remedy the more minor market-regulatory failures.

REFERENCES

1. Camdessus, M., and J. Winpenney (2003), Financing Water for All. Report of the World Panel on Financing Water Infrastructure.
2. Case, K.E. and R.C. Fair (1992), Principles of Economics. Prentice-Hall International Editions.
3. Chadwick, E. (1859), "Results of different principles of legislation in Europe: Of competition for the field as compared with competition within the field of service". Journal of the Royal Statistical Society, series A22, pp.381-420
4. Council of Europe (1998), The size of municipalities, efficiency and citizen participation. Local and Regional authorities in Europe. No 56.
5. Eurobarometer (2002), Consumer' opinions about services of general interest. Special Edition Report drawn up by the European Opinion Research Group for the European Commission "Health and Consumer Protection Directorate General.
6. EUREAU (1992), Management Systems of Drinking water Production and Distribution Services in the EC Member States.
7. EUREAU (1996), Management Systems of Drinking Water Production and Distribution Services in the EC Member States.
8. European Commission (2003), Guidelines for Successful Public-Private Partnerships. European Commission Directorate-General Regional Policy
9. Eurostat (2005), European Business. Facts and Figures. Data 1998-2002. Office for official publications of the European Communities. Luxembourg.
10. Eurostat (2004), Quarterly panorama of European Business Statistics. Office for official publications of the European Communities. Luxembourg.
11. Gustafsson, J.E. (2001), Public Water Utilities and Privatisation in Sweden. Working Paper ESPSU Public Service Conference. Brussels.
12. IWA (2003), International Water Utility Leaders Forum, charters on sustainable water management, governance and regulation. 1st draft.
13. Juuti, P.S. and T. S. Katko (Eds.) (2005), Water, time and European cities. History matters for the futures. Fifth framework. European Commission
14. Kotler, P. (1998), Marketing Management: Analysis, Planning, Implementation and Control. Prentice-Hall. New Jersey, USA.
15. Lannerstad, M. (2002), Water Supply and Sanitation in Sweden: A Public Trust. Stockholm Water Front Magazine. SIWI.
16. Lannerstad, M. (2003), "Sweden. Country Report" in Aqualibrium: European Water Management between Regulation and Competition. Edited by S. Mohajeri, B. Knothe, D. Lamothe, J. Faby. For the Directorate-General for Research Global change and ecosystems.
17. Lipsey, R.G., P.O. Steiner, and D.D. Purvis (1987), Economics. Harper & Row Publishers. New York. USA.
18. SUEZ (2005), Interview of Marco Schouten with Jacques Labre and Jean Pierre Maugendre (SUEZ). Paris. September 30, 2005.
19. Thomas, D.A. and R.R. Ford (2005). The crisis of innovation in water and wastewater. Edward Elgar Publishing Limited. UK.
20. VNG, UvW and VEWIN (2005), Samenwerken aan water. Folder of the Vereniging van Nederlandse Gemeenten (VNG), the Unie van Waterschappen (UvW) and the VEWIN. June 2005

Chapter II

Policies Regulating the Water Supply and Sanitation Sector in Nine European Countries

David Aubin and Frédéric Varone

INTRODUCTION

Even not so much liberalised or privatised the organisation of the water supply and sanitation sector changed much in the past two decades. A comparative study of the different organisational arrangements at national level was deemed appropriate to highlight the trends in the evolution of water management. It brought the contextual elements within which water actors' strategies took place and from which scenarios were designed. Precisely, the ambition of this chapter is *to present common trends and differences in the current administrative regulatory frameworks that govern the WSS sector* at national level in EU countries. This formal and institutional analysis participates in the common attempt to identify the future trends in the evolution of the sector, in particular towards (more?) liberalisation. Rather than presenting a necessarily brief and narrow description of each national arrangement governing water management, a comparative policy analysis was conducted from synchronic analysis of the legislation at national level. It turns around the following research questions: Is there a trend towards harmonisation of the regulation and liberalisation of WSS services across EU countries? Do we observe rather an international convergence or a national or sector-based path dependency? *What are the main similarities and differences in the regulation of WSS services observed between the selected countries?* Some broad tendencies come out from the apparent impressionistic picture of institutional arrangements governing water supply and sanitation, notably in terms of choice of policy instruments, organisational arrangements between public actors, suppliers' statuses, financial flows or liberalisation practises.

This presentation of the water legislation and regulation in the EU Member-States uses the *Comparative Policy Analysis* as theoretical background. Applying this approach consists in examining *"how" and "why" governments pursue particular courses of action or inaction in the water and sanitation sector* (Heidenheimer et al. 1990). It aims at providing a broader empirical knowledge based on *how public policies are designed in the WSS sector*, i.e. with an emphasis on the content of policy (Linder and Peters 1988; Schneider and Ingram 1990; Dryzek and Ripley 1988; Bobrow and Dryzek 1987). The paragraphs below describe the main theoretical concepts and methodological tools mobilized for describing water policies adopted and implemented by nine EU-Member-States.

Various public policies are conducted within the WSS sector. A *public policy* is defined as a series of intentionally coherent decisions or activities taken or carried out by different public –and sometimes private actors–, whose resources, institutional links and interests vary, with a view to resolving in a targeted manner a problem that is politically defined as collective in nature. This group of decisions and activities gives rise to formalised actions of a more or less restrictive nature (outputs) aimed at modifying the behaviour of social groups presumed to be at the root of the collective problem to be resolved (target groups, e.g. polluting industries) in the interest of the social groups or the

 ISBN: 1843391139. Published by IWA Publishing, London UK

environment who suffer the negative effects of the problem in question (final beneficiaries, e.g. producers and consumers of drinking water, fishes). Thus, the term public policy refers to a large number of legislative and administrative activities aimed at the resolution of real problems. Most modern legislation is only effective when the political, administrative and social actors involved in the different institutional arrangements are involved in decision making. The desired effects are only attained, however, in the aftermath of a group of complex decisions. It is this set of decisions and activities that is defined here as a "public policy". *Four main public policies are distinguished between drinking water production and distribution, sewerage and sanitation that cover the whole chain of the WSS sector.*

Seven specific elements of the four WSS public policies are systematically scrutinised: policy *problems*, policy *objectives*, policy *instruments*, *institutional arrangements*, *target groups*, policy *outputs* and *outcomes* and, the *model of causality* of the public policy. A public policy presupposes the formulation of a political-administrative programme, i.e. the selection of objectives, instruments and procedures to be implemented in order to resolve the problem under consideration. This suggests that a policy is based on a – usually implicit, partial or indefinite – model of causality. The *model of causality* designates the actors' behaviour held responsible of the problem (causal hypothesis) and points out the instruments most appropriate to influence this (target groups) behaviour (intervention hypothesis). This kind of analysis makes easier to discern the links between the different actors and the ways in which they are altered in the aftermath of public intervention and illustrates the *dynamics of the public intervention.*

In order to analyse the *policy problem* relevant to the public sphere, it is necessary to adopt a constructivist approach (Berger and Luckmann 1966; Cobb and Elder 1983; Steins and Edwards 1999; Swyngedouw 2004). In fact, no objective fact constitutes a problem in itself. All social reality should be understood as a historical construction, situated in time and space. It is never a question of denying the objective conditions that constitute a problematic situation (for example, the existence of an elevated level of effluent), but of stressing that these established facts only represent one of the dimensions that constitute a social problem. *In order to simplify the coding of the policy problems defined by the EU-Members States in the WSS sector, we classify them in a limited number of categories: water availability and supply, drinking water demand, quantity problem, quality problem, conflicts between drinking water production and agriculture and industry.*

Each public policy includes a definition of a more or less explicit goal, on the basis of which the public intervention may be conceived. The *policy objectives* define the status to be attained by the adopted solution which would be considered as satisfactory. They describe the desired social status in a field of action once the public problem is resolved. *The three components of sustainable development are mobilised as types of policy objectives within the WSS sector: environmental objectives (surface water and groundwater quality/quantity, prevention against pollution, wetlands protection, etc.), social objectives (good quality, supply to all, price affordability, etc.) and economic objectives (cost-recovery, infrastructure financing, etc.).*

The *policy instruments* define the modalities of intervention or measures planned to fulfil the objectives of a public policy. These elements define the level of interventionism and the type, scope and quality of the proposed public intervention. Policy instruments can take a number of forms, usually classified in four categories. The *prescriptive* mode is based on bans, general obligations (e.g. to limit discharges of hazardous substances or to report regularly on tap water quality) and the allocation of various rights, which may be the object of sanctions (e.g. fines) in the case of non-compliance. The *incentive* mode works on the basis of financial payments aimed at influencing the behaviour of target groups by means of the "price signal". The incentive may be of a negative (e.g. tax, incentive levy) or positive nature (e.g. tax relieve, subsidy) with the intention of (re)distributive effects. The *persuasive* mode uses an information strategy to convince target groups of the proposed objectives (e.g. campaign on rational water use). The last intervention mode involves the *self-regulation* of target groups. In this case, State intervention consists in supporting the organisation of target groups and the "private" definition of rules of conduct (e.g. professional standards, code of good practice). *This empirical study applies to these four categories of policy instruments for describing the content of the intervention mode within the production, distribution, sewerage and treatment policies.*

The *target groups* of a public policy are made up of people and organisations whose behaviour is judged politically to be the (in)direct cause of the collective problem that a given policy aims to resolve. Consequently, the target groups' decisions and activities are the subject of concrete state intervention. The policy in question will impose obligations on them or grants them rights. The legislator and/or administration reason that, as a result of such measures, the target groups will alter their behaviour and that the collective problem can be resolved or at least mitigated. *The main target groups identified in WSS policies are the water suppliers, the final consumers, industries and farmers.*

The *institutional arrangement* of a policy represents the structured group of public and para-state actors who are responsible for its implementation. It results from decisions on the distribution of formal competencies and the allocation of various resources (i.e. financial or human) to the administrative units involved. *In order to simplify the empirical analysis of the policy networks, the study focuses mainly on two dimensions: the government level (national, regional, local) and the (public, private, mix) status of the administrations, para-state bodies (independent regulatory agency) and suppliers involved in the WSS sector. This allows an identification of the main challenges for the vertical and horizontal coordination within the policy networks.*

Policy impacts must also be considered. The formal implementation acts of a policy are defined as *outputs*. They are the set of end products of the political-administrative processes which, as part of the scope of its implementation, are individually aimed at the members of the defined target groups. The policy *outcomes* are defined as all of the effects – desired and undesired, direct and indirect, primary and secondary, etc.– desired and undesired, direct and indirect, primary and secondary etc in relation to the public problem to be resolved that are attributable to the changes in the behaviour of target groups, which are triggered in turn by the implementation acts (outputs).

The Comparative Policy Analysis Approach presented above is applied systematically for analyzing the WSS policies in *nine EU Member-States*. Each national case study required a collect of information and presentation of findings structured according to a common framework, insisting on policy elements and organisational arrangements currently in force (synchronic analysis). Data collection is based on secondary and primary legislation, mainly legal text and completed with interviews of national experts. In a second phase information was synthesised within comparative tables in order to allow the comparison. The countries compared are *Belgium (B), Germany (D), Spain (E), England & Wales (E&W), France (F), Italy (I), the Netherlands (NL), Portugal (P), and Sweden (S)*. These countries were selected given their representativeness of the water policies conducted in Europe. Thus, by taking into account the most different institutional arrangements, water situation and political systems, the comparative study seeks to isolate the relative influence of the different dimensions on the evolution of WSS policies.

The organisation of the chapter follow the traditional presentation of a policy design, beginning with the problems related to pressure on the water resource and policy objectives, and then detailing the policy instruments used as well as the organisational arrangements that supports or result from policy implementation. This leads thereafter to the observable impacts of public intervention and ends with more general perspective on the similarities and differences observed in this comparison of the different national water management structures.

DEFINITION OF OBJECTIVES IN WATER SUPPLY AND SANITATION POLICIES

Pressure on the resource regarding water availability, pollution and demand, is differing much between countries. Although these differences in social problems raised should be reflected in policy objectives, the case is not so clear.

Policy Problems Observed and Outlined

The identification of the main (objective) problems or issues to be solved by the WSS policies is based on *actual water uses and rivalries between users*. An assessment of the relative pressure of drinking water abstraction on the water resource comes in complement. At first sight, we observe a great disparity in the availability of the water resource. From Belgium (1,619 m³/capita/year) to Sweden (20,200 m³/capita/year), the European countries have unequal endowments in water resources (see table 1). Paradoxically, *water is not scarcer in Southern countries than in Northern ones*. Belgium, Germany and UK have less water resource available per capita (respectively 1,619, 2,218 and 2,694 m³/capita/year), than Italy, France[1], Spain, and Portugal (with respectively 3,040, 3,265, 2,797 and 7,281 m³/capita/year) (Eurostat 2003a), a fact partly explained by the presence of the Alps and the Pyrenees (i.e. melting of ice and snow). These data, aggregated at country level, also *hide situations of local scarcities* as much in Southern countries as in Northern ones where important quantitative pressure can be put locally on the resource[2].

The total water withdrawals complete the view on the disparities observed between countries (Eurostat 2003a). In absolute terms, the most important volumes of abstractions are located in the most populated countries (32,323 mio m³/year in France and 40,590 mio m³/year in Germany). In Southern Europe, Italy withdraws 55,702 mio m³/year, while Spain withdraws "only" 26,053 mio m³/year. At the other extreme, the Netherlands has a low level of abstraction (4,655 mio m³/year), a position confirmed when looking at the level abstraction per capita (300 m³/year). Belgium is not under "water stress" (Gleick 1993), when looking at abstraction per capita (730 m³/capita/year) rather than at resource availability (1,619 m³/capita/year, i.e. under the threshold of 2,000 m³/capita/year). It even scores better than most countries (495 m³/capita/year in Germany and 553 m³/capita/year in France). In relative terms (per inhabitant), Southern countries obviously withdraw much more water than Northern ones. Spain has a relatively low level of abstraction (656 m³/capita/year), compared to Italy (976 m³/capita/year) and Portugal (1,102 m³/capita/year).

---

[1] France has a large opening on the Mediterranean Sea.

[2] It is worth mentioning important intra-country differences such as in Belgium, where Flanders has low and depleted water reserves for drinking water production, or in Spain where the rising importance of water transfers increases water availability in some regions.

Table 1: Drinking water supply and demand

	Countries								
	B	*F*	*E*	*NL*	*S*	*I*	*P*	*D*	*UK*
Density (inhab./km²) (2001)	337.1	108.8	79.8	473.7	21.6	192.2	112.1	230.6	241.3
				WATER SUPPLY					
Resource availability (m³/cap./y.)	1,619	3,265	2,797	5,788	20,200	3,040	7,281	2,218	2,694
Withdrawals (m³/capita/year)	730	553	656	300	303	976	1,102	495	300
Withdrawals (mio m³)	7,443	32,323	26,053	4,655	2,688	55,702	11,090	40,590	15,884
				DRINKING WATER CONSUMPTION					
Total (mio m³)	559,4	4,000	3,784	1,248	730	2,100	5,785	4,859	5,472
Per capita (l/day)	113	165	174	129	162	236	184	129	134

- Source: (Eurostat 2003b)

In general, water is mainly used for the cooling of power plants (for more than half of total abstraction), except in Southern countries where irrigation takes the most part of water (82% in Spain, 75% in Portugal and 48% in Italy) (Eurostat 2003a): this means that this water actually used is also consumed (which is not the case for cooling activities in which water goes back directly to the resource). Abstractions of water intended for human consumption represent varying shares of the total abstraction from one country to another, from 7.86% in Portugal to 37.67% in the United Kingdom. In average it turns around 15%. Thus, the pressure of the WSS sector on the resource should not be overestimated when compared to industrial or agricultural uses[3].

A direct link cannot be established between absolute availability of water and withdrawals per capita. A country that has more water available does not necessarily withdraw more from the natural environment. The Swedish has the more important availability (20,200 m³/capita/year), but withdraws far less (303 m³/capita/year) than a Portuguese (availability of 7,281 m³/capita/year and withdrawals of 1,102 m³/capita/year). Furthermore, no immediate correlation exists between population density and anthropogenic pressure (ratio abstraction/availability). The Netherlands do not necessarily have a high ratio because of a high population density, as is the case in Belgium or Italy, and Spain faces high pressure with a low population density. The natural conditions (resource availability) and the share of irrigated agriculture in water withdrawals seem to be best explanatory factors of the situations observed between countries.

Concerning supply in drinking water, two third of drinking water is produced from aquifers. This rate represents 85% in Italy. All countries partly use surface water to produce drinking water, but surface water is used in majority only in Spain and the UK, with respectively 75.83% and 70.34% (Eurostat 2003a). Concerning demand, the volume of domestic consumption is very dissimilar, varying from simple to double in relative terms (in litres per capita). Belgium has the lowest rate of consumption with 113 l/day, while Italy has the highest with 236 l/day. Self-supply partly explains the variance (e.g. with watering from drills) as well as uses for tourism that increase the volume[4].

When looking more precisely at (qualitative and quantitative) problem pressure and policy problems that the drinking water sector meets, the main rivalries in resource appropriation occurs with agriculture. *Pollution from agriculture has much impact on water quality*, especially intensive farming that requires fertilizers and pesticides that leak into the ground or rivers. This pollution pressure is pointed out as a major one in every country. Manure disposal causes particularly acute pollutions. In certain regions (e.g. Flanders in Belgium, most parts of the Netherlands, Baix Ter aquifer in Catalonia and Brittany in France), intensive pork and poultry husbandries produce manure surplus that is spread on lands not for yields improvement but economic reasons. A second problem with agriculture is the *over-exploitation of the resource* (e.g. in France, Spain, Italy and Portugal), with severe threats of salinity and desertification in Spain.

Industrial pollution of *surface and groundwater remains equally important in every country*, in particular discharges of hazardous substances in spite of generalised permit schemes. Concentrations of residues in surface and groundwater keep high levels, even at the tap, and are found in the environment, notably in fish, birds and mammals. Nutrients are a second source of industrial pollutants, even if much effort of reduction was consented in particular in the food sector.

Rivalries are quite similar between countries, even if the hierarchies of policy problems differ. In Spain, the most acute problem against drinking water supply concerns water quantity with over-abstraction of farmers in particular in

[3] It is true however, that domestic consumption is an important pollution source (e.g. discharges of nitrogen) and that most of industrial withdrawals are for cooling purposes.

[4] "The amount of water used per capita per day is calculated using the total population of the country. As a result, the quantity of water used per capita is underestimated for countries where a significant proportion of the resident population is not connected to the distribution network. The quantity may also be overestimated in countries where a substantial amount of water is used for tourism" (Eurostat 2003b: 3).

aquifers (this is true also in the Lille aquifer in France). In other countries, agriculture once again poses the problem but in qualitative terms with excessive disposals of fertilizers and pesticides. Industrial pollution comes close behind.

Policy objectives

In an analytical perspective, policy objectives are classified according to the three dimensions of sustainable development (ecological, economic and social). As a result, it appears that *ecological and economic objectives are converging*, while social objectives are dispersed. The quality of the water resource as well as improvements in efficiency and productivity gains are the leading objectives in most countries. These objectives already fit with the Water Framework Directive (WFD). In the social field (e.g. price affordability or improvement of the standard of living) however, disparities are numerous, a situation also reflected in the absence of a true social dimension in the WFD (see also EUROMARKET 2003).

The European countries are *converging on ecological objectives* with a priority on ecological protection everywhere. The objective of a *good status* for both surface and groundwater, as stated in the WFD, is already transposed in all the countries studied. This convergence on ecological objectives is not so recent. For long, Spain has been concerned by the restoration of polluted groundwater or the protection of surface water and dam reservoirs for drinking water production purposes, Sweden about the protection of ecological areas, in particular for water abstraction, and Germany has aimed at maintaining or restoring the ecological balance of waters.

Ecological objectives have not been developed with the single aim to protect nature. In many countries, they are strongly correlated with the need to *secure drinking water supply* (e.g. in Belgium, Sweden, Spain and Germany). In part, they conform to the 1975 directive on the protection of surface water intended for the abstraction of drinking water (European Communities 1975), which set quality standards for raw water used in drinking water production (see chapter 3). In order to reach these standards, Belgium, France, Germany and England & Wales notably regulated manure disposal and intensive catering far beyond the requirements of the 1991 Nitrates Directive (European Communities 1991b). In these countries, the concentration of nitrogen and phosphorus in aquifers has attained worrying levels and is hard to tackle. In sum, the European law contributed to the development of ecological objectives long before the WFD was enacted, but it is certainly not the only condition explaining this development.

Besides this convergence in ecological objectives, the *protection of water ecosystems is not so widespread.* In fact, the protection of wetlands is not systematically referred to. Wetland protection only seems to be a preoccupation in Belgium, France, Spain, Germany and England & Wales, in spite of the international law that has developed since the Ramsar Convention of 1971. Some other particular objectives are pointed out in different countries. One can also notice the Dutch specific objective of "giving space to water" (*Ruimte voor Water*), also recently transferred in Flanders (Belgium). This consists in developing a water management plan that respects the natural flow of water, in particular in freeing the major bed of the river, as a result of coordination with land-use planning. The objective is not strictly environmental, but closely tied to flood protection.

Convergence on the economic objectives is also observed, with a consensus on the principle of *cost recovery*. The final consumer should pay all the costs incurred by WSS services directly through the water bill. The cost recovery principle is going to be implemented (in line with the WFD) by the means of the full-cost pricing, i.e. the consumer would pay all the related costs in his water bill[5]. In practice however, the principle is never implemented strictly as countries maintain subsidies, at least on investments. Ensuring a secure and reliable provision of drinking water is also an economic (and social) objective, particularly emphasised in Belgium and Portugal but for different purposes. While Portugal aims at improving the connection rates to drinking water supply and sanitation, Belgium aims at preserving the quality of raw water in order to minimise drinking water production costs. In the same vein, Spain's absolute priority in water policy remains oriented towards economic development, notably with plans for large-scale water transfers and low water prices for farmers.

Social objectives are less considered than ecologic and economic objectives, with no convergence in formulation between countries. Presently, social objectives are defined with the three criteria characterising services of general interest (see chapter three): quality (plus continuity and reliability) of the service supplied, access to all at any point of the national territory and affordability (access to all through affordable pricing). The objective of *provision of a water of a good quality* made compulsory within the EU as of the early 1980s (European Communities 1980) has social ramifications. Other social objectives are more specific to countries. The objective to ensure drinking water supply to all is only referred in four countries (Portugal, Germany, Belgium and Italy) (indistinctively by access through connection or price). *Access to the service* is not anymore a problem, as connection rates to drinking water supply are close to 99 %

[5] While up to now, mainly financial costs are included in the water bill, the WFD requires an "adequate contribution of the different water uses… to the recover of the cost of water services" including also environmental and resource costs. – see article 9.

everywhere, except in Portugal and Spain (Eurostat 2003a). In Portugal especially, the objective of access to all remains tied to offering a connection to drinking water in isolated areas.

Concerning affordability, there is no convergence at all between the countries. Where social objectives are defined, each country developed its own correction mechanisms to maintain the affordability of drinking water in a context of price increases. To supply drinking water to all is explicitly mentioned as an objective in Germany, England & Wales and Belgium, as well as the prohibition of supply cuts (limited to the Brussels-Capital Region in Belgium). Affordability is also targeted with mechanisms of solidarity, i.e. the constitution of *solidarity funds* financed by a withholding on the consumer's bill and intended to make up for the water bills that people cannot afford (e.g. in France, Belgium and England & Wales). Disparities are also important concerning *social tariffs*. Flanders (Belgium) is unique in offering the first cubic meters for free (15 m^3/capita/year). Otherwise, some social tariffs exist in Spain (e.g. reduced licence fee for the Poor in Barcelona), Sweden, Germany (e.g. income support from social services) and England & Wales (e.g. affordability measures for water bills). However, these measures are dissimilar and often developed at the local level either by the municipalities or the suppliers themselves.

Ecologic and economic objectives are usually mixed together. Economic principles, such as cost recovery, polluter pays or resource users pay, legitimise cost increases necessary to invest in resource protection (protection perimeters, sewage collection and treatment infrastructure, etc.). Sometimes also ecological goals support or dissimulate more economic goals related with development of supply or the construction of drinking water treatment plants. Objectives purely centred on the protection of ecosystems are more limited, e.g. the protection of wetlands. In addition, ecologic objectives legitimise more economic objectives that principles. This combination of converging ecologic and economic objectives in the countries studied attests of *an early assertion of the WFD in the logic of the national water policies already developed.* However the hierarchy of objectives differs from one country to another, with a majority putting an emphasis on the economic objectives. Here we find again the traditional division between southern and northern EU Member States, northern countries being more aware of environmental protection (The Netherlands, Sweden and Germany) (Sbragia 1997). We observe in fact that "southern" countries (France, Italy, Portugal and Spain) as well as Belgium put a lesser emphasis on environmental protection than on economic aspects of drinking water management. All in all, a large extent of policy objectives is covered in the different countries quite uniformly, except perhaps the protection of wetlands and social issues.

POLICY INSTRUMENTS

The policy instruments define the modalities of intervention or measures planned to fulfil the objectives of a public policy. Public policies are often conducted with a mix of policy instruments (policy mix), i.e. a series of instruments of different nature targeting different social groups. First, policy instruments used in the WSS sector are successively presented for each water policy sub-sector (i.e. production, distribution, sewerage and treatment). Second, the convergence observed is further investigated through the design of a standard profile of policy intervention in the WSS sector inside the EU. Finally, the target groups impacted by the different sets of instruments are compared.

Policy Mix within Each WSS Sub-sector

Different instruments are used in the WSS sector. For each sub-sector, we precise the categories of instruments used (prescriptive, incentive, informative and self-regulative), insisting on the similarities and differences between countries.

Access to the resource and drinking water production are essentially regulated by prescriptive and incentive instruments. First, abstraction permits and protection perimeters were set up everywhere, even if implementation deficits are noticed for abstraction permits (Belgium, France and Portugal) and protection perimeters (Belgium, France, Spain, Italy and Portugal). Second, the prohibition of manure spreading, completed with an authorisation scheme, is not widespread. This disposition is not compulsory in Spain and Portugal and poorly implemented in France and Italy.

Drinking water production is regulated by technical standards everywhere (prescriptive instruments). Concerning incentives, abstraction is taxed, except in Sweden, but implementation fails in Portugal and Italy (symbolic fees only). Concerning incentives, the protection of the perimeters around catchments is only subsidised in Belgium and Germany, but the overall drinking water production remains much subsidised everywhere under the form of direct subsidies to water suppliers or loans at non-commercial rates that contributes to investment plans in infrastructure. Operation costs are sometimes subsidised, particularly in Italy and Portugal where public authorities almost totally finance drinking water production. To the contrary, there is practically no subsidisation in Germany, even if some Länder financially stimulate investments in infrastructure on a case-by-case basis. Informative or self-regulative instruments are not mobilised, except with the codes of conduct for good agricultural practice (as expected in the 1991 Directive on nitrates (European Communities 1991b) and the promotion of drinking water production from surface water in Belgium (Flanders) and the Netherlands (only forecasted there).

In the *drinking water distribution* sub-sector, all four categories of instruments are used. Above all, prescriptive instruments concern the quality standards for drinking water that are high, compulsory and effectively implemented in all countries, since the enactment of the 1980 Drinking Water Directive (European Communities 1980). In some countries, further missions of public services are translated into prescriptive instruments, e.g. the obligation of continuity for the water supply service in France and Spain. The obligation to provide a minimal quantity of water for free (15 m^3/capita/year) is only implemented in Belgium (Flanders). The prohibition of unilateral supply cuts is also implemented in Belgium (Flanders), as well as in Germany and England & Wales. Water meters are compulsory everywhere, except in Portugal and England & Wales[6] and in some big cities, such as Antwerp and Amsterdam, where the reform process is underway. Public control on water prices is also widespread, except in the Netherlands[7]. Incentive instruments in the distribution sub-sector essentially take the form of public subsidies for investment and operation. They are effective in several places, even if they are less developed than in the production sub-sector. In fact, we observe exceptions in Portugal, Germany, and also in Sweden where they are very limited, and in England and Wales where there are no subsidies at all. The installation of a second water circuit of non-drinking water for industrial uses or some domestic uses is still not economically encouraged, the same for rainwater tanks, with the exception of Belgium. Incentive instruments are not always environmental-friendly. For instance, the ecological price structure, with a progressive water pricing for domestic uses that would make abusive uses more expensive (e.g. for filling-in a private swimming-pool), is not developed at all. In the field of informative instruments, information to the consumer about water quality has been made compulsory everywhere, as well as alert schemes in case of incidental pollution of drinking water. Localised and dispersed actions in favour of water savings (demand-side management) are developed everywhere, such as information campaigns or lessons at school. As for self-regulative instruments, we observe the widespread certification to ISO standards (ISO 9000 for the quality of the industrial process and ISO 14000 for the respect of environmental requirements in the process).

Concerning *sewerage*, the prohibition of direct discharges in surface water is implemented in most countries, except Spain and Portugal where the measures exist but implementation is partial. On the contrary, the separation of rainwater from wastewater is not much developed, except in Belgium, France, and Germany. Subsidies for investment and maintenance of the infrastructure are also widespread, except in Spain, England and Wales and Germany. Consecutively to the separation of rainwater, codes of good practice have been developed for the construction of sewers in France, Belgium and the Netherlands.

The *wastewater treatment* sub-sector is dominated by the prescriptions resulting from the 1991 Directive on urban wastewater treatment (European Communities 1991a). All the selected EU countries transposed the obligation to equip areas above 2,000 population equivalent with installations of collective treatment into their national law. In addition, most countries decided to equip all areas with individual or collective treatment plants. This prescription is planned in Italy and Portugal, but not implemented, and also only partially implemented in Belgium, France and Spain. Standards for effluents from the wastewater treatment plants are compulsory and implemented, except some problems of implementation in Spain and Italy. Sewage sludge disposal is restricted in Belgium, the Netherlands, Germany, and France[8]. Sensitive areas are already defined in many countries, according to the obligation of the 1991 directive (European Communities 1991a). Some countries, e.g. Belgium (in Brussels and Wallonia), even declared that their whole territories were classified as sensitive areas. Concerning incentive instruments, wastewater discharges are subject to taxation in all countries, despite some implementation problems in Spain. Investment in infrastructure is highly supported by public subsidies, except in Germany under the reserve that Länder can launch individual initiatives, and in England and Wales where there are no subsidies at all. The existence of subsidy schemes for individual treatment plants is only mentioned in Belgium, France and Spain.

The great similarities observed between the instruments used to conduct national (or regional) policies in the water supply and sanitation sector create favourable conditions to draw a standard policy mix of interventions within the sector.

Standard Profile of the Policy Mix

Instruments converge, as attests the *standard profile of public intervention in the WSS sector* drawn in table 2. Certainly, European legislation is responsible of some of the common features, but actually most of these features are not the result of a compulsory harmonisation. Some prescriptive (e.g. abstraction and discharge permits) and incentive instruments (e.g. taxation of withdrawals and discharges) are common to the countries without any harmonisation being

[6] In England and Wales, consumers have a right to demand a meter being installed, but metering is not compulsory.

[7] In the Netherlands, the suppliers indeed can set the tariffs, although the final word is for the public shareholders.

[8] A specific European Directive concerns sewage sludge in agriculture (European Communities 1986), but its scope is limited to concentrations of heavy metals.

required by the EU. Other elements of convergence are a widespread involvement of public money in the sector and an underdevelopment of information and self-regulative instruments.

Table 2 presents the common policy instruments applied in the national water policies. Many instruments are common to the countries (i.e. all the grey cells), but only a few result from EU compulsory measures. First of all, common instruments are found in all the sub-sectors of the WSS sector. These instruments are mainly prescriptive (i.e. mechanisms of prohibition/ authorisation, compulsory rules, regulation of practises, etc.) and incentive (i.e. economic instruments, such as tax and subsidies, loans at non-commercial rates, financial penalties, etc.) ones. The more noticeable convergence concerns the abstraction and discharge permit schemes and taxation schemes also related to abstraction and discharge. Despite the absence of any EU programme establishing such policy instruments, all the countries adopted them.

In complement, other prescriptive instruments are common to the countries: contracting obligations with the consumer (i.e. signing of a contract between the customer and the supplier, eventually referring to general and particular regulations), installation of a water meter, and obligation to equip all areas. In addition to taxation schemes, mechanisms of public subsidies on investment and operation are also generalised (except in Sweden and Germany). Not only wastewater treatment is subsidised, but also often the investment and operation in the networks of production and distribution of drinking water (in Belgium, France, the Netherlands and Portugal). Implementation is made through direct subsidies or loans at non-commercial rates. As such, public intervention in the WSS sector is similar from country to country.

Table 2: Instruments common to the water policies of the selected EU countries

	Access/ Production	*Distribution*	*Sewerage*	*Treatment*
Prescriptive instruments	Withdrawing permit Protection perimeter Technical standard for production	Quality standards Contracting with the consumer Installation of a water meter	Prohibition/permit for discharges	Obligation to equip urban areas above 2000 p.e. Obligation to equip all areas Compulsory standards for effluents
Incentive instruments	Abstraction charge Subsidies on investment and/or operation	Subsidies on investment and/or operation	Subsidies on investment	Tax on discharges Subsidies on investment and/or operation
Informative instruments		Information to the consumer about water quality Alert system Rational use of water		
Self-regulative instruments	Code of good agricultural practice	Individual rain water collection ISO standards		Individual treatment

Legend:
- Instrument common to the Member States
- Requirement of the European Law
- Only partially implemented

Nevertheless, some of the common instruments result from a harmonisation of legislation at the EU level anyway. EU standards are well developed, particularly in the distribution and wastewater treatment sub-sectors. The 1980 and 1998 Directives on the quality of water intended for human consumption set up high quality standards for drinking water, as well as programmes of control. The 1991 Directive on urban wastewater treatment complains countries to equip all urban areas above 2,000 population equivalent with secondary treatment, and tertiary treatment was made compulsory in sensitive areas, where the concentration of nitrogen in surface water is already high (see chapter three for more details). The obligation of information to the consumer about tap water quality, the implementation of an alert system and the code of good agricultural practice eventually result from EU requirements, as well as the compulsory standards for the effluents of wastewater treatment plants. Furthermore, some instruments are now required in the EU directives, but have not entered into force yet (e.g. technical standards for drinking water production).

Target Groups of Policy Action in the WSS Sector

The similarities between the target groups as well as the instruments are great. Actually, instruments often pre-determine the targets groups identified (or respectively, see Schneider and Ingram, 1997). Some great categories of target groups are recurrent: Economic actors likely to pollute the water resource (surface water or groundwater); water suppliers; local governments and; drinking water users. In fact, policy programmes in the WSS sector do not often enter much into details concerning the identification of the target groups.

When looking at particular sets of actors, farmers are pointed out as target groups of *resource protection measures* in 7 out of nine countries (except Sweden and UK). Farmers still exerts a great pressure on the resource (see above). However, full coverage of their polluting activities, in particular manure management and spreading, is far from being realised or effective. Systematically, water producers are target groups of resource protection policies, being compelled to declare withdrawals and in most cases pay abstraction fees. More and more households, as any other landowners (such as industries) are also perceived as a threat for the resource (e.g. domestic use of pesticides).

In the *water supply sub-sector* (production and distribution), the water suppliers are the main target groups (e.g. drinking standards and monitoring). Sometimes the local authorities are also targeted, in particular when they have a competence in price setting. Concerning supply also, all kinds of customers (domestic and industrial) are regularly targeted, in particular in favour of reducing their consumption. As an exception, industries that were not customers yet are invited to connect to "grey" water supply networks (Flanders in Belgium). When social measures apply for the water bill payment, poor households are identified as more precise target groups.

Concerning *sanitation* (sewerage and treatment), competent authorities and suppliers are the main target groups (investments and technical standards), but certainly not the only ones. Industries and households are equally concerned. Industries must notably control their discharges in the public sewers and households are mainly submitted to taxation, though not always with the primary aim to influence their behaviour. Finally, the identification of target groups is broad and large but often incomplete in particular concerning resource protection and the financing of wastewater treatment, where a fine-tuning of the 'responsible groups' is necessary to an effective pollution reduction and a true application of the polluter pays principle (instead of the 'domestic user pays principle').

Despite some difficulties in implementation of water policies in some countries, we observe important similarities between the instruments used to conduct public policies in the WSS sector. These common features are partly due to the European law, but the phenomenon goes far beyond the strictly compulsory harmonisation process. Such similarities in instruments contrast with the differences among the "objective" social problems raised in the different countries. The availability of the resource, as well as the distribution of the uses, differ (e.g. importance of water abstraction for agriculture), as well as the major problems that water management faces (e.g. over-abstraction for agriculture). Even if differences in objectives are less noteworthy, the priorities between the countries also vary (e.g. objectives of prevention against the pollution or development of the drinking water production capacity or economic development are only mentioned in few countries). As a matter of fact, *problems and objectives differ between countries while the types of instruments are similar*. Such a finding certainly puts into question the effectiveness or reasonableness of the important policy transfer or learning processes between the European countries (Bennett 1991; Radaelli 2004).

INSTITUTIONAL ARRANGEMENTS

In order to implement a public policy, public actors, i.e. politicians and civil servants, organise themselves and decide who does what. The institutional arrangement precisely results from this decision. It defines how the public actors are organised, at which level implementation decisions are taken, and what are the relationships between the authorities and the suppliers. These relationships are presently described through a comparison of the suppliers' status, financial flows taking place within the sector and the liberalisation practises.

Organisation of the Public Authorities

In the present part, a particular focus is made on the political and administrative arrangement governing the WSS sector. The *division of competencies between the different (public) actors of implementation in the water policies differs much from one country to another*, given their specific polity and legal and administrative history, and this even in case of similar problems to be solved (e.g. financing needs, compliance to the EU requirements or modernisation of the sector). However, some particular trends can be highlighted with the comparison of the national cases.

The *Ministers of the Environment* and their administration play a leading role in the organisational arrangement, particularly at national level. They are precisely those that adopted the EU water directives in the Council of Ministers. In most countries the Ministry of Environment concentrates competencies in water management, prepares water legislation and supervises the whole WSS sector (e.g. in Spain, Sweden, Italy and Portugal and UK). Sometimes they share these competencies with the Ministry of Public Health (e.g. Spain and Germany) or any ministry in charge of foodstuff issues (e.g. Ministry of Agriculture in Sweden), notably about drinking water standards. In the weakest position, the Ministry of the Environment is limited to a coordinating role between several ministers (e.g. in France). An involvement of the Ministry of Public Works is also frequent, especially in issues related to construction and investments in infrastructure (e.g. in Italy) and in the management of waterways (where surface water is often withdrawn for drinking water purposes) (e.g. in The Netherlands, France, Germany and Belgian regions).

Attempts to decentralise water management are numerous, in particular in central States (e.g. Italy with the ATOs). In federal countries water management is often a regional competence, either exclusive (e.g. in Belgium) or shared/delegated (e.g. in Germany). At the federated level, the Ministry of the Environment monitors both the resource and drinking water quality. In such cases, the tradition of decentralisation is long lasting and not special to water. In contrast, decentralising arrangements, more specific to the water sector can be observed in centralised States. The closer to federalism is Spain where the autonomous communities implement the legislation and can decide additional environmental measures. They are involved in price setting (approval of municipal decisions) and subsidy investments. In Italy, the water laws empower Regions against municipalities as Regions design the Optimal Territorial Areas (ATOs) and delegate water supply and sanitation for each area to a single supplier, a former competence of the municipalities (Goria and Lugaresi 2004). In contrast, in France, national ministries delegate implementation tasks to deconcentrated administrations (*préfet*, DRIRE, DDASS, DDE, etc.), a situation equivalent to the one of the Provinces in the Netherlands. Exceptions are the French Water Agencies wholly dedicated to water management and that are financially autonomous from the central authorities. In fact, two tendencies are identified in the decentralisation process. In a top-down perspective, national (or regional) administrations want to adapt implementation to local situation, even if regionalisation is a much larger political issue (e.g. in Spain and Italy). In a bottom-up perspective, decentralisation responds to needs of rationalising water supply, especially in presence of small municipalities (e.g. in Italy and Portugal). Thus, decentralisation in the water sector tends to converge at the regional level.

A multiplication of specialised agencies is observable. In most countries, the organisational arrangement of the WSS sector encompasses number of agencies, not necessarily fully devoted to water management. The most obvious one is the *environmental protection agency*. Economic regulators and basin agencies are less frequent, but could expand in the coming years. First, environmental agencies are frequent at national level. The British Environment Agency (OE) falls under this category as well as the Portuguese *Instituto da Água* (INAG), the Swedish *Naturvårdsverket*, the Italian *Agenzia Nazionale per la Protezione dell'Ambiente* ANPA) or the Flemish *Vlaamse Milieumaatschappij* (VMM). These agencies usually collect information about the water resource and prepare protection plans to be adopted and implemented by the governments (see box 1). Some other agencies are specialised in the *supervision of drinking water supply and sanitation*, e.g. the British Drinking Water Inspectorate (DWI), the Portuguese *Instituto Regulador de Águas e Resíduos* (IRAR). They usually depend on the Ministry of the Environment, set the technical standards and monitor their application. The third type of agency is the *economic regulator*, poorly developed in the water sector. The English Office of Water Services (Ofwat) or the Scottish Water Industry Commissioner, which regulate water prices and the activities of the (private) suppliers, constitute notable exceptions. Other private sector participation reforms did not come with such an organisational arrangement (e.g. Spain or France where the establishment of an advisory *Haut conseil des services publics d'eau et d'assainissement* failed). *River basin authorities* represent a fourth type of agencies. Their common feature is more often their territorial division rather than their competencies. In Italy, the Water Basin Authorities promote the rational use of surface and groundwater within basin plans (and sometimes interfere in the Regional affairs about the design of ATOs), a similar end as the Spanish *Confederaciones Hidrographícas* and *Administraciónes Hidrographícas* which make up and follow the river basin management plans. In contrast, the French water agencies (*Agences de l'Eau*) are rather financial redistribution agencies. Along this line of financing also, some countries committing in a remedial policy created some kinds of *meta-operators* (not agencies at all) managed by the competent authority, either national or regional government (e.g. SPGE in Belgium, AdP in Portugal) (see box 1). As such the organisational arrangement in the water sector is characterised by a diversity of specialised and relatively autonomous public actors of implementation.

Box 1: Regionalisation and Remedy (Belgium)

Belgium has been a federal State since 1993. Water management is, to some exceptions, an exclusive competence of the Regions (i.e. Flanders, Wallonia and Brussels-Capital). While initially responsible for public hygiene, municipalities progressively devoted their competence to the Regions, which now concentrate policy-making and organisation of supply and sanitation. Regional competencies are shared between the regional governments and environmental agencies in Flanders (VMM) and Brussels (IBGE). The agencies monitor the environmental quality of rivers and aquifers and make remedial plans that point out the investment needs in wastewater treatment. The government adopts the plans and implements them either directly in Brussels or through a specialised regional company in Flanders (Aquafin). In Wallonia, institutional design differs a bit with SPGE that command the remedial policy in the field of sanitation. SGPE is a mixed company held by the Region (40%), water suppliers (35%) and financial institutions (25%). This meta-operator concentrates the public financial means within the sector, prepares investment plans and distributes money to the suppliers. Quality control of drinking water remains a direct competence of the regional administration that sets the standards and monitors the suppliers. Municipalities are mostly shareholders of water suppliers and holders of the sewage system. Water supply is operated mainly by regional companies (i.e. VMW and SWDE) and multi-municipal associations (e.g. TMVW, Aquasambre and CIBE). Organisation of sanitation differs between the regions, belonging to a mixed regional company in Flanders (Aquafin), to multi-municipal associations in Wallonia (e.g. IBW and AIDE) and to the Region in Brussels-capital with delegated operation either to a public (CIBE) or private company (Veolia).

The main victims of all these institutional innovations are of course the municipalities. In a recent past, they were still the main public authorities in charge of water supply and sanitation. They initially lost their regulatory authority on the sector as the legislation concerning drinking water quality and wastewater discharges developed (at national level). More recently they also lost organisational competencies as the (often regional) authorities decided to take a reorganisation of the sector in charge, compelling the local suppliers to merge or join regional companies (e.g. in Italy with the ATOs set up by the Regions). Nevertheless, those put aside were often consenting victims as the invitations to reform were discharging them of huge spending and liabilities in exchange to a modernisation of their networks. In complement, where they keep some sovereignty in the matter, their behaviour is much criticised in particular when they privatise or delegate the service operations. However, where the municipalities remained competent and were able to group together to set up multi-municipal associations or public limited companies, they succeeded in modernising their service and increased their power and influence on the sector (see box 2).

Box 2: Municipal Autonomy within Länder (Germany)

Germany's political system is characterised as a federalism of execution. Federal legislation is passed at federal level with legislative details adopted by each land individually. However, this arrangement leaves much competence to the Länder in water management (both resource and drinking water management). The Länder implement federal framework laws tailoring them to the regional conditions. Technical agencies (*Wasserwirtschaftsämter*) assist them in implementation. Municipalities organise and operate the water supply and sanitation services with much autonomy from the Länder. Given the regional law, they are free to choose between several organisational arrangements to provide the services, e.g. direct labour (*Regiebetrieb*) (1.3% of the suppliers), semi-autonomous municipal agency (*Eigenbetrieb*) (36.7%), multi-municipal associations (*Zweckverbände*) (16.5%), municipal enterprise (*Kommunale Eigengesellschaft*) and delegations to private enterprises (13.6%) (BGW Water Statistics 2000). While direct labour is possible, semi-autonomous municipal agency has become the most common suppliers' status that allows separate bookkeeping and decision autonomy while formally tied to the municipal council. Mixed companies are excluded, but private interests can be involved in delegations of management tasks or plant operations or even concession agreements (1.7 % of suppliers). Although engaged in a strategy of modernisation of the WSS sector, the German authorities should not push privatisation ahead in the coming years, but the municipalities perhaps.

Status of the Suppliers

The organisation of the WSS services is still dominated by *public and municipal management*, except few notable cases (e.g. France, Spain and England & Wales). This part provides an overview of how water supply and sanitation services are organised, notably given their status and level of operation.

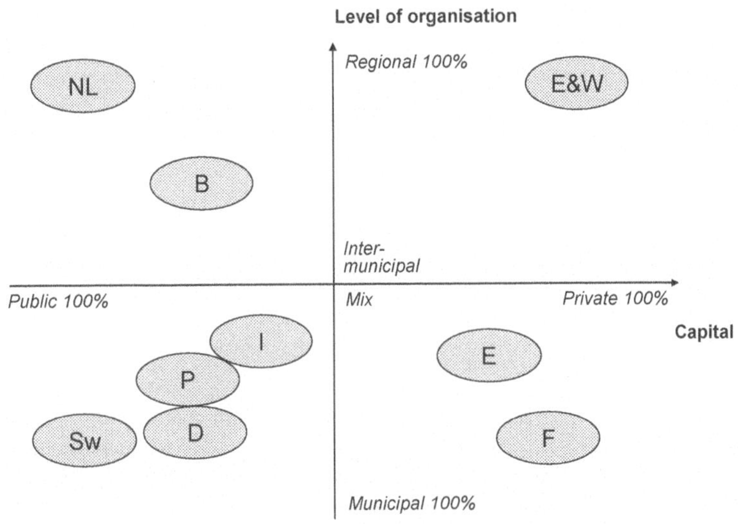

Figure 1: Dominant types of public and private actors in selected EU countries

The *municipal level dominates the organisation of the WSS sector* (see figure 1). Water and sanitation suppliers are in majority organised at local level and depend on municipalities, be they municipal or inter-municipal companies. However huge differences are noticed inside each country from one sub-sector to another. For instance, while the German *Regiebetrieb* (municipal service) is quite common for sewerage (around 70%), it is exceptional for drinking water supply (1.3%). Sweden is the only country that mostly counts municipal companies or services (see box 3). The dominant organisation is a combination of both municipal and inter-municipal companies or services (e.g. in France, Portugal, Spain and Italy). Of course, these organisations remain controlled by the municipalities. Only the Netherlands and England & Wales are fully organised at regional level[9]. Belgium combines both levels of organisation. In general, we observe *no significant differences in organisation between the sub-sectors (production, distribution, sewerage and treatment), except in Belgium.*

Box 3: Clean Water, High Levels of Services, at Low Price (Sweden)

The 289 Swedish municipalities (16,000 inhabitants in average) are responsible for the planning, construction and operation of water supply and treatment facilities. Usually, a single supplier operates both activities. The WSS sector traditionally belongs to the municipalities. 85% of the suppliers are still under direct municipal control and the remaining 15% are established as regional public limited companies (PLCs), operating more independently and sometimes involving private capital. Actually, in 7 out of 48 PLCs, the municipalities delegated the service to private companies through short-term management contracts (4-7 years) and one is fully privatised (Norrköping). Needless to say that the 1970 Water and Wastewater Law, that prohibits profit-making within the sector, discouraged further private sector involvement. This did not impede modernisation and related decreases in staff, by almost 50% within the last 10 years. While about 99% of the service costs are directly charged to the customers, some rural municipalities still subsidise the tariff. The debate on private sector participation in WSS services started at the local level in the 1990s around the questions of a lack in financial resources and needs in efficiency improvements. This process of opening to private capital would probably be stopped soon. After a demand from the Social-Democrats in March 2002 to conduct an investigation about the privatisation process in the Swedish economy, the Government stated that water supply and sanitation were "public tasks". In the meantime, the multiplication of public PLCs should continue.

Public ownership is widespread, and private participation exceptional in most countries, despite formal (or legal) introduction of some competition and private participation everywhere (see table 3). The Netherlands remains the single country in which private sector involvement is actually marginal. Conversely, England & Wales is the only place where all the WSS services are fully privatised. The situation is more contrasted in France. Private management dominates in both supply and sanitation, but under the control of municipalities that choose to either directly provide the service or

[9] A regional company has a different legal status than an inter-municipal company. Usually, it does not come out municipal reforms and generally operates on a larger territory. The public shareholding of the regional companies is not restricted to the communes, but involves the State or federated entities (e.g. the Region in Belgium). As such, municipalities have less control on regional companies than on inter-municipal companies.

delegate it to a (100%) private supplier. In Spain, different arrangements occur: 100% public, mixed or 100% private (see box 4). The other countries are led by the public sector, but open to private sector involvement (e.g. Portugal, Germany and Italy). It results from the comparison that *private participation increases when suppliers are organised at a higher level than the strictly municipal one.* The notable exception to that affirmation is France and Spain. The fact that delegated (private) management is operated by big private companies that are organised at national level with decentralised operative units, explains the difference (e.g. *Générale des Eaux* (Veolia) manages the French municipal services directly from regional offices). Thus municipal contracts are managed collectively with shared resources. Private sector involvement is marginal in many countries.

Actually, the dominant organisation mode in the WSS sector is public and municipal. Private involvement is important in some countries, but not necessarily in all sub-sectors. While many reforms in the water sector opened access to private companies, private actors did not always respond favourably to the legal opportunities.

Box 4: Structural Changes through Private Sector Participation (Spain)

A clear trend towards more private sector involvement and increased water prices is noticed. In Spain, the organisational arrangement distinguishes drinking water supply and sanitation. About supply, the municipalities are responsible, but usually delegate the service to private companies, mainly in large cities, that already supply 50% of the population. Aguas de Barcelona (Agbar), a tributary of Suez, manages 836 municipalities (11.4mio people), followed by FCC and Bouygues-Saur. The common management mode employed to involve private capital is the concession: The private company finances and manages the infrastructure that nevertheless remains under public property. Also, the mixed company model (with 51% of public shares) is developing. Direct public management still concerns 45.5% of the population, mainly in small towns. Concerning sanitation, private sector participation (PSP) also increased much after the transposition of the 1991 EU Wastewater Directive. Once again, municipalities in towns usually delegate to private companies with concession contracts. Agbar contracted out with 396 municipalities (15.1mio inhab.). FCC and Bouygues-Saur follow. Municipalities directly manage the remaining part. While autonomous in organising delegation, the municipalities must respect the 48/1998 Law that sets general contracting clauses for concession, not enough to provide effective regulation of the practises. About financial flows, companies propose the tariff to the municipalities that decide after prior consent of the regional Price Commission. Even in a context of PSP, cost recovery is not effective given huge investment needs in sanitation and partial national and regional subsidies.

Financial Flows

As "What is scarce is not water but money" (according to B. Barraqué), financial flows constitute an integral part of the institutional arrangement governing the WSS sector. The issue is approached with three simple questions: *Who pays for what? Who collects? And how are the revenues of the collected charges used?* Financial flows encompass water charges of all kinds, usually the payment of the service (service costs), taxes, fees and subsidies. Although taxes and service costs could be considered as separate financing modes, the difference is not clear-cut in the WSS sector as many taxes are reallocated to resource management and investment in infrastructure. This situation constitutes an *exception to the general principal of non-allocation of receipts in public budget.* The collected charges are directly invested in operation and infrastructure. Usually, abstraction charges, effluent and treatment charges are taxes or fees, and use and sewerage charges are rather services costs.

Box 5: Regional Coordination for Improvements in Efficiency (Italy)

In order to reduce fragmentation (13,000 suppliers), the 1994 Galli Law foresees to regionalise water companies. Municipal utilities must aggregate into single territorial areas (*Ambito Territoriale Ottimale*, ATO), with a single supplier for drinking water and sanitation (*Servizio idrico intergrato*, SII). The ATO authorities (AATO), set up by the Regions, are entitled to administer an integrated water service at local level. They draw investment plans for 20 years, delegate the integrated service to a supplier (management contract) and control its activities. In parallel, municipalities remain free to organise the service and to group under conventions (*Convenzione*) or consortia (*Consorzio*). In fact, municipalities benefit from a safeguard clause to temporary maintaining the current supplier. This apparent incoherence hampers a smooth implementation of the Galli Law, but helps preserving the national water industry. In 2003, only 25 ATOs were effective on a total of 91 forecasted.

WSS charges are one of the lowest in Europe, despite critical investment needs. Water price setting was formerly controlled and price increases restricted, but the Galli Law replaced the measure with the principle of full cost recovery. A unique tariff should be applied within the ATO, set according to a normalised method of calculation (*metodo normalizzato*). Users pay the water charges to the supplier which in return pays fees to the municipalities for using the netwok (*canone*), to the Region for water abstraction fee and finances the ATO authority. The *canone* incites the municipalities to join the SII. However, price increases forecasted with the normalised pricing method are not sufficient to cover investment costs. The compliance to EU requirements is estimated to EUR 28bio over the next 12 years (EUR 11.2bio for supply, 12.2bio for sewerage and 4.6bio for treatment). Thus, the Italian State and the EU considerably contribute to the financing of the investments. Only areas that have implemented the reform are eligible to the EU structural funds.

All countries collect charges (services costs, fees or taxes) at all stages of the WSS chain, except Sweden that does not charge water withdrawals, perhaps because water is abundant there (see table 1). Although water charges are collected everywhere, financial mechanisms are truly distinct. Charges are levied at every step of the process, but not always at the same rates and to the same target group. *Charges on industrial effluents and domestic wastewater treatment are less developed than abstraction charges.* Often, the final drinking water consumer pays most of the water charges (including service costs), but he is not systematically the direct payer of the services (see box 5). The money collected follows complex circuits of allocation or redistribution inside the sector (see box 6).

Box 6: Water Pays Water (France)

Water supply and sanitation remains a competence of the municipalities, which organise the service (currently 13,500 suppliers). Services are either directly managed (21% of the population for supply and 47% for sanitation) or delegated to private companies whilst preserving the ownership of the public infrastructure. Municipal services must be financially self-sufficient and consequently municipalities are free to set their own ratting schemes. In case of delegation, they negotiate water prices with the companies. The price of the service must correspond to the investment and running costs of the service, possibly including the remuneration of the private company. In complement, 6 basin agencies (*Agences de l'Eau*) with civil personality and financial autonomy finance water management through the collection of pollution and consumption fees. With this money, they subsidy all investments in favour of (surface) water quality that municipalities, industries or farmers make (up to 60% of expenditures).
The supply and treatment services are mostly funded by the consumers' bills according to the principle of cost recovery. In average, 42% of the bill finances water production and supply and 31% sanitation. Additional 20% fund the basin agencies to invest in infrastructure. Another fraction of the bill (1%), the National Water Fund fee, helps small-sized municipalities to build water networks (supply and sanitation). On a yearly basis, the basin agencies collect and redistribute EUR 8.5bio (of which 80% comes from the consumers) as well as the National Water Fund (EUR 145mio). Counties (*départements*), regions and the EU also subsidise the investments in the sector.

Money collection is fragmented. Starting from the concept of cost-recovery is a way to explore the circuits of finance inside the sector. Suppose the consumer would pay the full cost of the service provision, transparency would require that she receives a single water bill, detailing all the costs related to her use: abstraction charges, production and distribution, sewerage and treatment charges. In fact, *except for use and sewerage charges, water costs are not directly attributed to the drinking water consumer and the drinking water supplier is not alone in collecting the charges.* Abstraction and effluent charges are most of the time directly collected by a public authority. In spite of this *fragmentation* and involvement of public authorities, most money collected is reallocated to the water sector[10]. Fragmentation of payments and financial flows in the water sector hide the full service costs.

A dominant design emerges from the comparison. In a large majority of countries, *abstraction, use and sewerage charges are collected. Effluent and treatment charges are less systematic*, but may also represent important financial sources in many countries. At the same time, it is true that public authorities still inject public money in the sector under the forms of subsidies (see box 7), an indirect way for the citizen/consumer to pay water (involving interesting equalisation mechanisms). Once back to the three initial questions concerning financial flows (Who pays? Who collects? Where does money go?), one sees that the final drinking water consumer is the main payer, revenues are mainly used for operations and investments, and *money flows from the consumer to the infrastructure through the drinking water supplier.*

Box 7: EU Support to Remedial Policies (Portugal)

Portugal is structurally underdeveloped in water supply and sanitation, with one of the lowest coverage rate throughout Europe. The country has engaged a remedial process since 1993, with the primary objective to increase connection rates. The reform consists in "re-centralising" WSS services. Currently 90% of suppliers serve 20% of the population. In the new management model, municipalities remain responsible for drinking water distribution and sewerage (low pressure or *baixa* activities), but bulk water production and transfers and wastewater treatment (high pressure or *alta* activities) are entrusted to multi-municipal public companies controlled by a State-owned company, *Aguas de Portugal* (AdP), that also allocates the EU subsidies to the water sector. About a half of the population is already subject to this organisation arrangement that should be progressively extended to the whole country.

[10] Profit margin should be subtracted from the reallocation of the collected money in cases of private capital involved (either as a share of collected money, either on a negotiated profit margin on the amount of public money invested in the infrastructure, as it is the case for Aquafin in Belgium).

Over the period 2000-2006, a national investment of EUR 5bio was planned of which one third came from the EU structural funds. EU subsidies amount to EUR 1.3bio in 2001-2006, one third of the investments planned by AdP in the multi-municipal companies (subsidised at 80% in 1993-1999 and 40% in 2000-2006). The remaining funds come from the customer's bill (through price increases) and AdP equity funds. In other cases, financing is made through the municipal budget.

Liberalisation practices

Legislation does not impede private sector participation (see table 3). Throughout Europe, there exist many legal opportunities to involve private capital in the WSS sector. *Competition for wastewater treatment* is possible in all the countries under scrutiny, except Portugal. This situation contrasts with the low involvement of private actors, as observed in the organisation of the sector (see figure 1).

Table 3: Liberalisation practises

Competition in:	*B*	*F*	*E*	*NL*	*S*	*I*	*P*
Production	No	No	No	No	Yes	Yes	No
	25 y contract (W)	20 y max.	-	-	3-5 y	For the market Competitive bid or direct assignation	-
Distribution	No	Yes	Yes	Yes	Yes	Yes	Yes
	5y contract (W)	For the market (few competitors in practice) 20y max.	For the market 25y max. (up to 50y if construction)	No	3-5y	For the market Competitive bid but also direct assignation	For the market Competitive bid (20-30y)
	-	Laws but no regulatory body	Legal void for obligations regarding suppliers	Benchmark.	At municipal level	National and ATOs levels Price regulation	National regulator (recommend.)
Sewerage	No	Yes	Yes	Yes	Yes	Yes	Yes
	-	For sewer maintenance 20y	For sewer maintenance	No	3-5y	For the market Competitive bid but also direct assignation	For the market Competitive bid (20-30y)
	-	Laws, but no regulatory body	-	Benchmark.	At municipal level	National and ATOs levels Price regulation	National regulator (recommend.)
Treatment	Yes	Yes	Yes	Yes	Yes	Yes	No
	For the market (FL). Unlimited (lease termination of 20y)	15-20y contract	For the market 25y max. Up to 50y if infrastruct. built	No	3-5y	For the market Competitive bid but also direct assignation	-
	-	Laws but no regulatory body	No laws like in France	Benchmark.	At municipal level	National and ATOs levels Price regulation	-

If opportunities for competition are widespread, the modalities of tendering and contracting are not necessarily similar. In fact, contractual settings differ (see table 3). The mechanisms for competition are different from one country to another. Competition is mostly occurring in the format of *competition FOR the market*. We observe little possibility for competition IN the market. However, competition in the market for large industrial water supply is possible in the

UK, the Netherlands, as well as Germany. Furthermore, actual competition is sometimes restricted, when the local authorities can still proceed to direct assignations (e.g. in Italy) or when only a very select number of competitors submit offers (e.g. France). The *length of certain contracts* is also a barrier to true competition. The length of contracts is rather long (around 20 years), except in Sweden. In Belgium (Flanders), the delegation contract of wastewater treatment to a mix company, Aquafin, had not initially been subject to a public tender and it was not limited in time, with a lease termination of 20 years after a party had denounced the contract. In Spain, the contract can be effective for 50 years when the suppliers commit themselves to invest in the infrastructure. Contracting can also occur without liberalisation or privatisation. When the existence of a contract is mentioned while there is no competition, this means that public companies signed a "management contract" with the public authorities (as it is the case of the production and distribution sub-sectors in Belgium).

A systematic *regulation of practice* by an independent regulatory agency is absent in most places. Except in England & Wales (see box 8), the regulation of private suppliers remains limited (the absence of a regulatory agency that can take legally binding decisions, is particularly striking). In Portugal, the national regulator can only formulate recommendations (as well as the planned regulator in Flanders (Belgium)). In France, some form of regulation is made by the law, but no regulatory authority exists to monitor its enforcement. In Spain, there is even a legal void around the suppliers' obligations. In the Netherlands and Sweden, regulation is conducted by means of control at municipal and regional level control and some benchmarking activities (see box 9). In contrast, Italy organises regulation at national and ATOs levels, notably on price.

Box 8: A Cord Drawn Tight around Private Suppliers (England and Wales)

The administrative arrangement in England and Wales encompasses national regulatory authorities and (regional) private water suppliers. In England, water management is the responsibility of the Department for the Environment, Food and Rural Affaires (DEFRA), respectively the Wales Office in Wales. The three regulators, i.e the Office of Water Services (Ofwat), the Drinking Water Inspectorate (DWI) and the Environment Agency (EA) have important oversight powers. EA monitors river and environmental pollution and DWI monitors the quality of drinking water. DWI takes enforcement actions if standards are not being met and investigates incidents and consumers' complaints. Ofwat is a non-ministerial government department, led by a Director General appointed for a fixed term by the Secretary of State who is responsible for DEFRA. Ofwat is responsible for ensuring that the long term interests of the consumers are respected; that prices charged by the water companies are reasonable; and that the service-providing companies duly perform their duties. To this end, Ofwat steps in and compares the companies' performance against each other through a mechanism known as *yardstick competition*: an evaluation of the relative efficiency of each company in comparison to the most efficient company and price determination on the basis of the allowable price increase (*price cap*) for each regulated company. With the allowable price increase thus being fixed (every five years), each company has an incentive to further reduce its costs and to improve its efficiency beyond the one assumed when prices were set, in order to increase its profits. Thus Ofwat has the power to set overall prices and not approve company charges schemes. Ofwat also now has powers to fine companies for failures to comply with licence conditions and to apply for an enforcement order for more serious breaches.

Box 9: Benchmarking on a Voluntary Basis (The Netherlands)

In the Netherlands, benchmark is realised by the suppliers' association, VEWIN (*Vereniging van Exploitanten van de Waterleidingbedrijven in Nederland*). Established in 1952, VEWIN is the umbrella organisation for water supply companies in the Netherlands and promotes the interests of its members through political influence at national and European level, research and diffusion of information. VEWIN has been responsible of the benchmark since 1997, a benchmark that monitors four issues: drinking water quality, service performance, environmental performance, and finances (cost efficiency). Benchmark studies are conducted on a three-year basis. Different indicators are collected and compared. The benchmark study is used to increase the transparency and the performance of the companies and provides recommendations on how to improve the company's processes. More than 85% of the Dutch water companies participate. The benchmark studies can be joined voluntary and the results are not, until now, associated with sanctions or bonuses from the Government. As of 2001 this benchmark was made mandatory in a revision of the Water Supply Act (Bressers and Kuks 2003). In fact, VEWIN initiative was closely tied to the legislative reform. In a context of debates around privatisation issues, the suppliers wanted to demonstrate that the current organisation of the water supply sector, based on public limited companies under control of the municipalities and provinces was effective and efficient, that public suppliers performed well.

It is still too early to evaluate the efficiency of the introduction of competition in the WSS sector. Liberalisation processes are too recent in many countries to be properly assessed. Moreover, the level of activity submitted to liberalisation is in some countries too residual to justify the setting of a complete administrative arrangement or independent agency for regulation.

Institutional arrangements in the WSS sector are complex sets of interactions between public and private actors at different levels and with various consequences. They are deeply rooted in political and administrative histories, even if the latest reforms tended to advocate standardised organisation models involving regional public/private companies regulated by decentralised and specialised agencies. However peculiarities of former organisations tend to persist and adapt to the legislative reforms. In spite of many attempts to concentrate and privatise the suppliers, the dominant design organising the WSS sector remains municipal and public. The cost recovery principle is neither fully implemented, but the final drinking water consumer is already the main source of revenue for investment and operation (and less professional users and tax payers). In terms of liberalisation, the law formally gave many opportunities to contract with private companies, in opening the competition for the market, but few are the candidates and high the reluctance. Anyway, the policies conducted in the countries reached some pretty good results, particularly in terms of connection rates to the tap and to treatment plants.

POLICY IMPACTS

Policy impacts are the effects of the public action. They cover both the outputs (administrative products) and final outcomes (induced effects) of the public policy. Measuring policy outcomes revealed notoriously difficult, as evaluation tools are either not uniform between the countries or not developed at all. Therefore, this assessment rather focuses on strictly measurable outputs.

The effective *protection of the resource* is particularly difficult to assess. It would not really be worth comparing the qualities of raw water in different places, as local conditions are too determining in the figures. As a result, this analysis is tied to outputs. Even there, precise data is not available. For instance, in France, if water abstraction permits are systematically given to water users, information about the effectiveness of the measure is not collected. Are the maximum levels of abstraction determined according to the available flow of the resource or according to the needs of the users? The answer is often negative. Environmental permits are scarcely revised in case of depletion of the resource. Protection perimeters face rather a slow implementation process. In France, once again, 12,786 protection perimeters have been put in place, but only 35.3% of water catchments are sufficiently protected. Such approximation on data complicates the resource management. *Data are developed for monitoring the (anthropogenic) water cycle, but not for conducing to resource management.* The monitoring of distribution, sewerage and treatment is more precise, while data is lacking about the implementation of protection measures related to the resource. This gives information on trends that is however not detailed enough to be a guidance tool in the policy process.

Impacts are more evident and positive when considering the drinking water cycle. The connections to the distribution network are widespread. Most countries are close to a rate of 100%, except Spain (57.8%, in 1998) and Portugal (90%) (Eurostat 2003b). Both countries are under remedial processes and benefit from EU subsidies in order to do so (see box 10). *Leakage rates are much more dissimilar.* A good rate attests of the good maintenance of the network and of a concern of the supplier for resource preservation. Only the Netherlands and (Western) Germany have distribution networks in good condition (see box 11). Other countries have a leakage rate comprised between 20% and 30%. Italy and Portugal face a particularly worrying situation with over 40% of losses in the network.

Box 10: Fast Recovery (Portugal)

Water supply and sanitation services had been underdeveloped in Portugal. Despite huge improvements in the past ten years, the connection rate to wastewater treatment remains one of the lowest in Europe (55% in 2000). Hence legislation is still focused on the improvement of connection to water supply and sanitation. The 1993 reform aimed at increasing the production capacity as well as the connection to water supply from 75% to 95% in 2006 and to sanitation from 15% to 90%. It intended to attain this goal in centralising the production of bulk water and wastewater treatment at regional level through multi-municipal companies controlled by *Aguas de Portugal* (AdP). An initial investment of EUR 4bio was planned. Until 2003, 16 multi-municipal companies were created (8 for supply and 8 for sanitation), covering 83 municipalities (over about 300) and 4.5mio people (about 45% of the population on a total of about 10mio) for supply, respectively 35 municipalities and 2.3mio people (23% of the population) for sanitation. Water production of AdP rose from 355mio m^3 in 1999 to 465mio m^3 in 2002. Even if the WSS sector remains municipal and fragmented in the countryside, Portugal reached connection rates of 90% for water supply and 55% for sanitation in 2000. Investments represented EUR 980mio between 1994 and 2000 of which EUR 405mio for water supply, and EUR 241mio for the single year of 2002. Results are less convincing for local or retail supply as well as sewage and attest of a lack of co-ordination between the *alta* and *baixa* activities (see box 7). Only 20 concessions were signed with municipalities to re-organise local supply according to the new arrangement. All in all, most problematic situations have been tackled, but there is still a long way to complete the reform.

Box 11: Objectives Reached at Low Price (Germany)

Major improvements have been made over the past decades in the ecological balance of water bodies, through stringent protection policies, the compilation of relevant information, as well as information policies. these policies led to notable improvements in water quality, increases in the number of water protection areas, and decreasing pollution levels from substances hazardous to water. Incentive instruments also contributed to these improvements.
Around 99% of the population in Germany is connected up to the public water supply system, to which a high level of drinking water quality is universally attested. With a connection rate to wastewater treatment of 93.2% (76.478mio people in 1998), the sanitation system is practically complete in Germany. Point source discharges without any treatment were reduced from 115mio m^3 in 1995 to 65.3mio m^3 in 1998 (BMU and UBA 2001). Revenue obtained through effluent charges is earmarked for a range of water management activities, such as investments in water pollution control and cost coverage for other quality improvement measures. Such outcomes were reached with controlled water prices, ranging from 1.27 €/m^3 to 2.33 €/m^3 (BGW water tariff statistics 2000). Concerning other impacts, the water supplied to the average consumer in Germany decreased by about 19% within the last decade. In 2000, per capita drinking water consumption amounted to about 129 litres per day (BGW n.d.). This reduction could be achieved due to both technical measures as well as changes in the behaviour of consumers. In the same period, a reduction in leakages of around 20% was attained.

Apart from water supply, *the population is widely connected to sewers*. Connection rates are close to 100% everywhere (with individual treatment included), except in Italy (86%). However, all waters collected are far from being treated in many countries. While some countries have high connection rates (above 85% for The Netherlands, Sweden, Germany and England & Wales), the problem of direct discharges has not been resolved yet after ten years of implementation of the directive on urban wastewater (European Communities 1991a). Belgium is the tail-ender with a connection rate to (collective) treatment of 38.1% (in 1998, estimated to 55% today). The country, as others which have low connection rates (Portugal and Spain), is developing a remedial programme. However, it should not be concluded too fast that countries with a rate below 100%, such as France (76.9%) and Italy (75%), have incomplete wastewater treatment and remedial programmes. In Italy and France, with large rural areas and a very low density of population in some places, individual treatment is widespread. The last aspect is the quality of the treatment made. It comes out from the study that *an important share of tertiary treatment is generally associated to a good rate of connection to treatment plants*. However, this rate is remarkably low in England & Wales. The absence of renewal of the infrastructure partly explains this situation.

Price is another impact of water policies, more precisely an outcome, a consequence of past and present public policies. When looking at price range differentials between countries, it is striking to observe that there are certainly *more price disparities within each country than between the nine countries of our empirical sample*. The lowest prices are measured in Spain and Italy with respectively EUR 0.42-1.66 and EUR 0.15-0.89 (data collected by the researchers from various sources). Belgium, France, Germany and the Netherlands have a similar price range around EUR 0.75-3.00. The relative convergence on prices is striking, given the huge disparities between countries concerning resource availability, density of population, degree of fragmentation of the sector and the legal and economic structures of the suppliers. For instance, England & Wales with a unique legal and economic structure has approximately a similar level of pricing as other European countries (EUR 1.40 per cubic meter). Nevertheless the manipulation of these figures must be careful, as these are only gross approximations. In most countries, price structures are extremely difficult or quite impossible to compare.

Data on policy impacts in the water supply and sanitation sector is incompletely collected and processed. As such an evaluation of the policies conducted can hardly bring satisfactory results if based on them. While connection rates to distribution, sewerage and treatment are known, information about the cost and price structure of WSS services, as well as about the effectiveness of protection measures on resource renewability are missing. It is observed however that networks are growing smoothly in lagging countries and at an affordable price for the consumer.

PERSPECTIVES

The comparative policy analysis made about the institutional arrangements of the water and sanitation sector in various EU countries aimed at revealing the main similarities and differences that existed between the national arrangements with a view to validate or reject the idea that these arrangements were converging or harmonising under the pressure of the EU policies. While convergence is evident in certain aspects, notably the policy instruments used, major differences persist (see table 4).

Table 4: Summary of the similarities and differences in water policy in selected EU countries

Dimensions of comparison	*Similarities on:*	*Differences on:*
Policy problems	DW demand per inhabitant	Availability of the water resource as well as the volume of water abstraction
	Pollution of drinking water reserves by agriculture Concentration of hazardous substances	Over-abstraction by agriculture Much local/regional rivalries
Objectives	Environmental and economic objectives	Social objectives
Instruments	Prescriptive instruments (e.g. quality standards for drinking water, abstraction and discharge permits)	Economic/incentive instruments (Taxes/ charges/ subsidies)
Institutional arrangement	Dominance of municipal and public suppliers	Experiences of involvement of private capital
	Pre-eminent financial mechanism: from the consumer to the water infrastructure, with a collection by the supplier	Levels of direct subsidies
	Partial opening to competition for the market (but, in result still very different)	Regulation and terms of the contracts allowing involvement of private capital
Policy impacts	Rate of connection to DW supply and sewerage	Rate of connection to a treatment plant

The *major differences* between countries are the availability of the resource and, in southern Europe, the tough *competition between agriculture and drinking water supply* for the abstraction of the resource. *Social objectives* really differ, and are even absent in some countries. *Economic instruments* fit more to national traditions of public finances and public intervention, as well as the conditions of private involvement in the sector. Also, the level of wastewater treatment widely differs from one country to another, but should progressively converge. All these differences seem to be invincible in a short-term perspective as much of them are deeply rooted in the natural and political characteristics of the countries. How could it be talked about convergence or discussed of the creation of a single European water market in such a context?

In contrast, the number of *similarities* between the different cases is impressive, given the initial situations (including the heterogeneity between the countries and the differences in size and population) or water availability. *Drinking water consumption per inhabitant is similar.* The pressure on the resource is mainly due to diffuse pollution by agriculture, and discharges of hazardous substances by industry in a smaller extent. In the conduct of policies, an equivalence on *economic and environmental objectives* is noticed as well as on prescriptive instruments in all the sub-sectors. In the organisation of the WSS sector, the *public and municipal organisation dominates*, even if an (at least legal/formal) opening of the sector to *competition* for the market is occurring, though limited. The drinking water consumer brings the needed financial resources, which are collected by the drinking water supplier and mainly allocated to the water infrastructure (construction and operation). Finally, *connection rates* to drinking water supply, as well as to sewerage, are close to 100% everywhere. Is it fair to talk about convergence of institutional arrangements in front of such similarities? Actually, convergence seems restrained to the conduct of policies. Decision-makers imagine intervention programmes similar in many respects (e.g. the proximities of objectives and instruments or the solutions proposed about the status and size of the suppliers or the financing of investments). Once implemented these programmes meets resistance from dominant actors in place and transpositions of radical reforms drag on or fail (e.g. in Italy or Portugal). In sum, convergence of policies refers more to policy design than to organisational arrangements.

The factors explaining this convergence are manifold. However, some constraints are particularly powerful to force *harmonisation*. First the *EU requirements* pressed structural reforms at national scale. Since the late 1970s, EU environmental policy, in particular in the water sector, has imposed certain sets of instruments and monitoring programmes. In this long tradition, the WFD turned up as a powerful tool not so much to harmonise instruments, but rather to harmonise the policy objectives. The evolution towards harmonisation is also due to *economic and financial developments*. Adaptation to EU law requires substantial investments, which in the context of cuts in public finances of the early 1990's pushed the public sector in the direction of alternative financing mechanisms. The current private involvement in the water sector is in part due to economic reasons, but not only. The third factor explaining similarities and differences are the *technological developments*. The rising pollution problems, combined with an increase in quality standards for drinking water and discharges, required the recourse to high technology equipments. Traditional (mainly local and public) suppliers did not manage such technologies. As a result, public authorities called for the expertise of the major (private) companies of the sector that were able to provide the necessary technology and related services. It makes no doubt that these developments pressed for change and provoked in some way an harmonisation of the water policies conducted in the countries. Once again, if policies converge, it does not necessarily means that the organisational arrangements are harmonising. The conditions for the creation of a single market in the water supply and sanitation sector are not all met yet.

REFERENCES

- Bennett, C. J. (1991), "What is policy convergence and what causes it?" British Journal of Political Science **21**(2), 215-233.
- Berger, P. and T. Luckmann (1966), The Social Construction of Reality: A Treatise in the Sociology of Knowledge, Anchor Books, New York.
- BGW (n.d.), Die öffentliche wasserversorgung in Deutschland, notice in writing 26.09.03.
- BMU and UBA (eds.) (2001), Environmental policy, water resources and management in Germany, Bundesministerium für Umwelt, Naturschutz und Reaktorsicherheit, Berlin, Bonn.
- Bobrow, D. B. and J. S. Dryzek (1987), Policy Analysis by Design, University of Pittsburgh Press, Pittsburgh.
- Bressers, H. and S. Kuks (2003), "Geintegreerde regimes en duurzaam gebruik van natuurlijke hulpbronnen". Beleidswetenschap **17**(3), 232-259.
- Cobb, R. W. and C. D. Elder (1983), The Political Uses of Symbols, Longman, New York.
- Dryzek, J. S. and B. Ripley (1988), "The ambitions of policy design". Policy Studies Review **7**, 705-719.
- EUROMARKET (2003), Analysis of the European Union's explicit and implicit policies and approaches in the larger water sector. Final report of work package 1, unpublished work.
- European Communities (1975), Council directive 75/440/EEC of 16 June 1975 concerning the quality required of surface water intended for the abstraction of drinking water in the Member States.
- ——— (1980), Council directive 80/778/EEC of 15 July 1980 relating to the quality of water intended for human consumption.
- ——— (1986), Council directive 86/278/EEC of 12 June 1986 on the protection of the environment, and in particular of the soil, when sewage sludge is used in agriculture.
- ——— (1991a), Council directive 91/271/EEC of 21 May 1991 concerning urban waste water treatment.
- ——— (1991b), Council directive 91/676/EEC of 12 December 1991 concerning the protection of waters against pollution caused by nitrates from agricultural sources.
- Eurostat (2003a), "Water resources in the EU and in the candidate countries". Statistics in focus, OPOCE, Luxembourg.
- ——— (2003b), "Water use and waste water treatment in the EU and in candidate countries". Statistics in focus, OPOCE, Luxembourg.
- Gleick, P. H. (ed.) (1993), Water in Crisis: A Guide to the World's Fresh Water Resources, Oxford University Press, Oxford.
- Goria, A. and N. Lugaresi (2004), "The evolution of the water regime in Italy". In: The evolution of national water regimes in Europe. Transitions in water rights and water policies (eds. I. Kissling-Näf and S. Kuks), pp. 265-291, Kluwer, Dordrecht.
- Heidenheimer, A. J., H. Heclo and Teich Adams C. (1990), Comparative Public Policy. The Politics of Social Choice in America, Europe and Japan. St. Martin's Press, New York.
- Linder, S. H. and G. B. Peters (1988), "The Analysis of Design or the Design of Analysis". Policy Studies Review **7**(4), 738-750.
- Radaelli, C. M. (2004), "Europeanisation: Solution or problem?" European Integration Online Papers *(EIoP)*, http://eiop.or.at/eiop/texte/2004-016a.htm.
- Sbragia, A. (1997), "Environmental Policy: The 'Push-Pull' of Policy-making". In: Policy-making in the European Union (eds. H. Wallace and W. Wallace), pp. 235-255, Oxford University Press, Oxford.
- Schneider, A. L. and H. Ingram (1990), "Policy Design: Elements, Premises, and Strategies". In Policy Theory and Policy Evaluation. Concepts, Knowledge, Causes, and Norms (ed. S. S. Nagel), pp. 77-101, Greenwood Press, New York, Westport and London.
- Schneider, A. L. and H. Ingram (1997), Policy Design for Democracy, University Press of Kansas, Lawrence KA.
- Steins, N. A. and V. M. Edwards (1999), "Collective action in common-pool resource management: The contribution of a social constructivist perspective to existing theory". Society and Natural Resources **12**, 539-557.
- Swyngedouw, E. (2004), Social power and the urbanization of water. Flows of power. Oxford University Press, Oxford, UK

Chapter III

Analysis of the European Union Explicit and Implicit Policies and Approaches in the Water Supply and Sanitation Sectors

Authors Serge Garcia, Laetitia Guérin-Schneider and Lise Breuil*

INTRODUCTION

In many countries, new dynamics such as the questioning of the role of State in economic activities have led to an unprecedented opening up to competition and private sector participation in most network industries (electricity, natural gas, telecommunications, postal services, urban transports, air and railway transports). This process is in some ways accelerated in Europe by the willingness to create a single European market, see European Commission (1999).

However, the water sector has been excluded until recently from the restructuring processes achieved in other sectors. The European Union (EU) has affected numerous aspects concerning the water sector in Europe for thirty years. Standards concerning both quality of water bodies and emission limits derive mainly from the EU water legislation. The Water Framework Directive (WFD) rationalises EU water policy around the single objective of good ecological status for all waters and introduces the concept of integrated river basin management. It considers water as "a heritage that must be protected" and water supply as "a service of general interest". It also suggests the use of economic instruments like the principle of cost recovery. Because of its environmental, social and equity aspects, water supply and sanitation (WSS) are very sensitive topics for public opinion and need to be debated in order to define - or not - a European policy.

The main challenge of this chapter is to see whether the current EU policy in water standards, resources management and liberalisation of network industries could constitute a general framework for analysing the possible evolution of the WSS sector with regards to liberalisation. In particular, the objectives of this chapter are:

- to analyse the role of the EU in the current liberalisation process;
- to point out the elements that could foster or discourage a trend of liberalisation of WSS services that is already ongoing in some countries.

* This chapter is based on Euromarket (2003) the Deliverable 1 of the EUROMARKET project. This chapter would not have been made possible without the very valuable contribution of all partners of the project.

Since there is empirical evidence of a liberalisation trend in the WSS sector in several European countries, the indirect effects of the EU water policies as well as its policy of liberalisation in other sectors must be addressed now with a comprehensive approach.

Hence, three fields have been identified, where existing European policies could have an impact on WSS services: 1) standards in drinking water and sanitation systems, 2) management of the natural resource (by way of WFD), and 3) liberalisation and services of general interest. Each of these three fields has been treated in glancing with the three following points: the review of the European policy and legislation (stressing on its dynamics), the assessment of the principles or notions emerging from the European policy and legislation (requiring a clear definition of these notions), and the illustration of these principles and their perspectives at the Member State level.

The remainder of this chapter proceeds as follows. The first part deals with the European Directives on drinking water and sanitation systems, and shows that technological changes, financial aspects and the emergence of new stakeholders are important driving forces for the liberalisation of WSS services but with significant differences among Member States. We then highlight the implications of the WFD. The WFD constitutes the European policy for the management of water as a natural resource, and makes explicit the possible linkages between the WFD principles and the potential liberalisation of WSS services. In a second part, we consider the European liberalisation policy in other network industries considered as services of general interest, and investigate to what extent the process observed and the concepts developed could be relevant in the water sector. As a synthesis, a last part brings closer the three approaches, underlining the synergies between explicit and implicit European policies, and highlighting important remaining issues and needed safeguards.

EXISTING EUROPEAN POLICIES IN THE WSS SECTOR

Taking stock of the liberalisation definitions

Privatisation and liberalisation are often confused, although they refer to two different processes. It is possible to liberalise a sector while conserving public ownership of the incumbent operator. Conversely, it is possible to privatise a public monopoly, while maintaining its special or exclusive rights. It is also possible to carry out these two processes at the same time or one after the other.

1. Privatisation

The non-governmental organisation EEB (2002) uses the term ‘privatisation’ to refer to the management of the operation of a service such as water, or part of it, that is devolved to a private entity. It thus includes sales; concessions, leases, or, any form of joint venture or partnership between companies and public authorities in the services; management contracts; and the sub-contracting of elements of service such as maintenance. In the EUROMARKET project, ‘privatisation’ is a more restrictive term. It consists of the transfer of assets ownership (including infrastructure) and management from the public sector to the private sector. For example in England and Wales the WSS system and management is under the sole responsibility of private, but fully licensed, operators.

2. Private Sector Participation (PSP)

The term ‘private sector participation’ refers to the delegation of the management of an activity to a private firm from the former public operator, while retaining public ownership of infrastructure. For example, in the water sector, with the concession or lease contract to private operators, the management of services is delegated to private operators, but the water supply system always remains publicly owned (French model).

3. Competition

The European Community's competition policy (Articles 81 to 89 of the EC Treaty, formerly 85 to 94) is based on five main principles:

1) the prohibition of concerted practices, agreements and associations between undertakings which may affect trade between Member States and prevent, restrict or distort competition within the common market;
2) the prohibition of abuse of a dominant position within the common market, in so far as it may affect trade between Member States[1];

[1] The first two principles may, however, be subject to derogations, particularly when an agreement between undertakings improves the production or distribution of products or promotes technical progress. In the case of state aid schemes, social subsidies, or subsidies to promote culture and conservation of heritage, are also examples of possible exceptions to the strict application of competition rules.

3) supervision of aid granted by the Member States, or through State resources in whatever form whatsoever, which threatens to distort competition by favouring certain undertakings or the production of certain goods;
4) preventive supervision of mergers with a European dimension, by approving or prohibiting the envisaged alliances;
5) liberalisation of certain sectors where public or private enterprises have hitherto evolved monopolistically, such as telecommunications, transport or energy.

There exist different forms of competition. Indeed, the general meaning of competition refers to competition in the market (where the final consumers may directly and at any time choose the suppliers of the good they purchase). When it is difficult to implement direct competition, other forms are feasible: competition for the market refers to bidding process for franchising or concession; comparative competition consists in comparing performance of different companies operating on different geographical areas but on similar services.

The difficulty of pursuing an effective competition policy lies in the fact that the Community must continually juggle aims that are sometimes contradictory, since it has to ensure that:

1) the quest for perfect competition on the internal market does not make European businesses less competitive on the world market;
2) efforts to liberalise do not threaten the maintenance of public services meeting basic needs.

4. Liberalisation

The term 'liberalisation' refers to a process by which competition is introduced in situations or sectors hitherto characterised by exclusive or special rights, or monopoly granted to historical operators[2].

European directives on standards: various influences among countries

The standards in the water supply and sanitation sectors in the EU are most comprehensively covered by European directives. At European level, there is a clear gap between the intense activity triggered around the implementation of the drinking and bathing water-related directives and the poor impact of the fish, shellfish and pollution-control directives.

A related side-effect of the legislation has been the fostering of further research for technologies to comply with standards and for improving the scientific validity of the standards, which in turn have highlighted new issues and approaches (Kallis and Nijkamp, 2000). However the impact of these standards in different Member States is quite different.

In some countries like the U.K, France and Italy, huge investments were required. This led to significant institutional reforms in water management (i.e. privatisation of the service, establishment of new monitoring bodies, etc.). Hassan (1995) argues that the high investment costs related to the European water directives have been a major driving force for the re-organisation of the U.K. water industry. In order to expand the financial capacity for conforming to the directives' requirements the water service was transferred from the public to the private sector (extending the financial base through shareholders).

In other countries, water management was not as strongly influenced by European directives. These countries already had well developed environmental protection structures: environmental standards were already high, and water management was already very well organised. This is the case for the Netherlands, Germany, Denmark, Sweden, and Austria. This partly explains why private sector involvement did not significantly increase with the implementation of EU standards in these countries.

Hypotheses on impacts of standards on private sector involvement

We have decided to test the following hypotheses in order to highlight impacts on private sector involvement:

1) EU standards lead to growing technological complexity, which results in an increasing professionalisation of the water sector. This can induce more private sector involvement.
2) Directives represented increasing costs on treatment. This increase in treatment costs encourages private sector involvement.
3) Directives put pressure on water resources management (change of scale, concentration, network expansion). This has an effect on private sector involvement.

[2] Historical operators are (was) publicly owned companies that exercise (d) a monopoly over certain public services (e.g. British Telecom, Deutsche Telecom, and France Telecom, in the telecom sector).

4) Directives lead to the involvement of new stakeholders. New stakeholders have divergent impacts on water liberalisation.

Many factors influence the development of national water management systems and their pattern of industrial organisation and technical change. Geography and hydrology, for example, constrain local development since they shape patterns of quantitative and qualitative availability, determine the territorial size and the technical complexity of water infrastructure, require to a larger or smaller extent appropriate institutional developments in the model of water governance, and so on. Hence, it is difficult to isolate the sole impacts of European legislation on national standards and water management. Nonetheless, we have focused mostly on the financial implications of the European directives, the impact of legislation on technological complexity, and the emergence of new stakeholders.

Improving the quality of drinking water and wastewater can be considered as one of the most ambitious objectives of the EU environmental policy. But it is also proving to be the most expensive. After having analysed the impacts of these directives in Italy, France, Germany and the Netherlands, one can test the hypotheses advanced above. One can observe the following trends regarding liberalisation and private sector involvement:

- *Hypotheses 1 and 2: Increasing standards led to technological complexity, increasing costs on treatment and changing cost allocation between actors.* In France and Italy, this clearly boosted private sector involvement (in Italy this is true for the second generation of directives). Municipalities were confronted to growing complex treatment systems that they were not always capable of understanding and financing, whereas private companies were financially stronger, had qualified personnel enabling them (especially for the French companies) to assume the design, construction, and management of the most complex systems. This private sector involvement is much less pronounced in the other two countries (Germany and the Netherlands) that had a more developed environmental legislation and infrastructure. However, financial impacts have not been without important consequences: private sector involvement has increased with the development of public – private companies in Germany, and a growing dependence on external financiers can be observed in the Netherlands. However, the Dutch government prohibits the privatisation of the water supply.
- *Hypothesis 3: Increasing European standards led to a change in the scale of services.* In the French and German cases, this hypothesis is not confirmed. In Italy and in Portugal, it was the national regulation and not directly EU legislation that pushed towards a more concentrated structure of water management. However, these national reforms, aiming at more rationalisation in the management, were clearly presented as a way to reach the new EU directives requirements. In the Netherlands, EU directives may have enhanced the move towards the concentration of drinking water companies.
- *Hypothesis 4: New stakeholders emerge with the evolution of water management.* It is important to consider the relevant actors playing a decisive role in the water management system. Some of these actors come from outside the water operator system bringing key resources for the development of the sector. It is possible to consider the following stakeholders: education and professional associations, external consultants, planning officers, suppliers of technology, public research institutions, and independent institutions. In France, the change in managing water resources, from a supply-side approach to a demand-side, has brought the emergence of a new stakeholder: the water consumer. The water price increase has generated a strong opposition from consumers, accompanied with the debate about the consequences of "privatisation" of water industry (market power) and the lack of transparency of municipal management. The public opinion shows a growing sensitiveness toward health protection and quality of tap water, calling for more sophisticated water treatment processes, prevention of pollution risks, and the protection of the resources. It requires an improvement in professional expertise. In the Netherlands, stakeholders participate directly or through representation to make good management possible. Consultants are increasingly used by the Water boards and water companies to help them apply these regulations. This trend is reinforced by the drive for more efficiency in the 1990s, which led many water companies to doing away with staff departments and hire consultants instead. This facilitates public water management (and has no impact on private sector involvement).

Highlighted channels

The relationship between the development of drinking water standards and urban wastewater treatment and their impact on private sector involvement in the EU is not straightforward. One cannot point out a unique trend towards liberalisation, considering each country's particularities. On the contrary, it is possible to pinpoint several channels through which private sector involvement has developed or not: technical complexity, financial pressure, regionalisation and the emergence of new stakeholders.

1) The increase of technical complexity of water management forces local water management systems to delegate decisions to professionals having the necessary expertise. This leads to the creation of specialised water management organisations, who may belong to private or public sector depending on national paths and traditions.

This does not necessarily mean that water operators will themselves be privatised: Italian and German municipally-owned water companies are well-known examples of water industry developed within the public sector, though with some sort of entrepreneurial autonomy. In other countries, the need for professional skills has led quite early to delegation of operation to private companies, instead of developing the same skills in-house (French case).

2) The second channel is the financial one. The increasing cost of water and sanitation services in the present phase is paralleled by the crisis of public finance, therefore ending the traditional way of financing through the public budget. Self-finance through full-cost recovery starts to become necessary. Tariffs and prices paid by consumers will serve as a cash flow for sustaining market-based finance, for which the water operator becomes the intermediator. This does not necessarily lead to private operation of services, since public companies might also have the capacity to access financial markets (e.g. the Netherlands). Given the high capital intensity of the water industry and the very long depreciation schedule of water assets, the financial reputation of operators (private or public) is a crucial aspect in order to reduce the cost of capital and finally the cost of water to final consumers. Even if remaining under public control, many water companies have engaged in enlargement of capital to new investors, buyouts and other financial operations opening their property to external investors.
3) The third channel regards the "regionalisation" concerning water operators and water regulators. Local water management cannot be considered anymore as a "simple" task within the frame of a publicly-controlled water plan; the water operator itself has to become "a planner", and therefore acquire both suitable geographical scale and vertical integration.
4) Finally, the fourth channel regards the need of water service management to interact with a larger number of stakeholders with conflicting objectives. In particular, there is a stronger emphasis to be put on the demand side rather than on the supply of new assets and technology. Growing awareness of users regarding increased water price and problems of quality linked with new standards could put pressure against any liberalisation process (including private sector involvement). However, necessity of transparency and claims for decreased water prices can lead to the introduction of more competition.

Water Framework Directive: structuring principles for WSS services

The EU has affected numerous aspects concerning the water sector in Europe for thirty years. Whereas standards concerning both quality of water bodies and emission limits derive mainly from the EU water legislation, the Water Framework Directive (WFD)[3] rationalises EU water policy. The single objective is the good ecological status for all waters. The WFD also introduces the concept of integrated river basin management. We focus here on an assessment of the underlying "intervention logic" of the WFD. Moreover, we aim to identify and analyse some challenges resulting from the linkages between the new resource perspective of the WFD and a possible liberalisation process in the water sector.

Principle and content of the WFD

The WFD initiates a reform of the EU water policy. This corresponds to the wish of the Member States to reduce the financial burden and the complexity of action due to the many fragmented EU-water related directives. The WFD considers water according to a 'resource approach'. All the EU water directives are grouped together and the objectives of quality and limitation of emissions are combined. Its objective is to reach a good ecological status for all waters as of 2015 through an integrated water management at (international) river basin scale. The WFD sets up guidelines and leaves a large room for manoeuvre to the Member States during implementation. It defines the issues at stake for water management in the Member States for the forthcoming years. According to the principle of decentralised action, the WFD defines detailed objectives, but leaves the Member States with the duty to set the mechanisms of implementation for reaching these objectives.

The WFD targets the preservation and improvement of the aquatic environment of the Community as the ultimate goal. The Member States determine the '*good status*' of all waters. When the 'good status' is already achieved, it has to be maintained. The Member States have the duty to protect, improve and restore the water bodies in order to reach the 'good status' as of 2015. These objectives are legally-binding (article 4).

In addition to the 'good status' requirements, the WFD determines organisational objectives. It organises the planning and development of knowledge on the resource at the scale of river basins with an objective of international cooperation. The territorial unit retained for the water management is the *river district* (article 3). The river district corresponds generally to the river basin or watershed. The river basin is a territory where all running waters, including

[3] Directive 2000/60/EC of the European Parliament and of the Council of 23 October 2000 establishing a framework for Community action in the field of water policy.

rainwater, flow into the sea at the same estuary. Examples of river basins (not only the river but also the territories around all its tributaries) are the Rhine basin, the Seine basin or the Danube basin. The river district is an administrative territory. Sometimes its borders do not perfectly fit with the watershed, particularly for groundwater.

The WFD proposes to set up an *integrated water management*. The directive is not explicit on defining this term, but some elements help to delineate the concept. First, integration means that water is considered as a resource (resource approach). Water policy is not sector-based anymore (e.g. belonging to transport, agriculture, public health or environmental policies). Second, integration means administrative coordination. Policies are conducted through a single guiding document (planning approach) for the whole river basin (territorial approach): the management plan. The management plan contains a description of the resource, broad policy objectives for the river basin, all-national and European water legislation affecting the river basin (combined approach) and a monitoring programme. It also describes the cost-effective set of measures that will be taken in order to satisfy the objectives for all water bodies in the river basin district. Third, integration means rationalising and co-ordinating all policy interventions in the water sector. Integration here is a meta-principle developed in the more operational principles.

The main instrument of the WFD is the programme of measures. A *programme of measures* is elaborated for each (or the national part of a) water district. It inventories and contributes to implement measures necessary to reach the 'good status' (article 11). The measures must fit the characteristics of the district presented in the management plan. The possibility to adopt national measures is maintained. The programme of measures integrates the requirements of national/regional and European legislation in a single document. It is composed of compulsory and complementary measures. The *compulsory measures* encompass the measures required by the European legislation, measures concerning the distribution and use of water (i.e. cost recovery, promotion of a sustainable use, prior authorisation and control of wells), the measures of control and authorisation of recharges of groundwater, a regulation of pollutant discharges, a prohibition of discharges of pollutants in groundwater, and the regulation of the hydromorphological conditions of the water body. It also includes a regulation and control against diffuse pollution, a prohibition of discharges of pollutants in groundwater, measures to eliminate emissions of priority hazardous substances and measures of prevention against industrial hazards. The *complementary measures* are left to the discretion of the Member States (facultative measures). They are introduced in order to reach the objective of a 'good status' for all waters (see above). They include all kinds of public policies affecting (directly and indirectly) water quality/status. If the measures of the programme are not sufficient to meet the objectives, they are revised and eventually supplemented by new complementary measures.

The Member States have to implement the WFD gradually, but according to a precise time schedule. However, they have to set new institutional arrangement up quite quickly. In fact, the first deadline was December 2003 with the transposition of the directive into national legislation and the designation of water district authorities. After that, the programmes of measures and control are progressively put in place and implemented. Good status for all waters in 2015 thus requires early action by all the Member States.

All in all, the WFD is (to be) understood as a directive on the ecological quality of water. It rationalises the EU water policy and coordinates national water policies following a resource approach (as opposed to a sector-based or problem-based approach). The reforms that are currently conducted in the context of its implementation still raise many questions. The challenges raised by the WFD are approached through three perspectives: the ecological perspective, the economic perspective and the institutional perspective.

From WFD principles to liberalisation issues

While economic principles play an important role in the WFD, it should be stressed that the directive does not call for a one-dimensional "economisation" of European water management and protection. This manifests itself foremost in article 1 of the directive: "Water is not a commercial product like any other but rather a heritage which must be protected, defended and treated as such". At the same time, the question arises if the WFD does entail any information on preferable market structures for the provision of WSS services in light of the discussions on liberalisation/privatisation of these services in Europe. Overall, the WFD does not specify how the WSS sector should be organised in the different Member States. The Directive can thus neither be considered as being *per se* in favour nor against a privatised or liberalised water sector. It leaves this point deliberately open.

The WFD is based on a number of main guiding principles. One may examine the relationship of some of the WFD principles to the market situation in which they are applied. It highlights issues that are of importance to different liberalisation scenarios as well as their possible impacts on future developments in market structures (regulation, environmental safeguards etc.). From ecological, economic, and institutional principles anchored in the WFD, we briefly summarise their linkage to liberalisation in Table 1.

Table 1: Summary of issues derived from the linkage of the WFD principles (and their potential) implementation with a possible liberalisation process

Implementation of the WFD principles as...		*Issues related to liberalisation of WSS services*
ECOLOGIC	**Integrated basin management**	Integrated management requires coordination of different geographical scales and different uses (energy, transport, agriculture, fishing), as they are all interdependent. It can be more difficult to reach with private actors.
	Good status Environmental protection, conservation	The preservation of the ecosystems is an objective having priority, whereas any utility in a competition context looks for economic profit. Reaching the good status requires a long-term planning and changes of the current practices. It is sometimes incompatible with market transactions that are mainly short-term decisions.
	Prevention and Precautionary	These principles are based on intergenerational responsibility that is difficult to take on by markets. They include additional risks that are not always taken into account in a competitive market, and that should be imposed with precise common calculation rules. The unknown risks accounted for by these principles should be included in the calculations, but that implies an increase of costs.
ECONOMIC	**Polluter pays principle, cost recovery principle (for water services as defined by WATECO)**	The WFD induces a monetary vision of water values (including social and environmental aspects) that may open to a market vision. The reporting requirements of the WFD will induce increased transparency on financial information (costs, prices etc.) thus enabling comparative competition. The recognition that the price of water should cover the costs (investment and operation costs) is a pre-condition for private sector involvement because it legitimates increases of price. If cost recovery has to include also environmental damages, there is a need of a precise and common methodology to assess those costs, so that all actors make the same hypothesis in a competition context. Indeed, difficulties to assign costs in complex and synergetic impacts, difficulties to value eco-social assets and uncertainties of forecasting make the cost-recovery principle difficult to implement in a market structure. There is also a need of a strong regulatory framework to avoid short-term planning by private actors in order to ensure long-term cost recovery; to avoid 'cherry-picking' by private operators in a liberalised market; to avoid restricted access to information for calculating actual costs and degrees of cost recovery.
	Cost-effectiveness of measures	There is an increased difficulty of reaching actor consensus on cost-effective set of measures. However, safeguards are needed to avoid sub-optimal outcomes due to an unequal distribution of bargaining power.
	Derogation if disproportionate costs	Increased information asymmetries hinder the assessment of alternative options. One can anticipate a surge in derogation due to private actors interests.
INSTITUTIONAL	**Public participation**	As public opinion usually fears for new private sector participation in the water sector, new avenues for large public participation could probably be a constraint for a liberalisation process.
	Management basin authority	The concept of WFD is by itself crossing countries borders by putting emphasis on the river basin. This multinational character breaks down national frontiers and as such might be a driving force for addressing also the drinking water and sanitation sector over geographical and administrative country borders. Currently within the EU the drinking water and sanitation sector is still very much bounded by the country borders. However, problem of co-ordination between basin authority, water service regulator and market competition regulator can appear.

THE FRAMEWORK CREATED FOR A LIBERALISATION POLICY IN THE WSS SECTOR

Our purpose here is to analyse the European policy in the liberalisation of the public utilities in general, in order to answer two main questions:

1) Why has the water sector not been treated like other network industries and what makes the water industry 'special' from a European point of view?
2) To what extent does the European liberalisation policy of network industries influence a possible European policy in the water sector? Will the water sector follow the same path as electricity or telecom, or will it follow its own original path towards a regulated liberalisation?

In order to answer these questions, we propose two analysis frameworks: one to describe the nature of utility networks and one to describe liberalisation process. In particular, we first offer a general analysis of network industries. We present their specificities as 'services of general interest' and analyse the recent liberalisation processes, considering both empirical evidence and doctrinal evolution. Then we look more specifically at the water sector, underlining its own specificities. Finally we discuss the extent to which the processes and doctrine elaborated for the network industries in general could also be relevant in the water sector.

Empirical Evidence and doctrinal evolution in the liberalisation process of network industries

Driving forces and extent of liberalisation across sectors and countries

Until recently, due to the specificity of network industries, national network markets in Europe were in the majority of cases characterised by national or local monopolies, where the incumbent was State or local authority-owned. However, since the mid-1980s, liberalisation has become an important trend in these industries, clearly calling into question the State's role as a single monopoly provider. These liberalisation processes are initiated with two main objectives:

- to construct a common market for each sector of services of general economic interest defined by each Member State (communications, transports, energy, but not water or sanitation).
- to encourage greater efficiency to the benefit of users of these services, by introducing competition in fields that had often been "protected" by monopoly situations or by exclusive rights at the local, regional and/or national level.

The liberalisation process in network industries was supported by significant changes, namely:

- technological development (particularly in telecommunications and electricity),
- diversification of needs and of demand : users have new requirements concerning services, quality, and the satisfaction of their diverse needs and desires,
- failure of public management and dysfunction of previous types of organisation and regulation,
- potential competitors as industrial strategies led to a growing search for competitiveness,
- growing influence of neoliberal ideology and virtues of competition proclaimed by the EU (internal market).

The impact of these driving forces varies, so that the liberalisation process in network industries has been taking place unevenly (see Table 2). Furthermore, another reason for this unevenness is the fact that these industries were not part of the 1992 internal market programme, and only later an (*ad-hoc*) approach has emerged.

As Table 2 illustrates, the liberalisation process in the European network industries varies greatly between sectors. The telecommunications sector on the whole is the most advanced in terms of liberalisation (especially for the business segment). It has mainly been driven by rapid technological change and the strong growth in demand. The European institutions reacted promptly imposing full competition by January 1998[4], both in the market, and on and between networks. At the other end of the spectrum, are the cases of water and railways (especially passenger transport), where there has been a major problem to introduce free entry and competition. In railways for example, the major problem has been mainly due to network compatibility problems, to strong anti-competitive habits, and to powerful labour unions[5]. Directive 91/440/EEC[6] on the development of the community's railways was a first step towards establishing a single railway market.

---

[4] Commission Directive 96/19/EC amending Directive 90/388/EEC.

[5] Pelkmans (2001b).

[6] Amended by the Directive 2001/12/EC of the European Parliament and of the Council of 26 February 2001.

Table 2: Main driving forces for liberalisation in six network industries

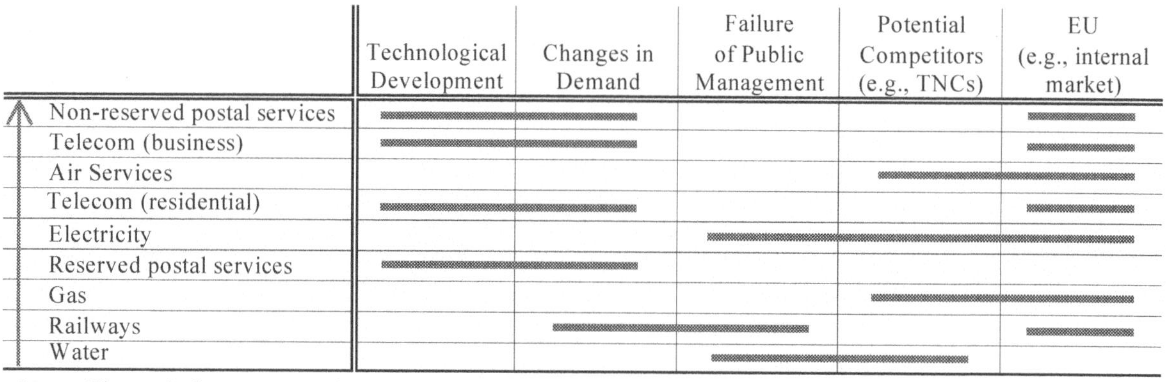

Notes: The vertical arrow means that sectors on the bottom are less liberalised than those on the top.

At the same time, and because liberalisation in network industries remains mainly a national issue, it also varies greatly across countries. The liberalisation process is more advanced in the northern European countries (e.g. the UK, which was a precursor in terms of privatisation starting from early 1980s, and Scandinavia) than in Southern Europe[7]. Table 3 illustrates the disparities found in terms of the degree of liberalisation (low, medium and high) both between sectors and across seven selected European countries.

Table 3: Degree of liberalisation in six network industries in selected European countries[8]

	Germany	*Spain*	*France*	*Italy*	*Sweden*	*UK*	*Switzerland*
Telecom	H+	H	H	M	H	H+	H
Air Transport	H	M	M	H	H	H	M
Energy	M	M+	L	L	H	H+	L
Railways	L	L	L	L	H	H+	L
Postal Services	M	H	L	L	H+	M	L
Water*		H	H+	L+	L	H+	L

Source: adapted from Bergman et al. (1998).
Notes: * refers to opening competition for the market.

A cross-sector analysis of the liberalisation process in network industries

Liberalisation is a means of establishing the conditions of market rules and competition, and its rational lies on the recognition that, in principle, competition is more prone to achieve efficiency than monopoly. Although it is difficult to analyse these different industries within a general liberalisation framework, we suggest the following framework:

1) the different routes to liberalisation: competition in the market, for the market, comparative competition and self-supply;
2) the possible means of implementing competition and liberalisation: private sector participation, unbundling and breaking down national borders;
3) the accompanying mechanisms (positive and negative) that can be observed in competitive situations: a customer driven approach and cream-skimming of the market[9].

[7] European Commission (1999).

[8] This table resumes the actual opening of markets per sector according to the available data. Nonetheless, one must take into account the opening of markets that is officially mentioned in laws and rules, but also examine what happens in reality to competition. For example, one often remarks that the German market concerning electricity is 100% open compared to the still protected French market, whereas in fact a bigger percentage of industrial consumers changed suppliers in France compared to Germany.

[9] "Cream-skimming" refers to a situation where new entrants tend to target the most lucrative markets.

The routes to liberalisation: introducing different forms of competition

1. Competition in the market

In the case of competition in the market (*ex-post* competition) operators compete for end users. Therefore, network compatibility (which may be achieved by standardisation), access and pricing[10] are fundamental to establishing a competitive environment in the market. Competition is normally fierce at the early stages – to persuade users to use the network. Due to switching costs, competition between operators after the first stage is more limited. The entry of a new operator into a mature network industry is likely to occur only in the event of a significant technological improvement.

Since the late 1980s, the European Commission has been trying to build an internal market and to introduce competition in the electricity and gas markets for generation and supply. However, these attempts have largely failed. The main impetus to liberalisation came from national plans in the U.K. It was not until the late 1990s that electricity and gas directives[11] were agreed upon. These called for competition in the market, although in practice energy networks remain fragmented and largely based at national level. Nevertheless, at least for electricity, there is already a long tradition of national co-operation, with frequent power exchanges through interconnections[12]. This suggests that the main constraint to full liberalisation may not be technical. Concerning access to the network, both directives set out several possibilities, regulated third-party access being the most commonly adopted[13].

With regard to air transport, there is a need to distinguish between charter, cargo and scheduled passenger traffic. The first two market segments are lightly regulated and benefit from a highly competitive market, while competition in the last is still limited by bilateral agreements that regulate world air transport[14]. The air transport sector has been gradually liberalised in Europe since the 1980s, following three packages[15] of liberalisation measures in 1987, 1990 and 1992 that enabled greater competition to take place in the market.

2. Competition for the market

Due to strong economies of scale or to the obligation to provide services of general interest (SGI)[16], some segments or markets cannot be opened to full competition. The solution in these cases is to institute competition *ex-ante*, i.e. competition for the market. For instance, water supply and sanitation are sectors where large-scale competition in the market is unlikely to develop because many segments are characterised by natural monopolies.

3. Comparative competition

When direct competition cannot be created, a method currently used is to compare performance of different companies operating on different geographical areas but on similar services. This is called yardstick or comparative competition. The comparisons can be made for segments of the utilities' operations and can cover a range of variables such as capital maintenance costs, operating costs, prices and quality of service.

4. Private supply

A definite good may be obtained from a public supply or instead may be privately supplied. Private supply occurs when one consumer (self-supply) or a group of consumers (cooperative supply) supply themselves rather than rely on the incumbent utility (Webb and Ehrhardt, 1998). When a user decides to get himself the good, he may construct his own network within a legal basis for his own final use. For instance, this is the case of a firm that needs a large quantity of this good but the connection to the public network is more costly than self-supplying. However, the main obstacle to such competitive service at a city scale is that operations, such as private abstraction, are prohibited by law in many countries.

---

[10] The exceptions are postal services and airlines, which are logistical and not physical networks, see Pelkmans (2001a).

[11] Directives 96/92/EC and 98/30/EC respectively.

[12] European Commission (1999).

[13] The options regarding access to the network are single-buyer system, and negotiated and regulated third-party access (TPA). TPA is where producers and consumers contract supplies directly with each other, see Bergman et al. (1998).

[14] Chicago Convention of 1944.

[15] Council Regulations No 3975/87 and 3976/87 of 14 December 1987; Regulations No 2344/90 and 2367/90 of July 1990; and Regulations 2407/92, 2408/92 and 2409/92 of August 1992.

[16] Services of general interest will be developed in the following section on the construction of a European doctrine.

Means of implementing competition

1. Private Sector Participation (PSP)

The private sector may be an important source of capital and new investment that are relatively scarce for many (developing) countries. It also brings management expertise, new technologies and knowledge, which may not be presently available.

Under the traditional public procurement model, public authorities can utilise the services of the private sector for design, construction and management of network utilities, with the award of individual contracts made on a competitive basis. The delegation of services poses a certain number of problems. First when the contract is written and then at the time of contract follow-up. The main stake is the clear definition of the responsibilities of each part but also the reduction of informational asymmetries.

There are a number of ways in which greater private sector participation can be introduced. It is possible to create three classes for outsourcing former public functions to the private sector. These mechanisms present opportunities for the private sector to participate in varying degrees in the maintenance, operation and management of infrastructure improvements. These categories are concession, leasing and intermediary management (services contracts, operation and management contracts). The main feature of concession is that the infrastructure is financed by the private sector. The private firm also operates the service. Leases provide a means for private firms to purchase the income streams generated by publicly owned assets in exchange for a fixed lease payment and the obligation to operate and maintain the assets. Moreover, public authorities can enter into service contracts with private companies for the completion of specific tasks. Service contracts are well suited to operational requirements and may often focus on the procurement, operation and maintenance of a limited range of equipment. Management contracts are used to transfer responsibility for asset operation and management to the private sector. These contracts involve both service and management aspects and are often useful in encouraging enhanced efficiencies and technological sophistication.

PSP does not necessarily mean liberalisation. A handful of companies, mostly based in France and England, dominate the water multinationals. The top two, Veolia and Suez capture about 70 percent of the existing water market share. The oligopoly nature of the market characterized by a strong concentration of companies prevents all forms of real competition.

2. Unbundling

Despite the fact that the need for compatibility and interconnection might suggest a pressure towards monopolisation and vertical integration, various segments may be exposed to some forms of competition. Therefore, there is the need for unbundling, i.e. for separating the network into its reserved (natural monopoly) and competitive elements. The definition and allocation of exclusive rights between these elements, as well as the prohibition of cross-subsidisation, become key issues[17]. For example, in the electricity industry, recent technological change enables the efficient generation of electricity at a small scale, decreasing considerably the economies of scale at the generation level, while transmission remains a natural monopoly.

3. Breaking down national borders

Another aspect concerns the breakdown of national borders, which can help the realisation of the internal market and in some cases is directly promoted by transnational corporations (TNCs). However the extent to which national borders have been broken down depends on the specificities of each sector and on the degree of liberalisation already achieved.

Accompanying mechanisms of liberalisation

1. Developing a customer-driven approach

As mentioned previously, the liberalisation process is about the opening up of network industries to competition. An increasingly competitive market also significantly strengthens the need to have a more customer-driven approach. This becomes critical in order to lower costs and to attract new clients (i.e. through marketing, reputation and trust). In the postal service, the main impetus to opening up some segments to competition[18] stems not only from the increasing potential for substitution of traditional mail products (created by rapid technological change), but also from the considerable pressure coming from new customer demands (in terms of quality and type of services). The ability of postal operators to respond to these changing demands is becoming critical for their success.

---

[17] Pelkmans (2001a).

[18] The 1997 Postal Directive (97/67/EC) engaged the sector in a process of "gradual and controlled" liberalisation, leading in practice to a highly segmented market only partially liberalised.

2. Cream-skimming the market

When entry is made possible, new entrants tend to target the most lucrative markets ("cream-skimming"). Prices tend to fall the most in markets where this selective form of competition is active[19]. The problem is that other segments of the market, which are essential to fulfil social objectives or for the proper functioning of the entire economy, may become under-supplied. Therefore, "cream-skimming" may justify the need to define the obligation to provide services of general-interest.

The Construction of a European doctrine: from deregulation to re-regulation

The liberalisation processes whose main characteristics were developed above led to important changes in ownership and in the market structure itself, representing major regulatory challenges. One can identify two main objectives for the regulation of network industries, depending on the way competition is considered: on the one hand, if competition is seen as the unique path towards efficiency, regulation should prevent the creation of artificial barriers to entry; on the other hand, if the limits of competition are highlighted, the regulation should control the quantity, quality and pricing of the service provided, and the obligation to provide SGI[20]. From the beginning of the liberalisation process in the 1980's, the construction of a European regulatory policy for SGI took into account these two sets of objectives – although sometimes contradictory.

Separation between operators and regulators and creation of regulatory authorities

The liberalisation process in network industries was based on the separation between operators and regulators: regulatory authorities have been created in telecom, energy and postal sectors in several countries (see Table 4 for United Kingdom, Germany, Sweden and (later) France).

Table 4: National regulators in network industries

	Energy	*Telecom*	*Postal services*
UK	Office of Gas and Electricity Markets (OFGEM), 2000	Office of Telecommunications (OFTEL), 1984	Postal Services Commission (POSTCOMM), 2000
Germany	in project	*Regulierungsbehörde fur Telekommunication und Post,* 1996	
Sweden	*Energimyndigheten*	*Post & Telestyrelsen (PTS)*	
France	*Commission de Régulation de l'Energie (CRE), 2000-2003*	*Autorité de Régulation des Télécommunications (ART), 1997*	in project

The main mission of these regulators is to ensure that the access to networks is open to new entrants so that competition is fair. This objective has been partially reached for telecommunication: thus, for example, the incumbent in Germany (*Deutsche Telekom*) operated only 47% of long distance communications in 2000, while *British Telecom* in United Kingdom still had 60% of the market in 2000.

Introduction of concepts that limit the negative effects of competition

There cannot be a complete liberalisation process in the network sectors, by relying solely upon European competition law. In these sectors, competition tends to be oligopolistic, to undermine previous tariff structures (that explicitly sought to cross-subsidise in favour of lower-income citizens), and finally to ignore positive and negative externalities (both town planning and impacts on the environment for example).

As a result of these conditions, European rules - the result of debates, actors' initiatives and social movements - have been based on implementing a controlled, organised and regulated liberalisation, with the emergence of European concepts, both sector-by-sector and cross-sector. Consequently, two main concepts were elaborated, to introduce a minimum set of obligations (Universal Service Obligations), and to affirm the specificities of network industries considered as Services of General Interest.

[19] European Commission (1999).
[20] European Commission (1999).

1. Services of general economic interest (SGEI)
The first reference to service of general economic interest dates back to 1957: the article 86 of the treaty of Rome mentions the term of service of general economic interest subject to particular Public Service Obligations (PSO) defined by each Member State, according to their history, traditions and national institutions.

Jurisprudence of the Court of Justice of the European Communities acknowledges, since 1993-1994[21], that services of general economic interest come under other objectives, missions and types of organisation than general rules of competition. Later, the new article 16 of the Treaty of Amsterdam (1997) refers positively to the SGEI (common values, promotion of social and territorial cohesion), but has no direct juridical effect and has a limited impact: it does not modify the imbalance of the treaty that mainly focuses on competition policy. Article 36 of the Charter of fundamental rights (2000), proclaimed during the European Council of Nice, also highlights the importance of SGEI for citizens and residents of the European Union.

2. Service of general interest (SGI)
The concept of service of general interest is developed in 1996 in the communication by the European Commission on the services of general interest (revisited in 2000) and later in a report at the European Council of Laeken (2001). Services of general interest cover market or non-market service activities, that are considered by national public authorities to be of general interest, and are subject for this reason to particular Public Service Obligations (PSO). It is important to notice that the scope of SGI and the content of PSO are defined by each Member State (for instance, some countries have included in the content of PSO for telecom the access to high-speed Internet, or the cover of the whole territory by mobile phones) However, one limit of the SGI is that this concept does not have any juridical foundation at a cross-sectoral level.

The difference between SGI and SGEI is clarified in the *green paper on services of general interest* published in May 2003. The distinction between services of an economic nature and services of a non-economic nature is important because they are not subject to the same rules of the Treaty. For instance, provisions such as the principle of non-discrimination and the principle of free movement of persons apply with regard to the access to all kind of services. However, the freedom to provide services, the right of establishment, the competition and State aid rules of the Treaty only apply to economic activities. Concerning the distinction between services of an economic nature and services of a non-economic nature, any activity consisting in offering goods and services on a given market is considered as an economic activity. However, this distinction between economic and non-economic activities has been dynamic and evolving, and in recent decades more and more activities have become of economic relevance. For an increasing number of services, this distinction has become blurred (CEC, 2003).

3. Universal Service Obligations (USO): a minimum set of obligations by sector
Along with the European Directives introducing competition in several network industries, the European Union was progressively led to define a universal service obligation in the sectors of telecommunications (1992 and 2001), postal services (1997), and recently electricity (2003), and also public service obligations for transports (article 73, Treaty of Rome, 1957).

The definition and guarantee of a universal service obligation ensures, for all users and consumers, the accessibility and the quality of services during the transition from a situation of provision of services under monopoly to that of markets open to competition. Universal service, in an environment of open and competitive markets, can be defined as a minimal set of services of a given quality and at an affordable price, to which all users and consumers have access to, considering particular national circumstances. The content of this minimal set varies across sectors and is also likely to evolve. In the telecom sector, the European obligation is limited to the access to traditional telephone. In the postal services sector, the universal service is more ambitious, as it implies the collection and distribution of letters and parcels everywhere and 5 days a week. This content imposes itself to each Member State, contrary to the Public Services Obligations that can be defined at a national level.

The *green paper on services of general interest* (2003) specifies that during the last two decades, the concept of universal service has developed into a major and indispensable pillar of the Community's policy on services of general economic interest. However, this is quite new, as this notion was considered as a transitory concept in 2000 and 2001 by the Commission.

4. Current debates on financing and organisation of PSO and USO
The *green paper on services of general interest* mentions that it has always been the core responsibility of public authorities to ensure that basic collective and qualitative needs are satisfied and that services of general interest are preserved wherever market forces cannot achieve this (CEC, 2003). Up to now, the crucial importance of this

[21] See the Corbeau and Almelo judgements of 1993 and 1994.

responsibility has not changed. However, what has changed is the way in which public authorities fulfil their obligations towards the citizens. Indeed, public authorities increasingly entrust the provision of such services to public or private undertakings or to public-private partnerships and limit themselves to defining public objectives, monitoring, regulating and, where necessary, financing those services. The green paper remarks that "this development should not mean that public authorities renounce their responsibility to ensure that objectives of general interest are implemented." (CEC, 2003, point 24, p. 8) The Commission intends to reaffirm this responsibility by stimulating a European debate on the political choices to be made concerning services of general interest at European level.

The national, regional and local authorities of each Member State are in principle free to define what they consider to be a service of general interest. Public authorities in each Member State retain considerable freedom to define and enforce public service obligations and to organise the provision of services of general interest. This allows Member States to define policies that take into account specific national, regional or local circumstances. But the absence of specific legislation can lead to legal uncertainty and market distortions: "in the water sector the absence of specific, relevant regulation has led to very different industry structures across Member States" (CEC, 2003).

Guiding principles in the construction of the European Doctrine

From the various contributions and evolutions developed above come out several principles which constitute the core of the European doctrine.

1. Transparency and proportionality of means with respect to objectives
As mentioned above, each State and public authority is responsible for defining what comes under services of general interest, within the scope of its competence. This responsibility implies that missions of service of general economic interest be clearly defined on a regular basis with transparency and that subsidies be accorded proportionally to objectives defined.

2. Consumers as new stakeholders
One of the justification of liberalisation processes and the introduction of competition rest on the efficiency gains from which consumers will benefit. This would imply that regular, comparative and pluralistic (as for criteria chosen and assessment methods) assessments be conducted. This would also mean that consumers and their representatives be regularly consulted and associated to the definition of the trends as well as to their implementation.

3. Subsidiarity
This principle is often incorrectly presented as giving priority to local levels in each circumstance. Some accordingly assert that the definition and organisation of services of general interest in the European Union should mainly come under States or infra-national authorities. In actual fact, article 5 of the treaty of the Communities specifies that one must assess *on a case by case basis* whether such objective must come under such level, and this is so for a matter of efficiency[22].

4. Neutrality
Article 295 of the treaty mentions the neutrality of European integration as regards property rights of operators, of nationalisations/privatisations: "This Treaty shall in no way prejudice the rules in Member States governing the system of property ownership".

New issues arising from the combined evolution of doctrine and practice

While EU internal market and competition policies have contributed to the opening of markets to competition in sectors so far protected by monopolistic situations and exclusive rights, it was stressed that other (non-doctrinal) driving forces have also contributed to the process. Nonetheless, the nature and degree of interactions between doctrine and practice towards the liberalisation of network industries are different depending on the sectors. For example, in the liberalisation of the telecommunications sector the main driving forces were the rapid technological change and strong growth in

[22] Article 5 of the treaty of the communities: 'The Community acts in the limits of the competencies that are imparted to the latter and of the objectives that have been attributed to it by the actual treaty. In the fields that do not come under its exclusive competence, the Community does not intervene, in accordance with the subsidiarity principle, unless the objectives of a contemplated action cannot be adequately achieved by Member States and can therefore be better accomplished at community level, owing to the dimensions or impacts of the envisaged action. Action of the Community does not go beyond what is deemed necessary in order to attain the objectives of the present treaty.'

demand while the EU doctrine surged as a reaction to these main drivers. On the contrary, in the case of railway transport, the essential driver towards liberalisation came from the European Commission and the necessity to develop Trans-European Networks, i.e. strong networks properly linking the different European regions into a single market characterised by freedom of movements[23].

Independently of these dynamics, the fact is that the combination of the opening of network markets to competition and the construction of a European Doctrine is in every sector raising important challenges in two areas: regulation and restructuring of utilities.

The need to guarantee independent regulatory agencies and better coordination with competition authorities

The traditional approach to regulation relies on the assumption that effective competition will develop in network markets, progressively weakening the role of regulation. Thus regulation would be needed only where natural monopolies prevent competition: in that case, regulation would try to do its best to reproduce competition (like pseudo-competition through comparison in water sector). However, the move towards competition in network industries like electricity or telecom has created a need for better coordination with newly created independent regulatory agencies. Thus, when market failures in network industries (specificity of the assets, high sunk costs) prevent complete competition mechanisms, the need to guarantee independent regulatory agencies remains an important challenge.

Despite the great regulatory variety existing between sectors, the profound transformations that can be expected in these industries will bring about sufficiently common characteristics to justify equivalent responses regarding regulation. This means that transectoral regulation could probably arise in order to deal with the common need to introduce further competitive mechanisms; to develop integrated European markets; and to regulate segments that remain out of the competitive market and possible PSOs (including their financial issues).

Restructuring and the development of multi-utilities

The change towards a more competitive market progressively leads to the restructuring of former vertically integrated network industries. The need to "unbundle" the supply-chain to permit direct competition in appropriate businesses is leading to structural changes in these industries. Furthermore, the reinforcement of horizontal mergers, take-overs and joint ventures aimed at achieving economies of scale through cross-sectoral integration of similar activities, is leading to a high level of concentration[24]. This is also increasingly occurring between sectors, generating new structures known as multi-utilities. This new strategic concept of utility companies, normally operating at a global level, must be distinguished from the prior concept of municipal multi-utilities, which in Germany, Austria and Switzerland have been delivering local "multi-utility packages" since the 19th century[25]. The rationale on the basis of municipal multi-utilities is mainly based on economies of scope, i.e. on the belief that the average total cost decreases as a result of increasing the number of different goods provided. For global multi-utilities, the motivation to operate in more than one utility sector is based on the expectations to benefit from economies of scale. Efficiency gains and cost reductions are also expected from grouping activities with similar risk and return dynamics. The emergence of these global multi-utilities makes regulation a more difficult task, mainly due to rising problems of asymmetric information.

To ensure that the long-term objectives of liberalisation are achieved, the threat that further liberalisation efforts may engender the dominance of a small number of oligopolistic players needs to be addressed. Common challenges to all sectors are to make sure new entrants have competitive access to the markets and that they are able to sustain investment in order to ensure efficiency and sustainability.

Evolution of the EU doctrine and practice for the WSS sector

While electricity, telecommunications, postal services and transport have undergone significant changes due to a European liberalisation process, the water sector has until now been treated differently, and was not subject to the European policy of liberalisation of network services. The objective of this section is to underline the specificities of the water sector, using the same framework as other network industries, and to analyse to what extent the liberalisation processes and the concepts introduced above are meaningful for the water sector.

---

[23] Under the terms of Chapter XII of the Treaty (Articles 154, 155 and 156), the EU aims to promote "the interconnection and interoperability of national networks as well as access to such networks".

[24] This concentration phenomenon has been (at least implicitly) encouraged by the EU Directorate-General for Competition who has tended to treat each national market as the basis for assessment of dominance, see Helm (2001).

[25] See Rothenberger (2002).

Specificities of the water sector compared to other network industries

Nature of the good: a merit good and a quasi-public good

1. A merit good

Water supply and sanitation services (WSS) have strong merit good characteristics, meaning its social benefits exceed the private benefits (given the strong externalities). Therefore, if water provision and sanitation are left only to the private sector, it is likely that they are under-supplied, especially for the low-income population. The public authority often has a greater understanding than consumers themselves of the benefits of clean water and proper sanitation to their health, well-being and to the global environment. Indeed, governments may limit the potential of water utilities to exclude customers (from either connecting or being disconnected) on the basis of national social objectives. If this occurs then water and sanitation services increasingly look like quasi-public or common pool goods (again the classification will depend on the supply-demand balance which may alter over time).

2. A quasi-public good

One difficulty to characterise the nature of WSS services is that it is often confused with the nature of water resources. On the one hand, water resources are subtractable when they are finite, and water infrastructures are subtractable when they are at the limits of their capacity. On the other hand, water resources are excludable when abstraction/discharge licensing and monitoring is in place, and water infrastructures are excludable when connection points are easily identifiable - as is the case with households. This implies that almost all water infrastructures are - depending on capacity considerations – theoretically private or tollable services. This is similar to the other network utilities. However, unlike some of the other network utilities, there is a high risk of market failure and a need for strong government involvement. Moreover, the important externalities and social objectives may, in practice, limit excludability. Hence when there is sufficient infrastructure capacity, water and sanitation services are possibly transformed into a quasi-public good.

Characteristics of the production: a natural and local monopoly and strong externalities

1. Natural and local monopoly

Whereas technological changes have allowed the distribution in the telecom sector to be competitive, the water sector has remained a natural monopoly. Contrary to electricity, the transport of water is very expensive (important irreversible sunk costs) compared to the value of the good in itself. Moreover, duplicating water networks (whether for raw water transfer or the delivery of potable water) is not realistic and is a real barrier to entry for possible competitors. The obligation of quality requires a relative proximity between water treatment and consumption place. Thus, it has traditionally been treated and even sourced locally. In the absence of any dramatic technological change, this situation is unlikely to change in the foreseeable future.

However, despite relatively high transport costs it is not uncommon in some Member States (the UK with its regional water companies and some large cities in southern Member States with local supply constraints) for raw water to be abstracted, stored and then transferred in bulk to the treatment works from over 100 km away. In southern Europe it is also not uncommon for these bulk raw water transfers to be managed by organisations other than the municipality responsible for drinking water supply. Indeed the raw water so transferred may be sold and used for several different purposes – agricultural irrigation, industrial cooling and potable water supplies. Such a regionalisation (and diversification) of raw water resources has helped to improve water availability, and hence customer service, in times of drought.

Whilst the water network utility has been justly described as the natural local monopoly *par excellence* (Littlechild, 1986), there are some trends towards regionalisation (both in terms of management and remote resource development - as potable water demands grow and local water sources are fully exploited) which are extending the geographic impact of this so-called local monopoly. That is the main difference with other network services (e.g. electricity, telecom) that were historically produced and supplied at national (rather than local or regional) level. Consequently, the management of WSS services varies greatly across countries, as there are numerous local (or even regional) solutions. This diversity of management modalities is a specific characteristic of water sector.

Similar arguments about the high cost of transporting foul sewage and drainage water to the treatment works pertain. Indeed the preference for gravity based sewerage systems further encourages the development of small local monopolies. However there will always be opportunities for neighbouring municipalities to construct treatment facilities jointly (thereby gaining economies of scale).

So again, whilst a local monopoly, cost pressures are encouraging the formation of more regionally based sewerage monopolies.

2. Strong externalities on health and environment

Water supply is unusual among network utilities because it has strong positive and negative externalities that are health-related. The provision of clean water provides enormous positive externalities to public health through the control of water-borne diseases. On the other hand, water is part of a cycle, and needs to be cleaned before it is discharged back into the environment: the unregulated disposal of raw sewage by water utilities can cause enormous negative externalities in the form of disease and environmental contamination.

3. Important asymmetries of information between providers and public authorities

One important feature in the characteristics of WSS services is the asymmetry of information between the (public or private) provider and the public authority in charge of controlling the sector (local or national): their relationship is dominated by an asymmetry of information on costs.

Social objectives

Drinking water, its supply and sanitation services are considered as prime examples of 'general interest' or 'public interest' activities. Indeed water is essential for the life and the organisation of various economic activities. Most countries guarantee a universal access to water and/or to a minimal level of service. Moreover, environmental considerations are increasingly taken into account. This is in addition to the risks that poor water quality poses to human health. It appears that water and hygiene education is necessary to improve and guarantee the population a high level of health.

Characteristics of the demand

1. Elasticity of demand

With ordinary economic goods there is a (negative) relationship between price and demand. The slope of the demand curve is called the price elasticity of demand. This is defined as the percentage decrease in demand resulting from a percentage increase in price. For primary uses of water, the elasticity is close to zero (inelastic) since these are essential needs of the user. For the most essential use of (drinking) water, few substitutes are available. For sectors such as industry and agriculture, demand for water is generally more elastic. This is because alternatives for water use exist in these sectors (e.g. introducing water saving production technologies, shifting to less water demanding products/crops). In urban water supply, price elasticity is low[26]. Poor consumers can often only afford to use small amounts of water, and any increase in tariffs will have little effect because they cannot do with less water. For richer consumers (that irrigate their gardens and wash their cars) their ability to pay is such that the need to save money on water is limited. When the price elasticity of demand for water is low, as is the case for urban water, the unregulated water (monopoly) provider may raise tariffs since users are captive.

2. Important asymmetries of information between providers and consumers

The 'perfect market' model requires that consumers and producers have all the information that they require in order to arrive at rational decisions. But this may not be so in the case of water. For example, the producer often has a greater understanding than consumers of the quality of the water than consumers themselves. In other sector such as Telecom, this problem (related to the quality of the good) is less predominant.

Processes visible in the water sector: their extent and their limits

Despite the fact that many parts of the water and sanitation sector are considered as natural monopolies, many of the features that are at the core of the liberalisation process of other network industries can already be observed to some extent in several countries. We discuss in next paragraphs their relevance for the water sector.

The routes to liberalisation and competition

Contrary to other network industries, competition in the market is difficult for many reasons: natural monopoly (very high sunk costs and extensive economies of scope and scale), transportation costs predominate, extensive universal service obligations, geographical, environmental and local constraints. With possibly one notable exception, competition in the market has therefore not flourished across Europe, or indeed the world. After briefly describing this

[26] See for instance Nauges and Thomas (2000).

experience, we present two main models that provide tools to foster competition for the management of the whole water or sanitation service: competition for the market (French model) and comparative competition (implemented in England and Wales).

1. Competition in the market: the experience of common carriage in UK

The separation of the management of infrastructure (mainly the pipe network, which is a natural monopoly) and the delivery of the commodity is one unique example of promoting competition in the water sector. In England and Wales, since the Competition Act 1998, common carriage[27] has been possible. It occurs when for instance one company agrees to carry water through its pipes for another company. However, no such arrangements currently exist on a competitive basis. The main issue to solve here is the condition (notably the price) of access to the network. Common carriage has been recently formalised in the national Water act 2003). However, it has been restricted (at least initially) to large water consumers (the threshold is currently set at 50 megalitres per year). This pilot experience in the UK is designed to encourage new (possibly more efficient) entrants into the water market. However, they will still have to become licensed (like the existing water companies).

2. Competition for the market

Competition for the market occurs when potential (public or private) operators bid competitively for what is called in France a delegation contract, otherwise known as a franchise contract. The winner of the competitive tender should be the bidder that proposes the best quality of service with the lowest price. The delegation contract can take various forms but the most frequent are concessions and leases. The competitive tendering[28] for private sector provision of water supply and sanitation is a growing mode of competition in numerous European countries (e.g. France, Spain and Czech Republic). This way of involving private sector financing and management is also strongly promoted by international lending agencies such as the World Bank, especially for developing countries.

However, there are important limitations on competition for the market[29]. First, competition for the market can only occur every 10 to 15 years because most contracts require a long duration to amortise the investments made by the operator (in the case of concessions or concessive leases). Besides, competition should be transparent and based on objective criteria (e.g. price, quality of water), whereas the delegation contract is a long-term relationship based on trust. Hence, there is a trade-off between transparency and quality of the contractual relationship[30]. Another significant problem is that only a small number of major private companies are involved in the international water market (oligopolistic competition). They are extremely powerful in their bargaining with municipalities and governments alike, and such companies can easily engage in collusion. Moreover, it is well known that there is an important advantage for the incumbent operator that arises from its detailed knowledge of the state of infrastructure and operating conditions, known as ‘first mover advantage’. Finally, the specification of contract is a difficult issue to deal with because of problems of asymmetric information (e.g. incomplete contract, difficulty to monitoring) and uncertainty (e.g. on future investment and revenue stream). Should bidding for WSS services become compulsory, it is important to introduce not just one criterion on the price of water, but also criteria on quality of service.

Alternatively, it is possible to observe increasing competition for some parts of the vertical chain of activities in the water distribution process (e.g. network maintenance, leakage location, metering, billing and vehicle maintenance). With simple service contracts, it is possible to contract out some specific tasks to the private sector while the monopoly provider (private or public) remains responsible for the overall management of the utility. These contracts are typically for short periods - from six months to two years - and take advantage of expertise of operators. This strategy of breaking up water and sanitation services into smaller business units leads us naturally to consider the phenomenon of unbundling.

3. Comparative competition

Indirect competition through yardstick or comparative competition is another way to improve cost efficiency and the quality of service. It is an approach that has been adopted by OFWAT (Office of Water Services) in England and Wales. Key indicators of company performance are collected and published. Unit and marginal costs are also used to set price-

---

[27] Common carriage occurs when one service provider uses another's assets, such as its pipe network or treatment works, see Ballance and Taylor (2001).

[28] This procedure is regulated by the European Council directive for the entire infrastructure sector that obliges water companies up for tender on a regular basis. See Directive 98/4/CE.

[29] See Ballance and Taylor (2001).

[30] In France, the principle of free choice for the local community is called *intuitu personae*.

cap formula and as the basis for periodic price reviews.[31] The objective of comparative competition is to 'mimic' the normal pressures of a competitive market environment. However, such comparisons require accurate information in order for the regulator to assess performance in a specific environment. Moreover, as the number of companies in England is decreasing because of mergers, the value of such an approach may gradually be eroded.

The implementation of competition in the WSS sector

1. Private Sector participation

Increasingly public authorities have been turning to the private sector for help in developing and delivering WSS services. They hope to take advantage of private sector wealth and knowledge to finance new investments and to improve the efficiency of existing services. As seen in the paragraph on competition for the market, the public-private partnership is carried out thanks to different types of delegation contracts.

2. Bundling and unbundling: scale issues

Unbundling refers to different industrial realities: vertical disintegration (between production and distribution of water), geographic disintegration or horizontal disintegration (between water and sewerage).

The objective of vertical disintegration is to separate the different constituents of the sector. For instance, the separation between water production and water distribution or between collection and sewage treatment is possible. Different firms can operate each of these parts of the utility. Another example of vertical disintegration happens when private companies operate customer centres specialised in billing, and handling complaints. Such operations may be consolidated with similar customer management facilities for electricity or telecommunication utilities, thereby benefiting from economies of scope.

Geographical disintegration can lead to indirect competition if a water regulator can compare the performance of separate companies. However, this may run contrary to claimed economies of scale. In many countries, the objective is to achieve the most efficient (minimum) size (in terms of average cost minimisation). Hence, the geographical service area is actually growing (e.g. intermunicipality in Belgium and France, optimal territorial basin in Italy (ATO, *Ambiti Territoriali Ottimali*...)).

Horizontal disintegration may destroy potential economies of scope. Is it efficient that the same operator is in charge of several urban services: water supply and sanitation, but also electricity and natural gas? For instance, a common billing could reduce costs. However, there is no clear evidence of the benefits of horizontal disintegration[32] and experiences differ among countries. In Italy and in some municipalities of France, water and sanitation networks are operated by the same firm. In Germany, the same utility also operates other urban services.

In any case, there is a need of strong coordination between the different activities. For instance, the production stage requires data from the distribution stage for the estimation of long-term demand. The necessity of coordination is also trivially crucial concerning the quality of water. Thus, when unbundling is performed, there is a risk of loosing the benefits of economies of scale, scope and vertical integration.

3. Breaking down national boundaries via operators

As water transportation is costly, international transfers of water are not frequent (one of few examples is the Rhone-Barcelona water transfer project). Thus, the breakdown of national boundaries in the water industry arises mainly via TNCs' and not yet via infrastructure. As noted by Finger and Allouche (2002), the TNCs presence in the water sector is highly concentrated with only four private companies and their subsidiaries controlling the privatised global water market. Table 5 shows the presence of these companies in every region of the world.

---

[31] See the report of OFWAT in 2001 "The role of long run marginal costs in the provision and regulation of water services". See also the annual reports of OFWAT: "Financial performance and expenditure of the water companies in England and Wales"; "Levels of service for the water industry in England and Wales".

[32] A paper by Hunt (1995) shows that there exist cost complementarities between water supply and sanitation on the one hand, and between water supply and environmental activities on the other hand. This means that there are significant benefits in terms of efficiency in joint operation of activities.

Table 5: Main global presence of TNCs

	Western Europe	Eastern and Central Europe	Asia	North Africa	Africa	South America
Suez - ONDEO	H	H	H	L	L	H
Veolia Environnement (VE)	H	H	H	L	L	H
SAUR international	M	M	M	L	H	M
Thames Water (RWE)	M	M	H	L	L	M
Aguas de Barcelona (Agbar)[33]	M	L	L	M	L	H

Sources: adapted from Silva, Tynan and Yilmaz (1998), Hall (1999) and TNCs web sites.
Notes: H, M, L are respectively for high, medium and low presence of TNCs.

A common strategy of these groups is to increase their presence in different parts of the world by acquiring or increasing shareholding in affiliate companies. However, there are sometimes difficulties in buying the shares of these firms because of national constraints (e.g, Veolia Environnement in France in 2002). One last limitation to the breaking down of national boundaries is that the international presence of TNCs is usually restricted to the big cities for reasons of financial profitability.

Other accompanying mechanisms

1. Segmentation of the market

The segmentation of the market has different meanings and implications according to the geographical level where it occurs. At an international level, competition through private participation is limited to large middle-income cities, where profitability is secured. To that extent, there is a segmentation of the global market. At the local level, the market is segmented between domestic and industrial consumers. Increasingly specialised private firms contract with industrial firms that have a large consumption of water, that need special quality of water, or that need their own sewage treatment facility. The last step of segmentation would be among domestic users who require specialised services. This has already been observed in telecom and electricity sectors.

2. Growing importance of customer service

The customer-driven approach has already been implemented with interesting results in the water sector. The concern for better communication with users originates in England and Wales with the creation of customer charters that define standards of service. OFWAT has set up regional Customer Service Committees (CSCs) in order to investigate customer complaints regarding water and sewage companies. Moreover, the performance of customers' service is measured by different indicators. This approach is expanding in other countries, notably France and Scandinavian countries, with the development of performance indicators that include user satisfaction.

European doctrine in the water sector

A sector that cannot be assimilated

As mentioned previously, the management of WSS services has been excluded until recently from the global trend for liberalisation. Thus, the concepts developed to limit the negative effects of competition were normally not necessary. However, most countries have developed a 'right to water' (see Smets, 2002) that guarantees the access to water for everybody (e.g. disconnection is forbidden in some States). This right to water is similar to a Universal Service Obligation that would include access and continuity of service. Furthermore, Public Service Obligations could include social aspects on participation of users – as prescribed in the Water Framework Directive.

For a long time, the European Union has not taken a decision regarding the nature of water and sanitation services. According to WFD (2000), "the supply of water is a service of general interest." However, the recent Green Paper on Services of General Interest (2003) considers water supply as a Service of General Economic Interest and "the provision and organisation of theses services are subject to internal market, competition and State aid rules provided that these services can affect trade between Member States". To date the UK alone – with publication of *the Competition Act*

[33] Agbar is owned 47.7% by Hisusa (owned 51% by Suez), 11.8% by Endesa, and 1.47% by SUEZ.

1998, application in the water and sewerage sectors (31 January 2000) – has effectively identified water supply and sanitation services as Service of General Economic Interest.

Recent publications from the European Commission on services of general interest and public-private partnership

Until 2003, the EU adopted the regulation relating to the Service of General Interest (SGI) on a sector basis (in particular for the large network industries). The EU degree of intervention depends on the nature of the SGI, which, on this point, are classified in three categories in the Green Paper on SGI of 21 May 2003:

1) the Services of General Economic Interest (SGEI) provided by the large network industries, which "have a clear Community-wide dimension and present a strong case for developing a concept of European general interest";
2) the other SGEI, "such as waste management, water supply or public service broadcasting, [which] "are not subject to a comprehensive regulatory regime at Community level" but are subject to the rules relating to the domestic market and competition;
3) and finally the non-economic services and the services without effect on trade between Member States, which are not subject to the rules relating to the domestic market, nor to specific Community rules.

The sector of water is thus excluded from the European policy of liberalisation, and subjects itself only to the general rules on public procurement and concessions.

1. White Paper on SGI – COM (2004) 373, May 2004

According to the consultation following the publication of the Green Paper, there is no agreement emerging on the question on the opening of the water sector at Community level. Even if, in its internal market strategy published in 2003, the Commission writes in the paragraph devoted to network industries: "one area where new action may be required is the water sector – which remains fragmented and where there are potential gains to be had from modernisation.", the White Paper is limited to announce an evaluation of the water sector, by recognising its specificities fully. Insofar as the Commission prefer not to present a proposal for a framework directive relating to the SGI, there is thus no direct impact for the moment on the water supply and sanitation services.

Following the Altmark Judgment of 24 July 2003 which affirms that the subsidies given to compensate for services related to the obligations of public service are not regarded as State aid and thus do not need to be notified, under certain conditions, the Commission proposes various initiatives which make safe the financing of the SGEI.

Nevertheless, the White Paper could have an indirect impact on the access to the information, which is mentioned within the guiding principle "ensuring consumers and users rights". This same paragraph evokes "the existence of independent regulators with clearly defined powers and duties". "These include powers of sanction (means to monitor the transposition and enforcement of universal service provisions) and should include provisions for the representation and active participation of consumers and users in the definition and the evaluation of services". The sectoral evaluation of services is also one of the new announced orientations ("regulators should also monitor market developments and provide data for evaluation purposes").

2. Green Paper on public-private partnerships and Community law on public contracts and concessions - COM(2004) 327, April 2004

If the White Paper on the SGI has only a limited impact on the management of the water services, insofar as it pushes back the decision to later, it is not the same thing concerning the Green Paper on Public-private partnerships ("PPP"), which explicitly calls into question the framework for the award of the service concessions.

The Community legislation distinguishes first the public procurement contracts, for which procedures of awarding resulting from recent directive 2004/17/EC apply (relating to the special sectors which are water, energy, transport and the mail service). On the other hand, the concession gathers sub-contracting subject to publicity, and the service concessions, which obey only the general principles of transparency, equal treatment, of mutual recognition and proportionality.

In its Green Paper, the Commission underlines the legal insecurity related to the absence of clear rules on the concessions, which are in addition often difficult to qualify. Proposing the new device of "competitive dialogue" set up in the last directive (2004/18/EC), for complex markets where the adjudicator is not able to define the technical means to meet its needs, the Commission suggests unifying the mode of awarding of all PPP of the contractual type. This proposal, which would have the merit to simplify the procedures, must be considered with prudence:

1) on the one hand, one cannot qualify "particularly complex" delegations of service in the water sector, where the communities as well as the private actors have very great experience, and where the needs are easy to

define; it would thus be necessary to plan to soften the conditions of recourse to this procedure as stated in directive 2004/18/EC;

2) in addition, it is important to preserve the freedom of negotiation during the conclusion of the contract between the community and the private operators, insofar as the two partners engage for long durations;
3) finally, it seems relevant to evaluate the implementation for the specific cases for which it was created (and in particular, complex public contracts implying at the same time the design, realisation, financing and the management of equipment on behalf of the authorities) before extending it.

In any event, the French principle of the *intuitu personae* would be threatened, with the reason which it restricts competition.

3. Directive "Services": discussions in progress

If the consultation on PPP is closed, the discussion on the Directive "Services" is currently in hand, and can have an impact on the procedures of attribution of contracts concerning the sanitation services. In its last version, the Directive "Services" excludes the water supply services from the principle of the country of origin (chapter III, article 17). However, without additional notification, the Directive includes the services of sanitation and waste management. The regulation (through procedures of authorisation and approval) of these two local SGI is necessary to control the negative externalities, which they can have on the environment and the public health. The draft Directive is today in the course of rehandling by the Commission.

Common trends in water and sanitation management and regulation in Europe

Despite the great diversity among European management of WSS services, one can observe three common trends in the evolution of water and sanitation services. These are: a shift towards market logic (with a larger private participation), a process of regionalisation (grouping of municipalities and increasing scale of WSS services), and the creation of new regulatory systems. There is a strong correlation between these three trends, as we illustrate further.

Regionalisation of WSS services management and a shift towards market logic

Regionalisation of WSS services management refers to an increase in the territorial scale of WSS services, often due to a grouping of municipalities. In Italy, as a result of the Galli law in 1994, responsibility has moved from municipalities to the ATOs (*Ambiti Territoriali Ottimali*), regional bodies of a larger geographical scale. This law also states that the delegation and competition for WSS services must become easier. However, private sector participation still remains relatively low. In Portugal, a 1993 decree-law reformed water policy by setting up a "multi-municipal" management arrangement and a framework for the system of delegation to public or private operators. In England and Wales, regionalisation dates back to 1973. The complete privatisation of these regional water and sanitation services was carried out in 1989. France and Spain are the EU countries where private operators are the most active through concession or lease contracts, while a trend of 'cross-municipal collaboration' has been strong in France since 1999.

Increasing the geographical scale of WSS services and introducing private sector participation could be two ways to address the same issue: the move towards professional and performance-orientated services. This look for performance implies increasing technical skills, access to information and access to financial resources. In some countries, where water supply is provided by public operators/companies, they come under private law (e.g. *Eigengeselschaft* in Germany), even though the capital remains in the hand of local authorities. This type of governance is more and more common in countries that have chosen the public management arrangement (e.g. the Netherlands and Scandinavia). Furthermore, all countries are taking into account the fact that water is a commercial product and should follow the cost recovery principle[34] adopted as a general rule in the WFD

[34] The cost recovery principle consists in setting a water price that covers not only operation and investment costs of water supply and sanitation but also the resource and environmental costs.

Regionalisation and professionalisation are thus often combined in several European countries. Figure 1 shows the trends of water management for some European countries during the last thirty years by correlating the geographical level of management (local, local consortium or regional) with the degree of private sector participation: one can observe that most countries are following the same path, although their initial point greatly varies.

Source: adapted from Guérin-Schneider (2001).

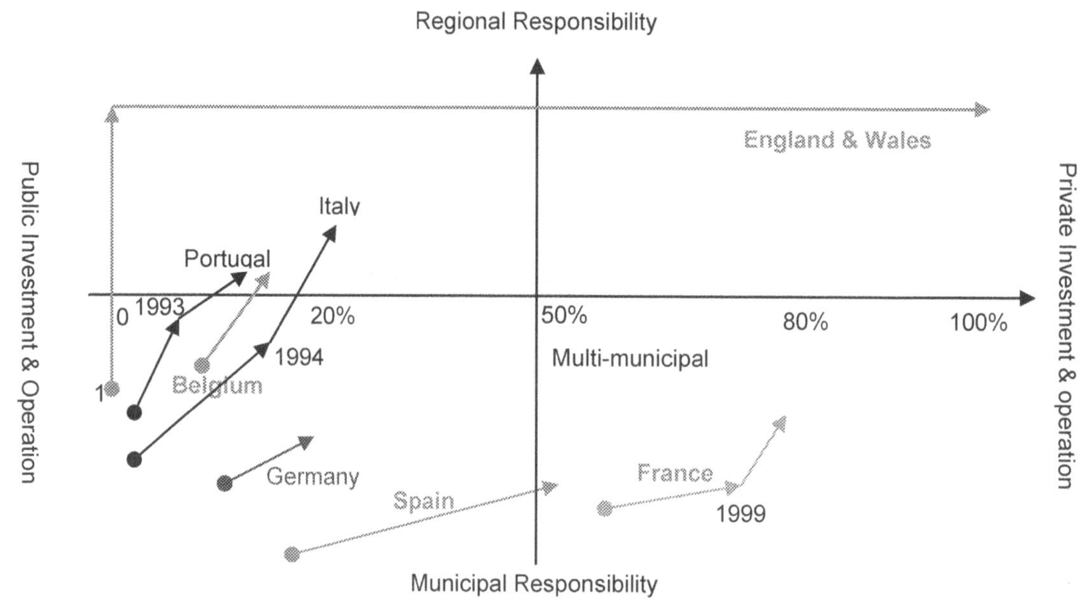

Notes: an arrow indicates a trend, the point at the beginning of the arrow is the initial date of observation (1973), a date mentions an important reform. The percentages on the horizontal axis refer to private investment and operation. Finally, the vertical axis refers to a range of situations: municipal responsibility at the bottom, grouping of municipalities within a "consortium" (France, Portugal, Italy for example) in the middle, and complete absence of municipalities for water management at the top (England only), replaced by regional authorities.

Figure 1: Trend for privatisation and regionalisation in some European countries since 1973

Regulation authorities Re

The water sector presents a great diversity of situations, ranging from the separation of operators and regulators with the creation of a national regulator (e.g. Office of water Services in England), through the complete integration of local operators and local regulators (e.g. municipal management in Germany), with some intermediate situations, where there is a separation between operator and regulator, but where the regulator remains at a local or regional level (delegated management of French model). However, as water supply is a natural monopoly, the missions of a regulatory authority for water (e.g. OFWAT) will differ from the missions of a regulator for telecom (e.g. OFTEL), see Littlechild (1988): whereas OFTEL has to organise a transition towards competition, determining the tariffs for different competitors for the access to network, OFWAT has to organise forms of comparative competition and competition for the market within a natural monopoly.

With the development of private sector participation, there has been a focus on regulation. Many European countries have tried to control operators, whether public or private, by way of regulation authorities. The most complete system is the Office of Water Services (OFWAT) in England and Wales. OFWAT monitors prices in setting a "price cap" formula. A system of quasi-competition ("yardstick competition") has been set up in order to compare the costs and performance of water companies. After a considerable increase during the first five years, water prices have fallen in real terms since 1995 and it is expected that they will soon reach their pre-privatisation level (see Byatt 2001). Moreover, it is now acknowledged that the water companies are much more efficient, both in terms of operation and maintenance, water quality and service quality. The main question about the success of the regulation remains the efficiency of asset management with the risk of under-investment on the long term and the capacity to enforce the public service obligation (see Defeuilley 1999 and Mac Gowan 1995).

In other European countries, regulatory authorities have been created, while the delegation of water and sanitation services has been developed and private participation facilitated. In Italy, the *Comitato per la vigilanza sull'uso delle risorse idriche* is an independent institution established by the Galli law that is in charge of driving the reform on the WSS services and the regulation of water prices. The regulator has also to represent the interests of users. A central objective in the reform was to decentralise the activities of the regulator to the level of ATOs with the setting up of regional authorities in charge of collecting and processing statistics. But until now, there has not been sufficient willingness to coordinate between all stakeholders. In Portugal, the *Instituto Regulador de Aguas e Residuos* (IRAR) began operations in 2000. IRAR is an authority under the supervision of the Ministry of the Environment. Its competencies focus on tariffs, quality, market access and investment. However, IRAR was initially only allowed to make recommendations and on the economic aspects of regulation. During the first year of implementation it was generally recognised that IRAR did not have sufficient powers to implement a genuine system of regulation. Things have changed since the nomination of a new Director and with the implementation of a benchmarking process with a list of compulsory performance indicators collected, analysed and published by IRAR. The first results should be published before December 2005. In France, a report by the *Haut Conseil du Secteur Public* (1999) recommended the creation of a regulation authority for water and urban services in charge of establishing technical standards, investment financing and price index rules. In June 2001, the government proposed a law to the Parliament that included the setting-up of such a regulation authority (*Haut Conseil du Service Public de l'Eau et de l'Assainissement*). The negotiations have been very hard, and the prerogatives of this authority have been progressively cut down. Since the 2002 general elections, the project has been shelved.

CONCLUSION

The objective of this chapter was to design a framework based on the analysis of EU policy on standards, resource management and liberalisation of network industries, in order to analyse water liberalisation scenarios. The EU has no clear role on the liberalisation of WSS services. First, the ultimate goal is not liberalisation. Second, EU policy in the WSS sector is somehow neutral in that respect. Indeed, there is no explicit liberalisation policy. However in the same time, there are some contradictory signs. On the one hand, the water sector is excluded from the general competition laws and still remains under the national and local responsibilities. On the other hand, some discussions within EU lead us to think that a future action in the WSS sector is on the agenda. The following analysis presents the drivers and the constraints underlined in the previous section in a schematic way.

A cross-analysis of the explicit and implicit European policies influencing the WSS services allow us to point out three topics that could relate to a possible liberalisation of WSS services:

- the emergence of new stakeholders, with on the one hand the development of (public and private) industrial companies in the water sector, (whose objectives are to widen their area of activity), and on the other hand, the suspicion of public opinion about the business of those companies in the water sector;
- national, European and international discussions on the nature of WSS services;
- an beginning evolution of political and juridical framework that allow to build innovative forms of public-private partnerships.

New stakeholders

The stringency of drinking water standards increase the technical complexity of water and waste water treatments, and forces local management systems to be professional, because technologies required imply a high degree of specialisation. Combined with the principle of cost recovery (from WFD), it leads to an increase of costs covered by WSS users. This increase in costs has two main consequences: on the one hand, it creates financing needs that appeal to external private financing; on the other hand, it increases prices, with suspicious reactions of users who do not want to pay high prices for water. Driven by the need of professionalism and the financial pressure, local management authorities have different options that can eventually be combined:

- they can delegate their services to a private operator who has the expertise and the capital;
- they can group with other local authorities in order to reach a critical size that allow the development of skills, the mutualisation of financial resources, and the access to financial markets;
- in the case of public sector, local authorities have to adopt an autonomous status that enables them to borrow and act as an enterprise.

Whatever the solutions chosen by countries are, according to their culture and their existing industrial structure, the result of these trends is that big companies, public and private, are developing a financial autonomy - often under private law status - and professional skills. They may be potential competitors if liberalisation is introduced in the WSS sector.

The nature of WSS services

The second causal chain is rather well separated from the first one, and it deals with the nature of WSS services. It includes the monopolistic structure of WSS services and their inclusion within ecosystems that impose ecological objectives. Thus, more regulation and safeguards are needed in order to avoid excessive rent from the monopoly and non-respect of ecological constraints. The nature of WSS services also includes on-going discussions on the qualification of WSS services as a Service of General Interest (SGI) or as a Service of General Economic Interest (SGEI). This qualification has strong consequences on the competition rules that must be applied to the water sector, and the latest green paper of the commission on SGI (COM(2003) 270 final) tends to indicate that water is acknowledged as a SGEI, thus calling for more competition.

The evolution of the regulation on public-private partnership and the rules for competition

The third causal chain is closely linked with discussions on the qualification of WSS services. Although EU has until now not submitted water to its general competition laws, the debate is now open[35] at European level. At a national level, several countries have introduced laws enabling concessions with competition for the market in the water sector. As seen above, the Commission published a green paper on Public-Private Partnerships in 2004 which is carrying a reform when a public authority decides to award the management of A service (including a SGI) to a third party.

Numerous questions concerning the rules of competition have no answer at this stage of debates at European level. Is there a choice to do between public contracts or concessions for the water services? If yes, it is maybe necessary to take care that the price be not the sole criterion for the choice of the operator and that other criteria such as quality of service be taken into account. The pure and simple abandonment of the negotiation is not desirable anymore because it allows for the adaptation of the contract by monitoring the quality of service and adjusting the schedule of conditions (because of incomplete contracts).

Figure 2 summarises the three causal chains that result from existing European policies:

1) The emergence of new stakeholders (public opinion and water industries) has contradictory effects on liberalisation issues, with two diverging lobbying groups;
2) The discussions on the nature of the good are still unsolved and their impact on liberalisation issues is still uncertain;
3) The regulation on competition and Public-Private-Partnerships - which is one important levy of EU - is currently evolving towards an increased liberalisation of WSS sector.

Hence, these channels do not exhibit obvious conclusions on the future liberalisation of WSS services.

---

35 See the recent communication of the commission on Internal Market Strategy (COM(2003) 238), where water is mentioned as "one area where new action may be required".

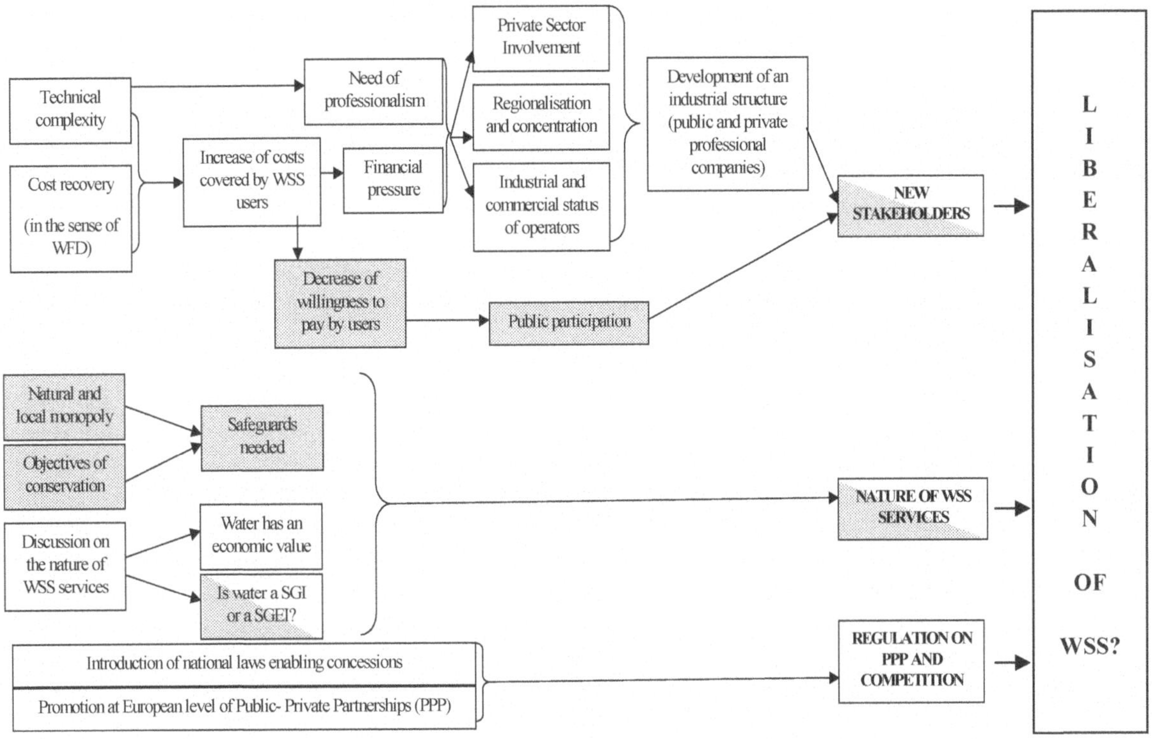

In grey appear factors that are considered as constraints for a liberalisation of WSS services.

Figure 2: Framework figuring drivers and constraints for WSS services liberalisation.

Many issues cannot be solved at this stage and will be further developed in next chapters of the book. Nonetheless, some of the safeguards needed in any liberalisation processes can be highlighted. In the perspective of an action or a policy of the European Union, different (non-exhaustive) questions are related to the channels leading to a liberalisation of WSS services:

- Is it necessary or desirable to precise the nature of the good 'water' and at the same time to introduce a new definition for a concept more specific to the WSS services on the pattern of concepts already existing such as SGEI or Universal service?
- Is it possible and/or desirable to introduce competition rules and/or derogation to strict competition (direct negotiation, best deal rather than cheapest deal, etc.)?
- In the case of a liberalisation process with the introduction of competition, which type of additional regulation process has to be set up? How have the problems of information and coordination to be treated?

This last question on the regulatory arrangements and the specific problems relating to the regulation of WSS services is crucial.

Planning a liberalisation process is not simply a matter of choosing the type of status of operators or the level of private involvement or the degree of competition. The reality and the consequences of a liberalisation process will depend on the regulatory authorities and mechanisms implemented to reach the best efficiency (i.e. to reduce costs, to invest and to improve the quality of service), optimal environmental objectives and a good social acceptance. Moreover, regulation objectives are not only there to protect the customers from the abuse of the power of monopoly suppliers and to avoid damaging interruptions to the service, but also to protect the new entrant competitors from the dominant position of incumbents.

In a regulation system, a very important factor is the availability of good information on operating practices and investments, but also on the environmental context (resource and quality). For instance, following the application of the principle of evidence-based policy, the WFD proposes to shape policies based on detailed information about the resource status and its socio-economic uses in each river basin. Collection of information at river basin scale should be systematic and compulsory. The obtained data set, given its final format, could act as a facilitator in a context of liberalisation or be "neutral" due to the high level of aggregation of information. In the second case, it could not be detailed enough to be useful to operators or potential investors interested in market entry. On the other hand, data collection may help to reduce information asymmetries between the regulator and the private or public operator responsible for the management of WSS services. Indeed, operators can use hidden information in a strategic way in order to increase their rents.

The WFD requires the designation of competent authorities at river basin scale. In most States today, water management is driven by territorial authorities with administrative boundaries. In a liberalisation context, the increase in the number of regulators (basin authority, water sector regulator) poses the question of compatibility between these different regulatory authorities and also with the local authority responsible for WSS services and even operators. Problems of coordination could act as a constraint in increasing transaction costs so that it reduces their efficiency. In another way, the enforcement process of integrated river basin management may stimulate the coordination between the different public administrations of water policy (e.g. administration of the Environment for the management of surface water, administration of Public Works for navigation and quantitative issues, administration of Agriculture for irrigation and manure spreading, etc.).

The competition rules but also the resource management (by way of WFD), and the setting up and the monitoring of environmental standards require the designation of competent regulatory authorities. In most countries, territorial authorities with administrative boundaries drive water management. Moreover, other authorities regulate the water services and operators (e.g. competition authority, public health agency, etc.). A crucial question is to know whether these regulation authorities falls within the competence of the European Union or whether these aspects remain in the context of national or local actions. Is the European Union the best level to enforce different rules of regulation or does the WSS sector come under the exclusive competence of Member States? The answers to these questions depend on different problems stressed above. On the one hand, there are important interactions and cross-border externalities that require coordination and justify the establishment of centralised rules at the European level. On the other hand, the need to have better information and to bring a rapid answer or action to any problem argues in favour of strong decentralisation and local decision-making. The solution is not black or white, and certainly some aspects will be managed at the Community level to achieve the objectives of the internal market, whilst other ones will be dealt with at national or local levels in accordance with subsidiarity principle.

REFERENCES

Ballance, T., and A. Taylor (2001), Competition and Economic Regulation in Water: The Future of the European Water Industry. IWA Publishing, London, UK.

Bergman, L., C. Doyle, J. Gual, L. Hultkrantz, D. Neven, L.-H. Roller, and L. Waverman (1998), Europe's Network Industries: Conflicting Priorities. Telecommunications. Centre for Economic Policy Research, London.

Byatt, I. (2001), "The Water Regulation Regime in England and Wales" *in* Regulation of Network Utilities: The European Experience. (C. Henry, M. Matheu and A. Jeunemaître, Eds), Oxford University Press: 79-93.

Defeuilley C. (1999), "Competition and public service obligation: regulatory rules and industries games." Annals of Public and Cooperative Economics **70(1)**: 25-48.

EEB (2002), A Review of water services in the EU under liberalisation and privatisation pressures, Special Report, 2002/012, Brussels.

Euromarket (2003), Analysis of the European Union explicit and implicit policies and Approaches in the larger water Sector. Deliverable 1. Euromarket Project.

European Commission (1999), "Liberalisation of Network Industries: Economic Implications and Main Policy Issues." Reports and Studies. N° 4, Directorate General for Economic and Financial Aspects.

Finger, M., and J. Allouche (2002), Water privatisation. Transnational corporations and the re-regulation of the water industry. London, Spon Press.

Guérin-Schneider, L. (2001), Introduire la Mesure de Performance dans la Régulation des Services d'Eau et d'Assainissement en France: Instrumentation et Organisation. Ph.D. Dissertation, ENGREF, Paris

Hall, D. (1999), The Water Multinationals. Report number: 9909-W-U-MNC. Public Services International Research Unit, University of Greenwich.

Hassan, J. (1995), 'The impact of EU environmental policy on water industry reform', European Environment **5(1)**: 45-51.

Haut Conseil du Secteur Public (1999), Quelle Régulation pour l'Eau et les Services Publics? La Documentation Française, Paris.

Helm, D. (2001), "The assessment: European Networks – competition, interconnection, and regulation." Oxford Review of Economic Policy **17(3)**: 297-312.

Hunt, L. C., and E. L. Lynk (1995), "Privatisation and Efficiency in the UK Water Industry: An Empirical Analysis." Oxford Bulletin of Economics and Statistics **57(3)**: 371-388.

Kallis, G., and P. Nijkamp (2000), "Evolution of European Union water policy: a critical assessment and a hopeful perspective." Journal of Environmental Law and Policy **3**: 301-335.

Littlechild, S. (1986), Economic Regulation of Privatised Water Authorities. Department of the Environment, London.

Littlechild, S. (1988), "Economic Regulation of Privatised Water Authorities and further reflections" Oxford Review of Economic Policy **4(2)**: 40-68.

Mac Gowan F. (1995), "Utility reform in the UK - The role of regulation and the impact on public service." Annals of Public and Cooperative Economics **62(2):** 169-184.

Nauges, C., and A. Thomas (2000), "Privately-operated Water Utilities, Municipal Price Negotiation, and Estimation of Residential Water Demand : The case of France." Land Economics **76(1)**: 68-85.

Pelkmans, J. (2001a), "Making EU network markets competitive", Oxford Review of Economic Policy **17(3)**: 432-456.

Pelkmans, J. (2001b), European Integration – Methods and Economic analysis. 2nd Edition, Financial Times Prentice Hall, Pearson Education Limited.

Rothenberger, D. (2002), Integrating Energy and Water Supply on company level - Municipal Multi Utilities. 12th Stockholm Water Symposium, http://www.cirus.ch/downloads_d/stockholm_vortrag_rothenberger.pdf.

Silva, G., N. Tynan, and Y. Yilmaz (1998), "Private Participation in the Water and Sewerage Sector - Recent Trends." Finance, Private Sector and Infrastructure Network - Public Policy for the Private Sector. Note n° 147. World Bank.

Smets, H. (2002), Le droit à l'eau, Académie de l'Eau.

Webb, M., and D. Ehrhardt (1998), "Improving Water Services through Competition" Public Policy for the Private Sector, Report 164. World Bank, Washington, D.C.

LIST OF EUROPEAN UNION COMMUNICATIONS

Commission Directive 96/19/EC of 13 March 1996 - amending Directive 90/388/EEC with regard to the implementation of full competition in telecommunications markets.

COM 2000/60/EC. Directive of the European Parliament and of the Council of 23 October 2000 establishing a framework for Community action in the field of water policy.

Commission of the European Communities (1996), The Services of General Interest in Europe, COM(96) 443 final, Brussels.

Commission of the European Communities (1997), *Fourteenth annual report on monitoring the application of community law* (97/C332/01), Brussels.

Commission of the European Communities (2000), The services of General Interest in Europe, COM(2000) 580, Brussels.

Commission of the European Communities CEC (2003), Green Paper on Services of General Interest, COM(2003) 270 final, Brussels.

Commission of the European Communities (2003), Internal market strategy - Priorities 2003-2006, COM(2003) 238 final, Brussels.

Commission of the European Communities (2004), White Paper on Services of General Interest, COM(2004) 374 final, Brussels.

Commission of the European Communities (2004), Green Paper on Public-Private Partnership and Community Law on Public Contracts and Concessions, COM(2004) 327 final, Brussels.

Council Directive 91/440/EEC of 29 July 1991 on the development of the Community's railways.

Directive 96/92/EC of the European Parliament and of the Council of 19 December 1996 concerning common rules for the internal market in electricity.

Directive 97/67/EC of the European Parliament and of the Council of 15 December 1997 on common rules for the development of the internal market of Community postal services and the improvement of quality of service

Directive 98/30/EC of the European Parliament and of the Council of 22 June, 1998 concerning common rules for the internal market in natural gas.

Directive 2001/12/EC of the European Parliament and of the Council of 26 February 2001 amending Council Directive 91/440/EEC on the development of the Community's railways.

Directive 2004/17/EC of the European Parliament and of the Council of 31 March 2004 coordinating the procurement procedures of entities operating in the water, energy, transport and postal services sectors.

Directive 2004/18/EC of the European Parliament and of the Council of 31 March 2004 on the coordination of procedures for the award of public works contracts, public supply contracts and public service contracts.

Chapter IV

Analysis of the Strategies of the Water Supply and Sanitation operators in Europe

Authors Patrícia Luís-Manso, Matthias Finger and Jeremy Allouche*

INTRODUCTION

The main aim of this chapter is to analyse the business strategies of European water supply and sanitation (WSS) operators. The concept of strategy is not consensual in nature and has significantly evolved during the past decades, influencing the development of business strategy theories. Strategy herein pertains to the definition of aims and objectives of an operator, on the basis of the recognition of opportunities and threats, and the response - including implicit and explicit (re)actions - for attaining these objectives. Both opportunities and threats come from external and internal forces to the operator, meaning it is important to consider the evolving environment, as well as the capabilities of the operator itself.

In literature, it is usually considered that one can distinguish four main theoretical approaches to business strategies (Whittington). Firstly, classical theories advocate a formal type of strategy aiming at profit maximisation, and focuses on internal characteristics of the organisation. Secondly, evolutionary theories look at the external context to the firm (namely to markets), and consider the reduction of costs as the most appropriate strategy to survival. Thirdly, processualist theories refute the perfect rationality of the economic agent and the perfect competitiveness of the markets, and see strategy as a tool to strengthen internal competences of the firm. Finally, the systemic approaches focus again on the external environment to the organisation, namely on the society, to explain the possible strategic options. These theories have however some limits when applied to analyse the strategy of operators in the water sector.

The strategies of European water operators have not yet been thoroughly studied. This chapter aims at presenting a general analysis of the different types of operators, as well as their corresponding strategies. In doing so, one can distinguish two main difficulties in coming up with a systematic and rigorous analysis of water operators; firstly their number and diversity; and, secondly, the specificities of the sector. As a matter of fact, it is estimated that there are roughly more than 30'000 water operators in Europe, providing water supply and/or sanitation services. These operators are very much diverse, ranging from small non-autonomous public entities to large multi-utility TNCs, as well as

* This chapter is based on Deliverable 3 of the EUROMARKET project. This chapter would not have been made possible without the valuable help and contribution of the University of Zaragoza (Prof. Pedro Arrojo and Isabel Almudi) and TU Delft (Prof. Aad Correljé and Delphine François).

encompassing regional public, mixed and private organisations. The strategies of water operators change from case to case and in fact, different types of operators react differently and are exposed to varying realities and dynamics. The second difficulty pertains to the main specificities of the water supply and sanitation sector, which prevent that theories from other sectors can be generalised (Schouten and Dijk, 2005).

This chapter is structured into three parts. In order to understand the various business strategies of WSS operators, the first part aims at establishing a typology of operators, as well as their distribution across Europe. The aim is twofold: firstly, it is to cluster the different types of operators so as to enable a general and comprehensive analysis of these operators. Therefore, the initial assumption is that the strategy of operators is mainly influenced by physical and business characteristics, rather than particular historical or geographical features. Secondly, it is to illustrate the distribution of water operators across Europe according to our typology.

The second part focuses on the analysis of water operators' strategies. It starts out by presenting a framework for analysing operators' strategies. First the main theoretical approaches to business strategy are presented in more detail including their limits when applied to analyse the strategy of operators in the water sector. Then, we aim at identifying and synthesising the main features directly affecting the strategy of water operators in a coherent structure pertaining to its accurate analysis.

As a next step we identify the main strategies per type of operator, based on a quantitative (the European map) and qualitative (case studies) analyses. Ten operators corresponding to different typologies and to their distribution across Europe were selected, so as to elaborate detailed case studies on the basis of document analysis and interviews (Euromarket, 2004). The objective of the case studies is to develop an in-depth analysis of the most significant types of operators.

Finally, in lieu of conclusion, we present a general overview of the main strategies in place.

TYPOLOGY AND MAP OF OPERATORS ACROSS EUROPE[1]

In order to come up with a possible generalisation of the strategies of WSS operators, the first step is to establish a general typology of operators rather than doing an individual case-to-case analysis. The typology of operators is defined according to four major characteristics, namely:

1) Ownership: operators can be public, private or mix;
2) Size: it is used the variable "served population[2] by operator" to classify operators as small (serving a population below 100,000 inhabitants), medium (serving between 100,000 and 10,000,000 inhabitants), and large[3] operators (serving more than 10,000,000 inhabitants);
3) Scope of the market: it defines whether the operator works on a regional, national, or international level; and
4) Segments of the market: it relates to the types of services provided by the operator, namely water supply, sanitation or multi-utility (e.g., water, urban waste, energy) services.

As emphasised in the introduction, this typology is based on the assumption that the strategy of operators is mainly influenced by physical and business characteristics. It therefore does not consider geographical, cultural or historical factors that may shape the formulation of business strategies.

The next step aims at providing a general picture on the distribution of the different types of water operators across Europe. Considering the fact that different sources were used, as well as the lack of detailed data for some countries and typologies, the data presented in this section is just an estimation of the European map of water operators.

Traditionally water services were directly provided by local public entities. This type of operator is still dominant across Europe however other typologies have emerged in the past twenty years. Today there are more than 30'000 water operators in Europe. At a first glance on the data (figures 1 and 2), some first conclusions can be drawn. Firstly, in terms of water services provision, the majority of European population (about 55 per cent) is supplied by public operators, followed by private (about 35 per cent) and mixed owned (almost 10 per cent) ones.

[1] Europe in the scope of this report comprises EU-15 and Switzerland.

[2] In the scope of this report, the total population stands for the total population who is served by water services.

[3] As it was already mentioned, these large operators participate in the ownership structure of other operators across Europe. Nonetheless, we consider local operators as the reference in calculating these figures, which means for example that even if a medium sized operator is owned up to 50% by a large operator (from another country), the operator is aggregated in the medium sized category.

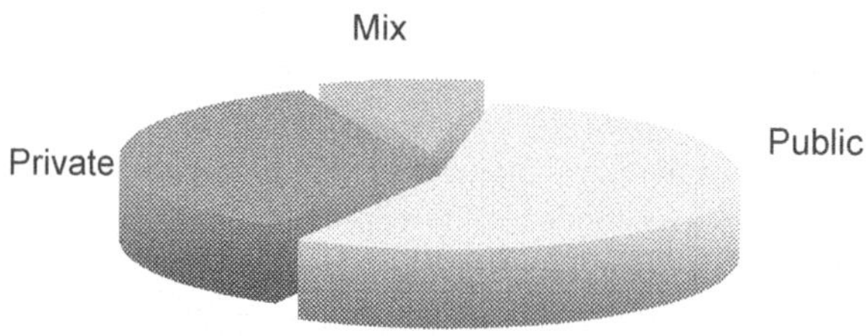

Figure 1: Type of ownership of operators per population supplied, 2003

Secondly, regarding the size of operators providing water services, medium sized operators supply almost 50 per cent of the population. This result is not very significant because, according to the defined typology, medium operators are those supplying between 100'000 and 10'000'000 inhabitants, which is a very large interval. Small ones that are the majority in nominal terms supply about 35 per cent of the population and, finally, large operators account for approximately 15 per cent of the inhabitants.

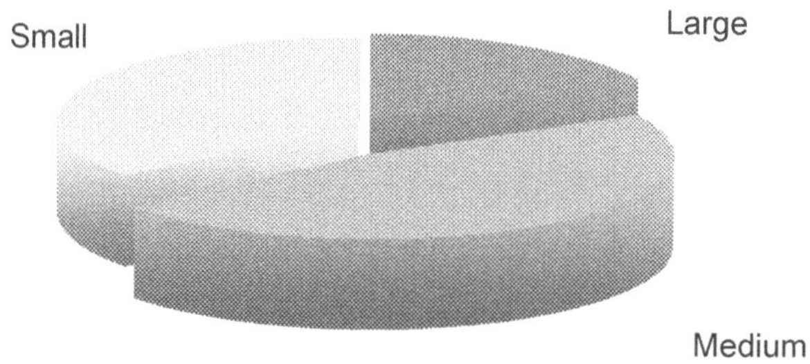

Figure 2: Size of operators per population supplied, 2003

Another interesting conclusion that can be drawn from the data (figure 3) is that there are no "public – large" water operators in Europe. All public operators are small and medium. Another conclusion is that small public operators account individually for the biggest share in terms of population supplied (about one third), followed by medium public and medium private operators (about 20% each). Yet probably the most interesting conclusion is that the participation of the private sector through medium sized operators is very significant (approximately the same as through large sized ones), which is especially due to the England and Wales case. This is especially true for water supply operators. It is nonetheless important to state that large private operators participate in the ownership of several small and medium private and mixed operators, and thus their influence needs to be analysed accordingly.

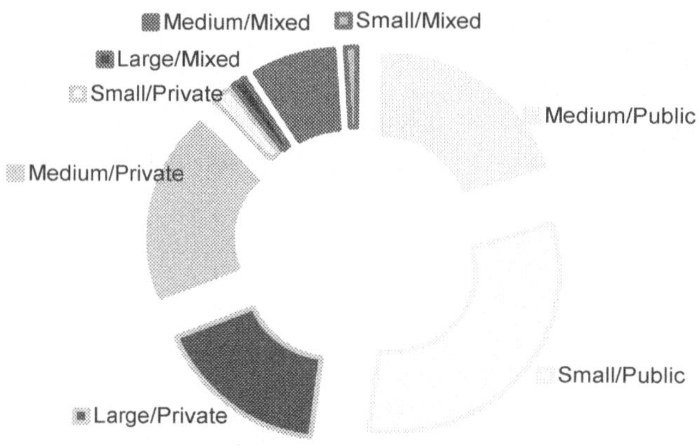

Figure 3: Partition of types of water operators in Europe, 2003

Clustering of the 16 countries

The analysis presented so far relates to a general picture of Europe in terms of operators' typology. It is important at this moment to present a more detailed illustration, namely with the aim of clustering some countries based on common characteristics (i.e., with respect to ownership and size). The first cluster that has been made is based on ownership, the second is based on size and the third and final cluster is a combination of these two.

In terms of population served per type of ownership of operators, public operators are providing water services to the majority of the population in 14 out of 16 countries. Only in France, England and Wales are the majority of the population served by private operators (although under very different modes of management). Private ownership is also considerably important in Spain, Denmark (due to private cooperatives supplying water especially in rural areas) and Italy. In contrast, private participation is almost neglectable in Switzerland, and it is even zero in Luxembourg and the Netherlands (although not for wastewater). Finally, in every country except for Spain, Italy and Greece, one form of ownership (almost always public) is dominant.

In terms of size, small and medium sized operators provide water services to the majority of the population. Only in France large operators are dominant in terms of population served, which can be seen as a result of the local model of delegated management to (private) companies. Apart from France, large operators are also present in the UK and Germany (although not in a dominant position).

Clustering based on ownership

Four clusters can be identified on the basis of the distribution of operators in terms of ownership:

1. France and England & Wales form the first cluster, because they are the only countries where private ownership is the dominant type (even though under different modes of management);
2. The second cluster is the mix cluster. This cluster is not based on mixed ownership, but based on the fact that these countries do not have a (real) dominant ownership structure. This cluster is formed by Italy, Spain and Greece.
3. The third cluster is formed by Denmark and Germany. These two countries have a dominant public ownership (i.e., public operators provide water services to about two thirds of the population), but less dominant than the countries in the last cluster (n.4).
4. The last cluster is the cluster where public ownership is dominant (i.e., it provides water services to more than 90% of the population), and it contains the following countries: Sweden, Finland, Ireland, the Netherlands, Belgium[4], Luxembourg, Switzerland, Austria, and Portugal.

Clustering based on size

Four clusters can be identified on the basis of the distribution of operators in terms of size:

1. France forms a cluster by itself, because it is the only country in Europe, where large operators are providing water services to the big majority of the population.
2. The second cluster is formed by Spain, England & Wales, Italy, Greece and the Netherlands where medium sized operators are the dominant size.
3. The third cluster is formed by countries where medium sized operators provide water services to (more or less) the same share of population as small sized operators. This cluster is formed by Austria, Finland, Sweden, Portugal, Belgium and Luxembourg.
4. The last cluster is the cluster where the small sized operators are dominant. Ireland, Germany, Denmark and Switzerland form this cluster.

Clustering based on ownership and size

Finally, it is possible to group the 16 countries into six different clusters combining these two typologies (i.e., ownership and size) (figure 4). The ordering of the clusters corresponds to their importance in terms of population, with the first cluster representing more population than the sixth.

---

[4] Belgium is considered in this cluster even if in terms of sanitation about 30% of population is served by mixed owned operators.

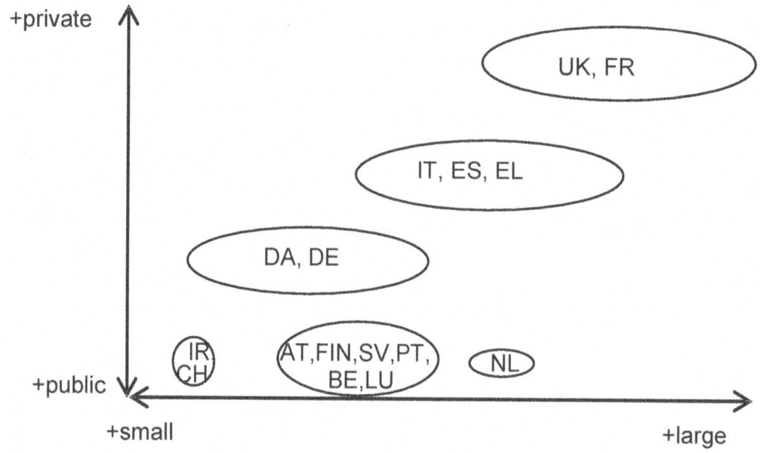

Figure 4: Clustering of European countries according to the typology of operators, 2003

First cluster

France and England & Wales, the only countries with the majority of the population being served by private owned operators, form the first cluster. Furthermore, the major water-related TNCs (RWE-Thames Water, Suez and Veolia) have in England and France their main domestic markets.

Second cluster

The second (largest) cluster is formed by Italy, Greece and Spain. Their water operators have many different forms, structures and sizes, but in general the majority of the population is served by publicly owned medium sized operators.

Third cluster

Germany and Denmark form the third cluster with dominant public ownership, but less dominant than the countries in the remaining clusters. Small and medium are the dominant size of operators.

Fourth cluster

The fourth cluster is formed by countries where the majority of the population is served by operators publicly owned, but equally divided into small and medium sized. Austria, Belgium, Luxembourg, Portugal, Sweden, and Finland form this cluster.

Fifth cluster

The fifth cluster corresponds to the Netherlands, which is the only country where medium sized publicly owned operators provide water services to the majority of the population.

Sixth cluster

Finally, Ireland and Switzerland form the sixth cluster, where small publicly owned operators dominate.

THE STRATEGIES OF OPERATORS IN THE WATER SECTOR

Framework for analysing operators' strategies

There is no universal model or theory of strategy capable of covering all types of strategies for all sectors. Strategy must be sensitive to social, cultural, economic, but also to sector-related features. Indeed, both the market structure (e.g., competitive or monopolistic) and the typology of the operators (e.g., type of ownership) have important impacts on the means and goals of strategy. The first step to analyse the strategies of water operators is to define the most appropriate framework for the analysis. After considering the main theoretical approaches to business strategies, we identify their limits when applied to the water sector and propose an adapted framework considering the specificities of the sector.

Theoretical approaches to business strategy

In literature, it is usually considered that one can distinguish four main theoretical approaches to business strategies, namely: (1) classical; (2) processualist; (3) evolutionary; and (4) systemic. It follows a list of the main authors (figure 5) as well as a short description of each approach.

Business strategy approaches	*Main authors*
- Classical	(Chandler) (Ansoff) (Porter)
- Processualist	(Cyert and March) (Mintzberg, Ahlstrand et al.) (Pettigrew) (Miles and Snow)
- Evolutionary	(Hannan and Freeman, 1988) (Williamson, 1991)
- Systemic	(Granovetter, 1985) (Whitley, 1990)

Figure 5: Main authors associated with the major approaches to business strategy

The ***classical approach*** finds its roots in empirical research and emerged in the 1960s. It is based on formal rational planning methods designed to maximise long-term profitability. This approach is very much influenced by economics (for strategy formulation) and military strategy (for strategy implementation).

The ***processualist theories***, which have "resource-based" strategy theories as the most important branch, postulate that rational planning is not effective because the processes of organisations and markets are rarely perfect. According to Mintzberg et al (1998), strategy emerges from a process of learning and adaptation. In fact, processualists both reject the perfect rationality of the economic agents and the perfections of competitive markets. The focus of this approach is the enterprise's internal resources and their capacity to generate the firm's competitive advantage.

Following the same path, the ***evolutionary strategy theories*** (emerged during the 1980s) state that environment is normally unpredictable to anticipate effectively, so future oriented planning becomes unrealistic. It is the market that makes the important choices and not managers (as Classicists support). Competitive pressures select out only the best operators - in terms of profit results - to survive in the long term, so the best strategy is to concentrate on maximising chances of survival today. The only real comparative advantage is relative efficiency; hence managers must concentrate on reducing costs.

In the 1990s, the ***systemic approach*** emerges, taking a relativist position and considering that the objectives and practices of strategy depend on the cultures and powers of the respective local systems. This means that strategists may pursue objectives other than profit maximisation if their social background frames other interests for them (e.g., personnel pride, managerial power). In the systemic view, the norms that guide strategy derive not so much from the human cognitive bounds as from the cultural rules of the local society. As a consequence, firms from different systems have different types of strategies (i.e., strategies vary according to social systems and, consequently, they vary geographically).

The limits of conventional approaches to analyse the water sector

The WSS sectors are characterised by very different types of operators, ranging from small non autonomous municipal services, whose main goal is the provision of a public service in a geographical limited area, to private trans-national corporations (TNCs) pertaining to profit maximisation and to the expansion of operations worldwide.

The majority of business strategy theories described above focuses mainly on private operators and on competitive markets, and considers profit maximisation as the natural outcome of strategy-making. The classical approach, which is still the most influential one, focuses mainly at industry-level, neglecting the potential social, economic and political direct impacts on operators' strategies. Processualist theories focus on the internal factors to the organisation, neglecting the direct influence of external factors, and evolutionary theories are based on the hypothesis of competitive markets. Even the systemic approach, which integrates sociological complexities and differences, and considers socially related goals for strategy making, fails to consider all the relevant factors and dynamics to analyse water operators' strategies.

The WSS sectors are characterised by certain specificities, which make a clear difference for the formulation of a strategy for this sector. Firstly, the WSS sectors display strong natural monopoly characteristics associated with the infrastructure network, limiting the scope and degree of competition in the market (Gordon-Walker and Marr, 2002). Secondly, given the fact that the transport of water is expensive due to high sunk costs and quality requirements, WSS services have traditionally been treated and even sourced locally (Gee, 2004). Finally, WSS have strong merit good characteristics, meaning its social benefits exceed private benefits (existence of strong environmental and health-related externalities) (Luís-Manso and Finger, forthcoming).

All in all, full-scale competition is unlikely to develop in the water markets because many segments are characterised by natural monopolies, very high sunk and transportation costs, as well as environmental and local constraints. Moreover, the most common operator in the sector is publicly owned and its main goal is related to the provision of a public service and not to profit making. For this reason, an adapted conceptual framework to analyse water operators' strategy is proposed in the following section in an attempt to surmount the identified problems.

Proposed framework for analysing operators' strategies

The contemporaneous business strategy theories when applied to the water sector thus raise two important problems: first, they do not adequately include all the issues of major importance for water operators; and second, they are based on full competitive market structure hypotheses and, consequently, they do not properly account for the existence of significant natural monopolies and externalities. This is explained by the fact that the majority of these theoretical works are based on empirical evidence, mainly in the United States, from sectors which are structurally very different from the water sector. Some studies nonetheless have been conducted in non-competitive sectors making use of these theories and producing relevant results (e.g., Brown and Iverson, 2004). However, we believe that it is essential to analyse the water operators' strategies in an adapted framework, considering the specificities of the sector.

The framework proposed in this chapter comprises two main axes: the operator's capabilities and resources (internal axis), and the evolving environment (external axis). The water operator's strategy has progressively become more complex, because the new environment, characterised by globalisation and liberalisation trends (Finger and Allouche, 2002), introduces new actors and dynamics (figure 6). From this perspective, the operator needs to take into account for the definition of its strategy:

1) its own capabilities and individual resources;
2) the fact it operates in a defined market, comprising consumers and probably other operators; and
3) the business environment (including the institutional and regulatory context, resources, other business sectors and civil society), which determines important market characteristics and affects its dynamics.

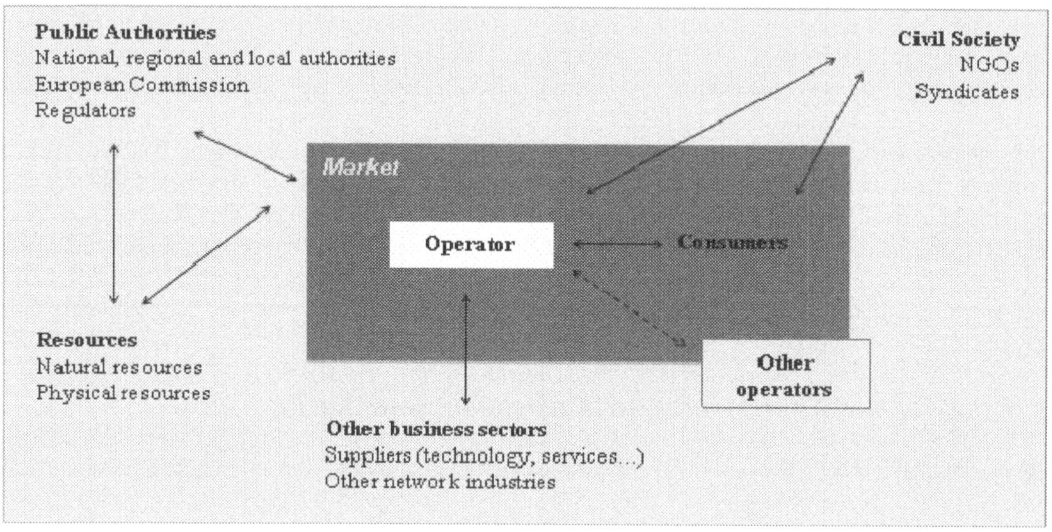

Figure 6: Market and business environment dynamics affecting the strategy of the operator

The main innovation of the proposed framework is the incorporation of specific features that directly constrain the range of feasible strategic options to the water operator. Thus, besides the conventional strategies in the market, the framework considers regulation, civil society and resources as essential elements to analyse water operators' strategy. These different features are most of the times interrelated and can be synthesised into the operators' strategic objectives. Without neglecting important dynamic phenomena, the operator's strategy may thus be broken up into: (1) strategy in the market; (2) strategy towards public authorities; (3) strategy towards resources; (4) strategy towards other business sectors; and (5) strategy towards civil society.

Strategy in the market

Strategy in the market includes operators' strategy towards consumers and other operators (if any). There are very important trends that affect operators' strategies in the market and that should be taken into account on a case-by-case basis. Probably the most important trend is related to liberalisation, i.e. the introduction of competition.

Until recently the WSS sectors were in the majority of the cases characterised by monopolies, where the incumbent was State or local authority-owned. However, liberalisation is clearly calling into question the State's role as a single monopoly provider. In fact, and despite the evidence that many parts of the sector are considered as natural monopolies, new dynamics are pushing towards the opening of markets to competition. In the industrialised countries the main driving force stems from financial pressures to respond to the needs of maintenance, to higher sewerage and water production costs due to increased pollution and water use, and to stricter environmental norms (Allouche, Luís-Manso et al., 2003).

In the majority of the cases, the main competitive challenge operators are confronted with is to get in the market. Partly associated therewith, also the pressure to enhance economic efficiency is increasing (as the England and Wales case show).

Strategy towards public authorities

Related to the issue of liberalisation is the emergence of new institutional and regulatory arrangements (Allouche, Luís-Manso et al., 2003). The most important functions assumed by regulators in the water sector are competition regulation, the regulation of scarce resources, technical regulation (related to the integrity of the network), and consumer protection. These functions are normally pursued at different levels - local, regional, national and international -, and influence considerably the way water supply and sanitation enterprises operate (see Box 1).

Box 1: The effects of Regulation – the case of the European Commission

A vivid illustration of the impact of regulation on operators' strategy is water quality standards and the European Union (EU) Urban Waste Water Directive (Council Directive 91/271/EEC of 21 May 1991). The purpose of the Directive is to stimulate the Member States to invest in the collection and treatment of urban wastewater in order to reduce nutrient inputs into European waters. The Directive includes minimum requirements for the treatment of urban wastewater, and for the disposal of sewage sludge. For example, it was estimated that the United Kingdom had to invest $60 billion in wastewater treatment between 1995 and 2005 in order to meet these standards. This amounts to about $1000 per capita, or about 0.6 percent of gross domestic product (GDP) to be spent on wastewater treatment alone over a ten-year period (Serageldin, 1994). At the same time, it was observed by Bowler (1999) that the regulator's downward pressure on prices is claimed by the water company directors to have limited the financial resources available to invest in sewage treatment and disposal systems. This in turn led water companies to take least-cost decisions, even though they may not be the optimum decisions as far as the government is concerned. In this example, one can therefore see the direct impact legislation had on operator's investment priorities and strategic choices. In this particular case, the operator was faced, on the one hand, with the regulator's (OFWAT) pressure on water pricing and, on the other, with stringent EU regulation requiring important investments for quality standards. The strategic choices of the operator are therefore much more restricted in the WSS sectors where there is a need for strong regulation due to the natural monopoly, and environment and health related characteristics of this sector.

Strategy towards resources

Strategy towards resources comprises natural resources and physical infrastructure. In this matter, the question of the ownership of the resources is central to understand the potential threats and opportunities to the operator. In general, natural and physical resources remain in public hands, so the type of relationship and the type of contracts agreed with the institution that controls the rights to use the resources are a key factor to operating the business. Finally, the degree of autonomy vis-à-vis the owner and, consequently, the degree and type of risk sharing are important issues for the operator, especially those related to the quality of the network.

One of the main competitive challenges in terms of resources regards the development of technology. The cases of desalination and water treatment techniques provide good examples of the strategic importance of technology developments in order to face, on the one hand, the increasing scarcity of water and, on the other, the increasingly stringent requirements of environmental standards.

Strategy towards other business sectors

In terms of operators' strategies towards other business sectors, one should distinguish suppliers (e.g., technology and service suppliers) from other network industries. The relationship established with suppliers may be analysed under the traditional business strategy's theories. Supplier markets are normally very competitive. In contrast, it is important to mention in more detail the importance of other network industries. In fact, despite the mentioned specificities of the WSS sectors, there are some important influences linked to the developments occurred in other network industries. This is especially important for multi-utility operators, i.e. companies ensuring at the same time the distribution of several services including, for example, drinking water, electricity, and natural gas. The main influences are related to the liberalisation process (more developed in other sectors than in WSS) and to the competition for the allocation of internal resources of the multi-utility operator among the different services.

Strategy towards civil society

Parallel to the liberalisation trend is the increased level of clients' demands. Being the service provided by public or private operators, consumers demand accessibility of water in sufficient quantity and quality. Because water is important both for human survival and for economic development, in many cases consumers are used to and demand subsidised prices. Nonetheless, the failure to meet basic needs for water is progressively leading to the recognition that water must be managed more efficiently and that giving it an economic value could be a reasonable tool to achieve the goal. The demands of consumers are supported and fed by civil society movements, which have increasing powers in the new era of globalisation, even though they are not homogeneous entities. Operators, with a special focus to TNCs, are increasingly aware of civil society movements due to the perception of their potential impact on the business. A growing number of TNCs engage in corporate citizenship programs with stakeholders to address social and environmental problems. Indeed, public (and even shareholder) expectations of corporations to deal with these problems in the communities where they operate have risen dramatically over the past decade, at the same time that the roles of national and local governments have been shrinking (Rondinelli and Berry, 2000).

Main strategies per type of operator

After identifying the main types of water operators and their distribution across Europe, the present section aims at presenting and analysing the main strategies per type of operator. This is both done on the basis of the distribution of the different types of operators (i.e., the European map) and on a selection of case studies. In fact, ten operators corresponding to different typologies and to their distribution across Europe were selected, so as to elaborate detailed case studies aiming at developing a deep analysis of the most significant types of operators.

At least one operator for the most significant types of operators was selected, namely: small public, medium public, medium private, large private, medium and large mixed operators. Figure 7 presents the list of the selected case studies[5].

	Public	Private	Mixed
Large		Suez (multi-utility)	RWE-TW (multi-utility)
Medium	ADP (multi-utility) Canal Isabel II (multi-utility)	Indaqua (WSS) Severn Trent (multi-utility)	Aquafin (sanitation) ACEA (multi-utility)
Small	TWM (water supply) Aguas del Prat (WSS)		

Figure 7: List of the selected case studies

Small public operators

In many countries, small, local public operators in the form of municipal services or municipal companies provide both water supply and sanitation services, and even other services such as electricity and gas. The objectives are normally formulated as follows: "an efficient supply of water of a reliable quality, at an adequate pressure, at a reasonable price".

[5] The detailed case studies are presented in Euromarket (2004). Analysis of the strategies of the Water Supply and Sanitation Operators in Europe. Euromarket Project.

Small, local public operators operate in a setting, which is highly affected by public policy and water management is considered to be a public task (see for example box 2). National, provincial and communal frameworks of regulation, including local implications of environmental permits, provide a whole range of economic, social and ecologic guidelines for the operation of the water companies. In fact, in a context with a larger number of small operators, conditions are more easily established by regulation, whereas in larger concessionary agreements conditions may be part of the negotiations on objectives. Specific legislation is directed towards the prevention of groundwater contamination in water extraction areas. This has direct implications on water quality norms as well as on economic conditions of water production and sanitation.

The financing systems involve a number of elements. An extraction fee is normally required for the production of water, to be paid to local or national government. The consumer price is based on cost plus a utility rate – the degrees to which revenues cover short-term or long-term marginal supply costs and include environmental externalities depend very much on the situation. Small users, as captive consumers, are supplied under supply conditions and tariffs defined by public authorities. Large users are sold water on a cost-plus basis. Smaller operators, as a consequence of their scale, may have high per unit cost for energy, levies and staff. Finally, the amounts of dividend transferred annually may be an important aspect to the shareholders. These elements may drive strategies geared towards integration and expansion, to reap advantages of scale.

According to some interviewees, the boards of these operators see no advantages in introducing competition in the sector, especially regarding small users, and are happy in maintaining the existing privileged relationship with the public authorities. Of course this secures stability in their generally good labour conditions. Related therewith, the objective to further develop a 'Socially Responsible Corporate Management' dimension is a means to underline the local connection with the water consumers and voters. By linking these aspects to their local world and creating a sense of connectedness with the population, as users and citizens, these small water operators seem to distinguish themselves from the large, anonymous, trans-national operating firms. This may be interpreted as a strategic move to fend off plans for restructuring and liberalisation.

Box 2: N.V. Tilburgsche Waterleiding-Maatschappij (TWM)

TWM is a municipal drinking water supply company with the objective to supply water in a sustainable way, meeting the traditional operating conditions for public works. TWM is characterised as a public owned company with a strong focus on product quality and cost price. TWM's main shareholders are local authorities but it is an autonomous company. By and large the company is happy in the relationship with public authorities and the strategy is to maintain these good relations.

The management of TWM thus perceives the current contractual arrangements as adequate. As the structure of ownership shows, the municipality is a key stakeholder with a direct influence in how the company is managed. Since 1960, The Netherlands has a system of non-transferable water production concessions, attached to the water companies. The production concession for groundwater extraction exists since 1898. TWM received these concessions for the municipalities of Tilburg and Goirle. The political intervention through the Municipality Council is reflected in the daily management of the company only weakly. Nevertheless, late 2003, the managing director of TWM had to withdraw in a conflict with the Municipal Council.

Level of compliance of the N.V. TWM with the laws and regulations is currently regarded as good, but not relevant for the competitive position in the market, as there is no competition in the drinking water market. The management of the TWM does not see any advantages in having a free market for drinking water in the market for small users, as profit maximisation is not an objective.

Gradually, competition is developing in the market for large industrial users of drinking water, which may choose to become self-reliant by abstracting their own water, or to go to other suppliers. The risk of losing large volume users in the long run is considered significant, with higher per unit supply costs and prices as a result of under-utilized capacity. So, water suppliers are developing separate strategies towards communal drinking water supply and sanitation and the service to large users.

It could be argued that depending on the quality of the drinking water supplied and the waste water released, the firms consider the Water Framework Directive a threat, or not. Indeed, when large investments are required, pressure may arise in the municipality to bring about a restructuring, away from public ownership. Depending on circumstances, consumer prices may rise, reflecting the possibly higher level of environmental and management costs to these firms. Also local governments may be pressing the firms involved to increase their scale and scope, by means of associations of municipalities or mergers to reduce supply and sanitation costs and to facilitate environmental strategies (see for example Box 3).

In addition, policies with respect to River Basin Management may – in addition to cost - become important drivers in the integration of operators into larger entities. The relevance to integrate water supply with water sanitation activities is often acknowledged as an effective and – low cost - strategy to enhance efficiency. So, in the future, these small firms may be incorporated into (public) regionally oriented companies or municipalities into multi-municipality companies, with common standards in respect of quality, hydrological and environmental issues. This may result into a diversification of the sources of water and in water 'products' by quality ranges and into the development of new activities in the area of integral chain management.

Box 3: Aguas del Prat Plc

Aguas del Prat Plc was set up in 1989 as a public entity to supply water to Prat de Llobregat - a city with a population of 65,000 inhabitants. The Prat de Llobregat council created this municipal service as a public management tool to improve the quality of the water service, consolidate the role of the public sector in providing basic services, and maintain autonomy in the management and planning of this area.

Faced with the growing possibility that private capital may enter the management of supply and sanitation systems, in 1999 the municipal companies of Prat de Llobregat, Manresa, Reus and Mataró created the Integral Management of Waters of Catalonia Company (IMWC). The setting up of IMWC has allowed for full use of synergies, an increase in individual representation and to offer their experience as a "public company" to other municipalities. The objectives of this grouping of municipal companies are:

- Define a common strategy and a common framework of action;
- Revaluation of the benefits and advantages of public management of services;
- Analyse and promote own culture of this type of management;
- Exchange information about the evolution of the water sector;
- Analysis of problems of the water service sector in this geographical area;
- Analyse future opportunities;
- Joint participation in market activities related to water cycles and the environment.

The need for additional capital and management expertise may eventually drive these utilities towards offering shares to private investors in return for cash. Indeed, the example of such a (partial) privatisation is provided by the local electricity and gas utilities in the process of restructuring, in which the local authorities' interest is clearly diminishing after having lost influence in day-to-day management and being confronted with a higher economic risk.

For the moment, one may be tempted to say that all these strategies are guided by the characteristics of the sector and its evolution and not so much by the size, segment or ownership. In fact, the first conclusion is that there is no real specific strategy for small local public authorities. However, one cannot forget that there is a sort of mission that small local public operators have towards consumers (with regards to quality of services), which in fact could be interpreted as a form of strategy to counter larger groups. Finally, the perception of profit is also largely absent or, in other words, different compared to other larger private operators.

Medium-sized public operators

Throughout the past decade, there has been a clear trend towards the concentration of small public operators in the form of multi-municipal companies, association of municipalities, and in some cases the creation of public holdings. The aim has been searching for the optimal size so as to exploit scale economies, as well as to allocate efficiently European funds.

Ownership, size and the legal form seem to influence the strategy of medium (i.e., providing the service to more than 100 000 inhabitants yet less than 10 millions) public operators. In terms of size, medium operators sometimes extend their operations outside their initial local markets[6] so as to benefit from scale economies. They do it firstly in adjacent geographical areas, and later in non-neighbouring markets including overseas. In fact, entities that have traditionally been publicly owned and managed on a municipal basis are becoming genuine commercial entities able to operate outside their initial regions. Nonetheless, for all of them, the initial market remains the most important in terms of population served and revenues. In general, the higher the degree of autonomy the bigger is the difference in terms of strategy in relation to small public operators.

The expansion of operations to international markets was restricted to developing and emerging economies and, in the majority of the cases, to countries presenting strong cultural and political ties to their domestic market (see for example Box 4). This strategic geographical orientation, which has followed the strategy of large operators in the sector, was to a certain extent backed by international financial organisations and the domestic countries' cooperation and foreign policies. This makes the withdrawal of operations in these markets in case of serious drawbacks much more difficult in political terms (contrary to large private operators). In fact, in the majority of the cases, the objectives of public enterprises are twinned with those of the (local) Government.

---

[6] This depends on the legal system. For example in Germany, the "principle of locality" restricts the actions of municipal entities outside their initial regions.

Box 4: Aguas de Portugal

Águas de Portugal (ADP) is a public holding company created in 1993. ADP works in all segments of water supply and sanitation sectors, and its core market is Portugal yet the Group also operates, through its subsidiaries, in Brazil, Cape Verde, East Timor and Mozambique.

The choice of creating ADP as a public holding company must be placed in the current context of extension of the systems, as well as an instrument to channel EU funds efficiently. The strategic choices of ADP are deeply correlated to the Governmental policies and strategic objectives. In fact, ADP was created to develop multi-municipal water and wastewater systems so as to rise in an efficient and sustainable manner the percentage of Portuguese households served by drinking water and wastewater systems. One can imagine that this model will evolve at the time most investments are achieved and EU funds are allocated. There is already a current debate about the future of ADP and several options implying various degrees of PSP are being considered.

The provision of water supply and sanitation services operated under a *de facto* monopoly is normally the core business of this type of operators. Nonetheless, some of these operators[7] are also competing for the provision of WSS services in systems open to some forms of competition (such as international bidding processes). In this case, medium and large private operators are perceived as the main competitors, yet the public ownership of the former may play a crucial role in obtaining the contracts. Amongst medium public operators' non-core business one can consider the provision of other environmental services, such as urban waste, and energy (see Box 5).

Regarding the scope of the market, these operators consider that size is not important and that their most important objective is to provide a public service to every citizen in their local area of operations. They have indeed a social strategy based on cross subsidies from heavily populated urban areas towards rural areas. This is made possible because their activity is considered a public service by the State, which has therefore granted the exclusivity to explore these systems.

The most important threats for medium public operators come from external factors, namely (1) the potential pressure to liberalise the WSS markets (e.g., from the European Commission); (2) the claims in favour of the implementation of the Treaty of Rome (in terms of non-discrimination of economic agents); and (3) the achievement of the objectives that were at the basis of the creation of these type of operators, such as the allocation of Cohesion Funds.

The establishment of stricter European Union environmental and public health requirements is in general welcomed by these operators, especially because up to now they have been complemented with EU funds for financing the required improvements. In fact, and in comparison to other typology of operators, medium sized public operators benefit the most from European credits and funds. Therefore, their main strategy towards the European Union is to comply with the required legislation, so as to obtain European funds to develop water supply and sanitation systems and the quality of their services.

Furthermore, these requirements have created new opportunities to get in rural and small town areas, which were facing difficulties to comply with the new requirements. In short, stricter requirements offer new opportunities for medium public operators to take advantages of their monopoly and economies of scale positions. Finally, their lobbying activities in relation to European institutions are usually developed through national governments.

Medium public operators are normally responsible for physical resources (i.e., networks and other infrastructures) and are committed to develop WSS systems. In sum, these operators are normally instruments of the State, provinces or municipalities in order to fulfil strategic objectives.

[7] So far multi-municipal associations have been excluded.

Box 5: Canal de Isabel II (CIII)

Canal de Isabel II is a public company depending on the Autonomous Community of Madrid. It employs over 1,700 workers and serves 5,000,000 inhabitants. CIII is a multi-utility with a firm basis in water services. The autonomous government of Madrid gave CIIII the exclusive concessions for network urban water and sanitation. CIII is thus contracted to operate the systems in 164 of the 190 municipalities in the Madrid region. In more than 100 of these CIII operates the full water cycle. CIII has a good financial record since the 1980s, when, after a period of drought, it was allowed by the government to invest in strategic infrastructures, financed by European funds, and to double its tariffs for water and sanitation services.

Given its position in the Madrid water management system, CIII's strategy to expand towards supplying other services, as well as to add new municipalities is basically dependent on its ability to convincingly show that it is able to supply high quality services at a reasonable price, to both its customers (and voters) and to the responsible politicians. Indeed, local political decision-making is the main determinant in this process, where contracts are not awarded by public tender.

In 2001, Canal de Isabel II started a new stage of diversification and internationalisation of its activities by developing new businesses. They focussed in Latin America, namely in Colombia, Venezuela, Dominican Republic and Ecuador, serving over six million (including Spain) people by the end of 2002. The energy sector is another sector of special interest for Canal de Isabel II. It has started operations in the Autonomous Community of Madrid, in areas of generation, distribution and commercialisation of electric energy and gas supply.

Medium-sized private operator

Private operators provide water services to about one third of the European population. Even though large private operators have a stronger visibility and lobbying power, medium-sized private operators indeed provide a bigger share of population than large ones. Medium-sized private operators are normally the product of the privatisation process of formerly regional public operators, such as in England and Wales (see Box 6), or the result of the diversification strategy of local private operators from other sectors (see Box 7).

This type of operators has some common strategic features with medium operators in general, namely in respect of the expansion of activities to other geographical markets so as to profit from economies of scale. For some of these operators, this is also a way of guaranteeing the sustainability of their businesses in the event of a competitive bidding process for their domestic market. From the point of view of these operators, when operating outside their domestic market, economic viability requires a minimum size which is sometimes at odds with the dispersion of the municipalities and their individualised management.

For medium-sized private operators, it is important that the EU ensures the respect for the non-discriminatory rules whenever concession contracts are awarded. In many cases, private operators join their efforts in associations aiming at lobbying for their interests.

Box 6: Severn Trent

Severn Trent was initially formed in 1974 as a regional, state-owned water authority based in Birmingham and responsible for water management and supply, and waste water treatment and disposal, in the catchment areas of two of Britain's greatest rivers - the Severn and the Trent. Since 1989, as a private operator, Severn Trent Water has invested heavily in replacing and repairing its assets and infrastructure. As well as complying with Government legislation on drinking water quality and the cleanliness of rivers, it sets its own internal standards on issues such as public health, leakage reduction and other sustainability targets in its catchment region. Severn Trent's mission is to be a leading supplier of environmental products and services for the benefit of shareholders, customers, employees and the wider communities.

Severn Trent International is focused on providing private sector management in Western Europe, where state-owned water authorities are being restructured. It operates in Italy, Belgium, Germany, Ireland, and Portugal. Furthermore, Severn Trent International management consultancy business works with funding agencies and governments to support sustainable development and capacity building in the developing world and Eastern and Central Europe.

In the event of competitive processes, the main competitors of this type of operators are medium public and large private operators. Medium-sized private operators have normally the advantage of knowing very well the local market and, besides that, of participating in the bidding processes in collaboration with local financial and credit institutions.

Finally, it is interesting to note that the anti-privatisation movements, which have increased in importance in the past decade, have somehow ignored this type of operators. And this is the more bizarre as in many cases these medium-sized operators are partially owned by the same large corporations that are the object of heavy criticisms. The reason probably lies on the closer relationship these operators can build with citizens when compared to large TNCs. And, in fact, medium-sized operators are highly aware of the important influence that the local civil society has on their businesses.

Box 7: Indaqua

Indaqua is a public limited company set up in 1994 with private shareholders (*Mota Engil Ambiente e Serviços*, *Teixeira Duarte* – two Portuguese construction companies - and Severn Trent Water International, which control more than 50% of the shares). Indaqua develops its main activities in the north of Portugal where the needs of the public water distribution service and urban waste water treatment are more sensitive. In this area, local town halls face the need to make large investments to broaden the cover of this service. This favours the concession of these services to private enterprise such as Indaqua. Its main sources of finance are: shareholders' capital (20 to 25%); and debt (75 to 80%) including financing arrangements from the European Investment Bank.
According to Indaqua, the economic viability of water management projects requires a minimum size, which is at odds with the dispersion of the municipalities and their individualised management (only 50 out of 330 municipalities have at least 50'000 inhabitants.
Indaqua and PAEB (Portuguese Association of Environmental Businesses) are lobbying to exert pressure on the national government so as to promote a model of progressive liberalisation of the market. Indaqua wishes this new management model to be based on the British model with an independent regulator and legislation that allows for the entrance of private capital in every segment of the water systems.

Medium-sized mixed operators

Medium sized organizations of mixed public/private capital in drinking water supply and water treatment have emerged the past decades in several countries. This, however, often happened in response to emergency situations, in which new control structures and private capital were needed in public organizations to catch up with lagging investments in the systems and weak operational and environmental performance (e.g., Box 8).

Box 8: Aquafin

Aquafin was founded in 1990 on the initiative of the Flemish Region to carry out accelerated work on water treatment in Flanders. It was therefore created by an emergency situation: Flanders was well behind compared to the neighbouring countries with respect to treatment facilities. Aquafin is responsible for the operation of wastewater treatment plants, pumping stations and trans-municipal pipelines. The total design capacity of the wastewater plants amounts to 4.9 million population equivalents. The responsibility for investing and maintaining the trans-municipal infrastructure is managed by governmental (Flemish) decree. Aquafin is confronted with changes in the legislation that cannot be influenced by the firm itself.
The strategic approach in dealing with the European Water Framework Directive is quite static and mainly consists of communicating the high investment costs that it implies towards the regional government. According to Aquafin, liberalisation appears to be deferred by the European legislator for several reasons such as health considerations. As politicians in Flanders are not aware of the financial implications of the European Water Framework Directive, liberalisation is not at stake for the moment. Therefore, Aquafin has no clear strategies to deal specifically with the process of liberalisation.

As ventures of public-private participation, local or regional authorities may retain significant amounts of shares, in addition to institutional (e.g., banks and pension funds) and other type of private investors (e.g., investors in the stock market). The strategy of these mixed local/regional monopolies is a combination of the orientations of the shareholder groups, each with its particular interest, inducing dynamism of a specific kind.

The public policy process establishes the geographical and institutional delineation of the tasks of these firms, in terms of their scale and the scope. In many cases, public authorities directly award concessions to the operators, so there is no competition at least in their domestic markets. Some of them compete in bidding for concessions outside their domestic markets mainly for two reasons: (1) they might fear to lose their concession contracts in the domestic market; or (2) to profit from their secure monopolistic position in the domestic market so as to increase their scale in new markets.

Box 9: ACEA

ACEA was established in 1909 as the electricity company supplying public and private lighting for the Municipality of Rome. It was only in 1937 that it became *Azienda Governatoriale Elettricità e Acque* (AGEA) with the addition of fresh water services. During the 1990s ACEA became one of the major Italian utilities and, since 1999, 49% of its capital is floating on the stock exchange. Independently from the type of arrangement, ACEA tries to establish partnerships with local authorities based on mutual trust rather than mere contracted obligations.
The Province of Rome and ACEA signed an agreement regarding management of integrated water services to come into effect on January 1, 2003. While acquiring the contract in its "captive" market in the area of Rome, ACEA started to engage in competitive bids in other regions, both as a partner in locally owned WSS operators and as contractor.

The financial framework involves a number of elements. The operator normally operates on a cost-plus basis and the government refunds costs, occasionally subsidized by the European Commission funds. Normally, the compensation is based on reasonable costs plus remuneration for the shareholders based on their investment of capital. The "plus" component is set at such a level that the financial results are appealing to private firms, as investors and as contributors of technology and knowledge. Depending on the circumstances, investments in the networks may be undertaken by the public authorities and employed by the firms at compensation. Conversely, the operator itself may invest and receive compensation from the public authorities.

As a consequence of this institutional context, these firms may want to establish themselves as efficient, effective and reliable operators, rather than a governmental organisation, or a pure for-profit firm. As a consequence of local or sector-related traditions and preferences, the networks and resources may be controlled through either direct ownership by the operator, or through a concessionary system in which the legal ownership remains with a public body. Public water policy determines objectives, standards, guidelines and priorities, in principle exogenous to the operator. The latter should be engaged with an efficient policy implementation and operation of the system from a technological point of view.

Whereas it is not party to the political negotiations, the operator still advises on the technical feasibility and costs of measures considered. As its remuneration and tasks are linked to the politically defined objectives, a main goal is to influence the investment priorities and the process of standard setting in respect of environmental goals. So, eventually, technical reports for policy-support are an important element in the relationship with the government. By convincing authorities that such objectives and standards should - and could - be enhanced, the requirement for additional activity and investment is ascertained.

In brief:

- The private shareholders and the operator have an argument to expand activities that are recognized as important by the public authorities, like environmental quality, security, etc;
- They will do so by convincing the authorities and the general public that this is feasible in an efficient manner;
- They may expand their services to other municipalities (normally neighbouring regions) not yet covered by adequate water supply or treatment facilities, offering low-cost off-the-shelf packages of solutions to municipalities or large industry;
- In a cost-plus regime, a main element of the strategy involves the communication of environmental, quality and services issues and technical skills, to justify quality improvement;
- Outperforming existing norms may provide a justification for, either, enhancing the norms further, or at least maintain the level of costs allowances – the "plus" being set as a proportion to costs.

Monitoring volumes, quality indicators and economic determinants of operations, through internal and external benchmarking and reporting systems, are used to convince the public authorities of the appropriateness of the operators' functioning and their being worthwhile their cost. The maintenance of a cooperative "partnership", based on quality and advice, is crucial to the role and the strategic position of the operator.

The European Water Framework Directive is expected to increase the importance of technological innovations and to enhance environmental standards. In addition, policies in respect of River Basin Management and Integral Water Management (Integral Chain Management) drive the integration of operators into larger entities, comprising several municipalities, and combining water supply with water sanitation activities. Such changes are particularly interesting from the perspective of this type of firms. It is through the need for technological developments and management approaches to the water cycle, that the importance of soft engineering techniques is stressed. In the future, small municipalities – lacking knowledge and access to capital - may seek recourse from such mixed capital regional operators. So, widening the range of services offered and deepening their quality and impact provides the main perspective for expansion to these operators.

These medium sized, regionally structured firms will have the scale to develop and use more advanced technologies. Moreover, and equally important, through size they will have the impact to be an effective, i.e. convincing, 'partner' to the local government in determining the outline of regional – or river basin level - water policies. Their size, and thus the scale at which technological solutions are applied, make these organizations an interesting objective for private technology and capital suppliers.

Public-private partnership is an option for the organization of the water sector, that is, with the advantages of both private sector involvement, like technology, capital, dynamism; and public assets, like low cost capital, reliance, high standards, legitimisation, etc. The limited freedom of action (strategy formulation and realization) of the firms can be perceived as a main weakness. However, the main strength of the mixed approach is the ability to focus on technological innovation, not hampered by short-term market conditions. Moreover, unlike what is often suggested, these firms might deliberately not be efficient in combining an ambitious public perspective on high environmental and security standards, with a common private appetite from the "plus" flowing from the "cost" associated, as the standards drive the "plus".

Large (private and mixed) operators

Large operators are emerging as very important actors in the European water market even though they only provide water services (i.e., water supply and sanitation services) to about 15% of the European population.

In terms of ownership, one can actually notice that there are no large public operators (i.e., providing water services to more than 10,000,000 people) in this particular sector, and all are privately owned, except one, (RWE-TW), which is mixed. However, if one compares mixed and privately owned large operators, ownership does not seem to influence their strategy, either at home or abroad. Finally, in recent years there has been a trend towards the increasing participation of financial institutions in the ownership structure of these large operators.

One could divide the major water supply and sanitation (WSS) large operators according to three markets: the French, the British and the German. These large operators participate in and, in some cases, control several small and medium operators in other European countries and in the rest of the world. French corporations are clearly dominating at the European level, namely Veolia Environment, Suez Environment (see Box 10), and to a lesser extent SAUR. The other major competitor is the German multi-utility group, RWE-Thames Water (see Box 11). In the UK, there are a few other large operators but they have not positioned themselves within the European WSS market as strongly as the other ones.

Box 10: Suez Environment

Suez is owned by the *Groupe Bruxelles Lambert* (7.2%), employees (4%), *Crédit Agricole* (3.3%), CDC (3.2%), Cogema (2.3%), and the General Public (74.6%). It provides water services for about 125 million individuals daily, which makes it one of the world's leading players in the sector. Since 1997, Suez started as a conglomerate but then refocused its activities to become a global industrial group providing services in energy and environment for companies, individuals and municipalities. Suez Environment (SE) is an operational unit and its strategy corresponds to the strategy of the Suez group for the specific sectors of water and waste. SE's main objectives are to strengthen its offer of environmental services at local level and to promote the public-private partnership model of delegated management, with a focus on operating performance and results.
Europe is by far the most important region for SE, once in 2003 about 77% of its revenues were generated in this geographical area. The strategy of SE has considerably changed since 2002. The collapse of the Argentine economy and the international stock market crisis induced the Suez businesses to refocus on its European base and more stable markets.

Although one can identify common characteristics in terms of strategy among these operators, there are also important differences most probably explained by the dominant management models and the water market restructuring policies in their domestic markets. Among the common characteristics of these operators is that they are listed in the stock exchange, which creates considerable pressures in terms of financial results. Indeed, all these operators have implemented the same type of business model pertaining to the provision of a public service yet remaining profitable.

All large (private or mixed) operators in the WSS sectors have international activities across different countries and even different continents. Nonetheless, for all of them, the European market remains the most important. Another important aspect is the prevalence in terms of importance of the domestic market, especially for operators created under the so-called British model.

During the 1990s, large water corporations in general focused on acquiring markets in developing and emerging economies. This strategic geographical orientation was to a certain extent backed by financing international organisations. However, after some serious drawbacks, the focus of these international operators is now increasingly shifting to the more stable and regulated markets in developed economies, namely in Europe, where financing is available.

These groups normally operate in several sectors, such as energy and the environment (i.e., water and waste), and are therefore able to propose specific combined solutions to their customers. The choice to provide multi-utility services highly depends on the characteristics of the markets, yet there is a general consensus that multi-utility packages work better for industrial clients. The combination of water services and energy (electricity or gas) is probably the most common type of multi-utility package. Even though the characteristics of the systems are very different, the commonalities lie in the similarity of the service and in the advantages of a joint front office and administration and contacts with public authorities and non-governmental organisations (NGOs). Another increasingly important type of multi-utility combines water supply and wastewater services – although the service contracts are normally separated the clients are the same, which enables interesting cost savings.

All large operators have developed a multi-utility approach but the French TNCs clearly distinguish themselves from the British and the German ones. It is only in the French case that water constitutes a dominant aspect in large operators' strategy of multi-utility development. Indeed, British large operators focus much more on water, sanitation and environmental services than utilities in general. For the German one, the WSS sectors represent a small share of its business compared to the other core businesses: electricity, gas and environmental services.

For large multi-utility groups, the WSS sectors are part of a diversification strategy whose purpose is to lower the risk of relying too extensively on one sector. Moreover, the WSS sectors are also seen as more stable compared, for instance, to electricity and gas since competition in the former sectors is less fierce, which therefore enables long term financial stability.

Box 11: RWE Thames Water

RWE is a German multi-utility group with a mixed ownership structure. RWE's traditional core business was electricity production and distribution but today RWE is a multi-utility group concentrating its activities in four core businesses: electricity, gas, water, and the environment. RWE provides water services through two of its seven management companies: RWE Energy and RWE Thames Water. RWE's key markets are Germany, the United Kingdom, Central Eastern Europe and North America.

The major goal of RWE Thames Water (RWO-TW) is to consolidate its third position in the world water market through business development within the existing portfolio of operations. Its key strategy is to evolve steadily the growth opportunities of its core business in the key markets of the United States (US), Eastern Europe and Germany. To face this challenge, RWE-TW relies on its expertise in the entire operational chain to generate extra efficiency gains in managing the systems. This is its real competitive advantage vis-à-vis its French counterparts, which are more competitive in bidding for contracts. RWE-TW strengths are related to managing investments and customer relationships. It is important to mention that these are related to the experience gathered in operating in their respective home markets.

European water markets differ considerably in terms of the degree and type of competition. There is a general trend towards the concentration of operators, although not amongst the largest ones. The water supply sector is considered to be more competitive than sanitation. In water supply, the main competitive pressures emerge at the moment of the bidding procedures to get in the market, while in the sanitation sector the pressure comes mainly from technological and pricing issues.

There is a growing belief among these operators that the reliance on private sector enterprises is increasing as more rigorous environmental and public health standards are being established (partly driven by the European Commission), and public sector entities reach the limits of their resources and expertise. They therefore argue that stricter requirements create new opportunities. In this context, according to their viewpoint, the European Commission should: (1) monitor the compliance to these requirements; (2) implement and monitor its competition laws and policies; and (3) monitor the proper application of the Rome Treaty rules of non-discrimination.

Large operators generally consider public-private partnerships the best solution to face the challenges of the sector. Thus, their preferences clearly lie on specific types of contracts, namely lease and operation and maintenance (O&M) contracts. However, the dominant model of management within the home-market of these operators again influences these choices.

Large operators normally provide water services to individuals, companies, and municipalities. They believe water quality is the most important and valorised aspect for the customer, followed by security of supply and, finally, pricing.

Regarding the scope of the market, these operators consider that size only becomes important outside their domestic markets where they do not have a strong physical or economic presence. Therefore, outside their domestic market, they only consider contracts in big urban agglomerations because fixed cost becomes too expensive for small contracts.

In the past years, several events have shown that citizens and civil society organisations could well have an important impact on these groups' overall strategy, leading to the setting up of several initiatives on ethical, environmental and social issues.

Finally, large enterprises make it clear that they are not interested in controlling or owning natural resources. Differences emerge in terms of physical resources with, on the one hand, large operators attached to the French model arguing they don't want to be the owners of the infrastructures but, for the duration of the contract, they accept to be responsible for their protection and maintenance. On the other hand, large operators attached to the British model consider the ownership of the infrastructure as beneficial because, according to their view, it enables higher rates of return on capital.

CONCLUSION

The current section presents the four main conclusions of the chapter, on the basis of the European map, detailed case studies and the results of a European wide questionnaire[8] elaborated in the scope of the project to illustrate some existing pan-European and national trends (Euromarket, 2004).

The **first conclusion** is that the traditional characterisation of operators as municipal non-autonomous public entities changed in the past two decades. We can identify three important trends:

(1) Small and medium-sized public operators remain central providers yet (medium and large-sized) private and mixed-owned operators became important actors;

(2) The evolution from non-autonomous to autonomous entities is a reality, in many cases accompanied by the delegation of management responsibilities to local or regional public operators, and sometimes to mixed or private companies;

(3) There is the regionalisation of operators, i.e. operators are pursuing a strategy of mergers and of alliances to increase scale. In effect, most operators believe the water industry in Europe will move towards more concentration of operators (yet not amongst the larger ones). Today, operators in the water sanitation sector are, on average, larger in terms of population served than water supply operators.

The **second conclusion** pertains to the recognition that several issues limit the efficiency and flexibility of the market, strongly conditioning operators' strategies. The most significant ones are herein recalled, namely:

(1) The creation of *competition in the market* seems rather limited due to the specificities of the sector and, therefore, in the event of liberalisation, the most common type of competition is *competition for the market*. This involves strategies that are focussed on competing for contracts through bidding. After winning the contract the service is offered in a monopolistic way.

(2) Long-term contract periods are a factor of inflexibility and inefficiency, yet they allow for stable planning horizons (required for the long-term investments inherent in the sector). Both the strategies of historical local operators and new operators are conservative to this respect, i.e., all operators aim at establishing long term relationships in what concerns the management of the services.

(3) In many cases competition and market options are limited to strategies of buying or participating in other companies. This is another strategy *for the market*, but in that case, without any competition with respect to the service itself; just a competition for the ownership of the operators.

(4) Over recent years the management of *European funds* for investments in infrastructures has been an important factor in certain countries, especially in the sanitation sector. These funds have usually been managed via local or regional institutions, being an essential part of the strategies of public operators (including municipal services).

The **third conclusion** is that the objectives, perceptions and strategies of water operators differ according to their typology[9] yet important features depend on the respective market liberalisation processes. Size, ownership and the liberalisation context in their domestic markets thus highly influence the way they operate and (re)act in relation to other stakeholders in the sector.

Where there are no significant liberalisation processes, operators present defensive strategies in terms of maintaining monopoly rights. They base their strategies on public service goals and a good relationship with their customers (e.g., users, councils, industry), but without taking into account their possible competitors. In fact, the perception of many of these operators is that the water sector is not very competitive. Most operators indeed consider that they do not have any competitors within their geographic area. In the few cases operators feel some competitive pressures, these are mainly related to pricing and technology and not to attracting clients, nor to gain markets. These are normally small publicly owned operators. It is increasingly common to find them in association with other small operators. They tend to emphasize their specialised technical and local know-how, and to implement long-term investment policies. They generally exert some political pressure for preventing liberalisation. One of their strong points is the flexibility of their strategies to adapt to changing public policies decided by local or regional authorities.

However, in some countries and for some segments (e.g., water supply in Spain), liberalisation processes have opened up (parts of) the market. Public operators complement their objectives of public services (with respect to users and local bodies) with strategies to fend off their competitors. For instance, small local public operators will stress proximity aspects and the local nature of water services in order to counter the arrival of new operators. For private

[8] This questionnaire was not used as the primary basis for the analysis, neither to construct the European map, due to the limited number of answers compared to the numbers of operators across Europe.

[9] See the annex for a summary of the main objectives and strategies per type of operator.

operators, their main objective is profitability in the long term, taking into account the competition of other operators. In this context, a specific and successful strategy of many operators is the Public-Private Partnership (PPP) strategy, based on sharing the risks and responsibilities between operators and public institutions. Water services remain under the final responsibility of the public institutions yet under the effective management control by the operator. In certain cases, this strategy allows the allocation of concession contracts by direct negotiations without the competition through public bid.

Operators in such markets are normally aggressive in terms of obtaining captive markets through long-term delegated management contracts. Especially the largest operators tend to take multi-utility approaches and to engage in long-term technological investments.

Even in markets witnessing such liberalisation processes, the water sector cannot be considered very dynamic in terms of new entrants. There are comparatively more new operators in the sanitation sector than in water supply. This can probably be explained by the fact that there have been more new opportunities in the sanitation than in the water supply sector following an increase in environmental and public health standards and concerns. Another reason for the lack of dynamics is the fact that competition in the water sector is very limited. In many cases, and due certainly to a certain search for stability, contract renewal are always to the advantage of the initial operator, and this is true whether one speaks of public or private operators.

Finally the **fourth conclusion** relates to a possible contribution of the empirical findings to business strategies' theories. That is the identification of cases where the framework, as well as the type of ordinary decisions (e.g., on pricing, investment and product/service specification), are largely fixed by a third entity in the form of regulation. One of the water operators' most important strategies hence relates to the management of regulation. In literature, the most common arguments state that operators under regulated environments can either take an adaptative/passive attitude or attempt to influence the public policy (i.e., to capture the regulator). Our statement is that water operators' strategies are being influenced and embedded in an interactive system, where they have an active role in defining the market and business environment. This means that the formulation of operator's strategies is regarded as a dynamic process with several (sometimes conflicting) goals. The ordinary decisions such as prices and investments may not be part of the spectrum of water operators' strategic choices. This is particularly important if we consider that in the water sector, regulation is not meant to be a temporary stage towards a more competitive environment. Considering water sector specificities, it will always be a regulated industry even though it might show some type and degree of competition.

ANNEX

Summary of the main objectives and strategies per type of operator

TYPE OF OPERATORS	
Small publicly owned	
Objectives	Efficient provision of Public Services to citizens at a reasonable price; Guarantee of the long-term quality of the systems; Application of sector-related public policies;
EU water-related legislation	Could be a threat related to pressure for restructuring;
Perceived competitors	No competitors;
Pressures	Shrinking public budgets; Increasing costs related to environment and public health standards;
Types of strategy	Defensive in the sense of maintaining the *status quo;* Political pressure for preventing liberalisation and/or participation of the private sector; Enforce sense of connectedness with the population; Long term investment policies; Specialised technical and local know-how; Association with other small operators.
Medium-sized publicly owned	
Objectives	Provision of Public Services to citizens; Guarantee of the long-term quality of the systems; Gather and allocate funds for investing in infrastructures; Application of sector-related public policies;
EU water-related legislation	Opportunity if complemented with funds; Compliance to guarantee continuity of funds;
Perceived competitors (if any)	Outside initial local market: medium and large public operators;
Pressures	Lobbying from private operators to open the market;
Type of strategy	Defensive in the sense of maintaining public management; Implement long term investment policies; Efficient implementation of public policies; Emphasize specialised technical and local know-how; Expansion outside initial local market.
Medium-sized privately owned	
Objectives	Provide a public service with profit; Efficiency in operation and management;
EU water-related legislation	Opportunity if non-discriminatory rules are applied;
Perceived competitors (if any)	Medium public and large private operators;
Pressures	Take-over from large operators;
Type of strategy	Aggressive in the sense of keeping captive markets through long term delegated management contracts; Motivate public-private partnerships; Pressure to raise transparency in regulation and bidding; Merging and strategy of alliances to increase scale; Multi-utility approaches; Emphasise specialised technical and local know-how, and reinforce local links with population.

TYPE OF OPERATORS	
Medium-sized mixed ownership	
Objectives	Application of sector-related public policies; Guarantee of the long-term quality of the systems; Improve efficiency so as to reduce costs and/or increase return on capital;
EU water-related legislation	Opportunity based on operator's technological and local know-how;
Perceived competitors (if any)	Outside initial local market: medium and large public operators;
Pressures	Lobbying from private operators to open the market;
Type of strategy	Defensive in the sense of maintaining the *status quo* in their initial local market; Motivate public-private partnerships; Emphasise specialised technical and local know-how, and reinforce local links with population; Convincing authorities about the need to expand activities.
Large-sized (private and mixed ownership)	
Objectives	Provide a public service with profit; Increase efficiency to raise profit margins;
EU water-related legislation	Opportunity (in the event of limited public resources and expertise);
Perceived competitors (if any)	Other large operators;
Pressures	Competitive pressures to get in the markets; Anti-globalisation and anti-privatisation civil society movements;
Type of strategy	Aggressive in the sense of obtaining captive markets through long term delegated management contracts; Motivate public-private partnerships; Pressure to raise transparency in regulation and bidding contracts; Long term technological investment; Multi-utility approaches; Integrated (vertical and/or horizontal) contracts; Development of international activities yet minimising financial and political risks (stable and regulated markets).

REFERENCES

Allouche, J., P. Luís-Manso, et al. (2003). "Water resources management in a globalised economy: towards a multi-level regulatory approach." Economia e politica industriale **116**: 217-232.

Ansoff, H. (1965). Corporate Strategy. New York, McGraw Hill.

Bowler, I. R. (1999). "Recycling urban waste on farmland: an actor-network interpretation " Applied Geography **19**: 29-43.

Chandler, A. (1962). Strategy and structure Cambridge M.A., MIT Press.

Cyert, R. and J. March (1963). A behavioural theory of the firm. Englewood Cliffs NJ, Prentice-Hall.

Euromarket (2004). Analysis of the strategies of the Water Supply and Sanitation Operators in Europe. Euromarket Project.

Finger, M. and J. Allouche (2002). Water privatisation. Transnational corporations and the re-regulation of the water industry. London, Spon Press.

Gee, A. (2004). "Competition and the water sector." Competition Policy Newsletter **2**: 38-40.

Gordon-Walker, S. and S. Marr (2002). Study on the application of the competition rules to the water sector in the European Community, European Communities.

Granovetter, M. (1985). "Economic action and social structure: the problem of embeddedness." American Journal of Sociology **91**(3): 481-510.

Gustafsson, J.-E. (2001). Public Water Utilities and Privatisation in Sweden. EPSU Public Service Conference, Brussels.

Hannan, M. and J. Freeman (1988). Organisational Ecology. Cambridge MA, Harvard University Press.

Lannerstad, M. (2002). Country Report Sweden. European Water Management between Regulation and Competition. AQUALIBRIUM.

Luís-Manso, P. and M. Finger (forthcoming). "The challenge of regulating the drinking water sector: lessons from Mozambique." Water International.

Mintzberg, H., B. Ahlstrand, et al. (1998). Strategy safari: a Guided Tour through the Wilds of Strategic Management. London, Prentice Hall.

Pettigrew, A. (1973). The Politics of Organizational Decision-Making. London, Tavistock.

Porter, M. (1985). Competitive Advantage. New York, Free Press.

Rondinelli, D. and M. Berry (2000). "Environmental Citizenship in Multinational Corporations: Social Responsibility and Sustainable Development." European Management Journal **18**(1): 70-84.

Serageldin, I. (1994). Water Supply, Sanitation and Environmental Sustainability: the Financing Challenge. Directions in Development. Washington DC, World Bank.

Whitley, R. (1990). "East Asian entrepreneurial structures and the comparative analysis of forms of business organisation." Organization Studies **11**(1): 47-74.

Whittinton, R. (2002). What is strategy - and does it matter? London, Thomson Learning.

Williamson, O. (1991). "Strategising, economising and economic organisation." Strategic Management Journal **12**: 75-94

Chapter V

Identification and Description of Plausible Water Liberalisation Scenarios

David Musco[1]

INTRODUCTION

Having looked at the formulation of European Union (EU) policies towards the water sector, as well as the current and likely future behaviour of major actors in the water market, we focus now on the methodology and the description of likely European liberalisation scenarios in the water supply and sanitation (WSS) sectors. By scenarios we mean coherent and credible stories about alternative futures. A scenario is therefore composed of two separate elements: the End State, which describes the situation at a particular future point in time, and the Storyline, which connects the present to the End state in a logical manner. Hence a scenario is both a description of the future and how we get to that future. In building these scenarios we recognised that there is not one uniform current situation for the water sector across the EU from which alternative future scenarios could develop. Instead, as the country profiles prepared in earlier phases showed, there are multiple national situations, which vary enormously across the EU. This made the scenario building exercise at a pan-European level particularly challenging. This chapter is divided into four sections. The first section introduces the concept of scenarios and scenario building. The second section identifies the initial Euromarket scenario outlines. And the third section describes the final detailed Euromarket scenarios that were developed by individual Euromarket consortium members. In the fourth section the author draws some general conclusions from the whole scenario building process.

[1] This chapter is based on Deliverable 5 of the EUROMARKET project. Special thanks go to Fiona Evans for all her help in preparing the guidance material on scenario writing and her assistance in organising the second scenario validation workshop. Thanks must also go to Andrew Nickson for his role in chairing both scenario building workshops.

ISBN: 1843391139. Published by IWA Publishing, London UK

SCENARIOS AND SCENARIO BUILDING

Scenarios

What are scenarios?

Whilst there is no single definition of scenarios, the following quotes are good approximations.

- Scenarios are coherent credible *stories* about alternative *futures*. Effectively they describe different paths (via a consistent set of events, trends and actor strategies) that lead to these alternative futures (Davis, 2003).
- A scenario is a *story* that describes a possible *future*. It identifies some significant events, the main actors and their motivations, and it conveys how the world functions (Shell website – http//www.shell.com/scenarios/).
- Scenarios are hypothetical sequences of events constructed for the purpose of focussing attention on causal processes and decision points (Kahn and Wiener, 1968).

Hence a scenario is both a description of the future and how we get to that future. A scenario is therefore composed of two separate, but inter-related, elements:

- The End State – which describes the situation at a particular future point in time.
- The Storyline – which connects the present (ie the "current" state) to the "end" state in a logical manner.

It is extremely important for the reader to realise that "scenarios are not projections, predictions or preferences, but alternative futures. They are purposely *challenging*, being designed to help us confront the assumptions we are making about the present and future. Scenarios are valuable because they stimulate *questions* rather than because they provide *answers*". (Sir Philip Watts, Shell website)

Why use scenarios?

A scenario can help us (via the end state) to see what the future will actually be like, and (via the storyline) how and why these futures may occur.

Scenarios are particularly useful in situations where changes in the business environment are recognised but are not well understood. They help to reveal to their users an understanding about what is important, how those critical factors will evolve in the future, and the linkages between the different factors. Scenarios can therefore expand our thinking, enable us to take on and explore new possibilities, and challenge conventional wisdom.

Assessing the liberalisation of Water Supply and Sanitation (WSS) services is therefore well suited to the scenario approach. The liberalisation process, by its political nature, is complex and poorly understood. The use of scenarios in this context will hopefully reveal some of the underlying complexities and help us to challenge conventional thinking in the area.

What are the main characteristics of scenarios?

Scenarios can be characterised on the basis of their time horizons, their geographic scale and their content details.

Time Horizons: The timescales for scenario building are normally judged according to the dynamics of the sector under investigation. The WSS sector is traditionally risk averse and quite slow moving. The project team decided to extend the original scenario time horizon to cover 15 years. This allowed for the development of more interesting and challenging scenarios. Noting that the scenarios were finalised in late 2004 the Euromarket scenario time horizon was therefore set at 2020.

Geographic Scale: Scenarios can be developed on different geographic scales – global, international regions (eg Europe), national (eg Member State), sub-regional or local. The Euromarket proposal states that the "likelihood of one single European water liberalisation scenario, or of different scenarios" will be assessed. The Euromarket scenarios (notably the end states) are focused on the European scale. But storylines have also been developed for selected Member States.

Content: Each EU scenario has been structured around an outline of the future European end state (structured via the end state template[2]), and a set of storylines (made of a series of future events, actor strategies and associated driving forces) - composed at both the European and Member State level (see below for a more detailed explanation).

Scenario Building

Stages to scenario building

An iceberg (see figure 1) can be used to understand the process of building scenarios (Lindgren & Bandhold, 2003).

At the peak of the iceberg are the observable events. Trends and patterns lie just below the waterline. This reinforces the point that observable events are not random, but are related to one another and give rise to trends and patterns in the events observed. Finally at the bottom of the iceberg there is the underlying causal structure – the driving forces behind the trends and events.

Scenario building can be conceived as a process of diving to observe the bottom of the iceberg. From an understanding of *historic* events and *historic* trends one can develop an understanding of the underlying causal structure. From this one can (hopefully) resurface to build the scenarios (by identifying the key driving forces and identifying the main *future* trends and *future* events that will characterise these scenarios).

The process of producing scenarios is just as important as the scenarios themselves (Davis, 2003). A dedicated scenario building process was developed specifically for the Euromarket project. It was designed to widen the project teams perspectives and to understand the key trends and events that are significant (that we, as a group, might otherwise dismiss).

The scenario building process can be conveniently split into five stages:

1. Define the problem and select the scenario building approach to be adopted (see introduction and discussion on scenario building below).
2. Build the Base (by understanding the market dynamics[3]).
3. Develop the outline scenarios.
4. Describe the scenarios in detail.
5. Validate the scenarios.

This chapter focuses on stages 1, 3 and 4. The preceding chapters have essentially built the base (ie stage 2).

Tools for scenario building

It is the adoption (and particular use) of specific scenario building tools (particularly those used to develop the initial scenario outline and then describe them in detail) that distinguishes the many different generic and proprietary approaches to scenario building. For an excellent summary of the various scenario building tools the reader is referred to Scenario Planning – the link between future and strategy (Lindgren & Bandhold, 2003) and Learning From the Future – competitive foresight scenarios (Fahey & Randall, 1998). The Shell website is also an excellent unrivalled source of information on scenarios and scenario building.

There is a large number of different scenario building tools. At least 25 different tools can be identified – although some are refined or elaborated versions of the classic techniques. These different scenario building tools can grouped under five method headings:

1. *Time-line methods* embracing tools that focus on the *past* (eg analogues, historical analyses, and trend extrapolation).
2. *Consequence methods* embracing tools that focus on the *future* (eg probability effects, future history, future event production, and consequence trees).
3. *Actor orientated methods* embracing positioning tools such as actor analyses and competitor analyses.
4. *Systems methods* embracing analytical tools such as STEEP[4] analysis, cross impact analyses (possibly adopting causal loop diagrams and four field analysis to simplify the results), morphological analyses, and system modelling.

---

[2] The template included 8 major categories to describe the future: nature of competition, market, operators, institutional arrangements, economic factors, social factors, environmental factors, and other factors.

[3] From different actor perspectives (the EU, the Market, Legislative/Regulatory Institutions, and the Operators) and from different themes (Social, Technological, Economic, Environmental, Political).

5. *Generative methods* embracing narrative tools such as future imaging, story telling (future history and story boarding) and future headline production.

The above tools tend to be applied at different stages in the scenario building process – identifying key driving forces (eg causal loop diagrams), picturing future outline scenarios/end states (eg scenario cross, future imaging), and describing in detail the selected outline scenarios (eg story telling, future event production).

Approaches to scenario building

A number of "off-the-shelf" scenario building approaches exist – both generic and proprietary. No single approach can be identified as best practice, especially considering the subjective nature of what actually constitutes a good story.

The only golden rule is that the scenario building approach adopted must be credible to the team building the scenarios and the key audiences who will eventually read the scenarios. In addition there must be a fit between the approach used to build the scenarios and the ultimate objectives of the work.

As no single approach is universally best it is sometimes appropriate to adopt, at least in the early stages of the scenario building exercise, a combination of approaches. Each approach used will generate particular insights. Hence the use of more than one approach to scenario building will provide an even wider understanding upon which to base future decisions. This concept was integrated into the Euromarket scenario building methodology (see below).

Internationally there is little consensus or consistency in the scenario building approaches employed. The influence of historical and cultural factors remains strong. National traditions therefore play an extremely important role in determining individual Member State approaches to scenario building (Henley Centre, 2001).

Types of scenario

There are three major types of scenarios – contrasted, trend based, and normative scenarios – as linked to possible (or plausible), probable, and desirable/undesirable futures.

Contrasted scenarios state an ordered set of possible events (irrespective of their desirability or undesirability), while normative scenarios take values and interests into account. Contrasted scenarios can be classified as either "deductive" or "inductive" (Fahey, 1998).

- *Deductive scenarios* relate to future end states that are deduced first (through an assessment of driving forces, or through future imaging techniques), with event pathways subsequently being developed to these future end states (a *future-backward* approach where plausible storylines are developed to fit pre-identified end states).
- *Inductive scenarios* relate to future end states that are gradually induced from linking chains of events into a plausible storyline (a *future-forward* approach where plausible pathways effectively predetermine the end states). This approach can also be likened to a sophisticated form of future trend based scenario.

The above classification will ultimately depend on the individual tools adopted to develop the scenarios.

Deductive scenario development can be further classified according to how the future end states are initially determined and how the associated storylines are then developed. There are a number of different proprietary approaches to the development of deductive contrasted scenarios (eg la prospective, intuitive logics, future mapping etc). And as expected each proprietary approach tends to rely on a slightly different set of scenario building tools.

- *La Prospective* tends to adopt tools from all of the above methods to develop and describe the scenarios. The French school particularly favours complex system methods (eg cross impact analyses to analyse the driving forces, morphological analyses to develop and describe the scenarios) and detailed actor orientated methods (eg actor analyses to validate the scenarios).
- *Intuitive Logics* tend to rely on the generative methods (eg brainstorming, future imaging) and simple systems methods (eg four field analysis, otherwise known as the scenario cross[5]) to develop the initial outline scenarios. Morphological analyses may be used to describe the initial outline scenarios in detail.
- *Future Mapping* approaches tend to rely on generative methods (eg future imaging) to develop the outline future end states. Consequence methods – particularly future event production – are typically used to describe the pathways to the possible end states.

---

[4] STEEP analysis assesses the drivers for change under five headings – Social, Technological, Economic, Environmental, and Political.

[5] Scenario cross consists of two using two key driving forces to create a simple 2x2 matrix. Four different scenarios will come out of the corners of the cross.

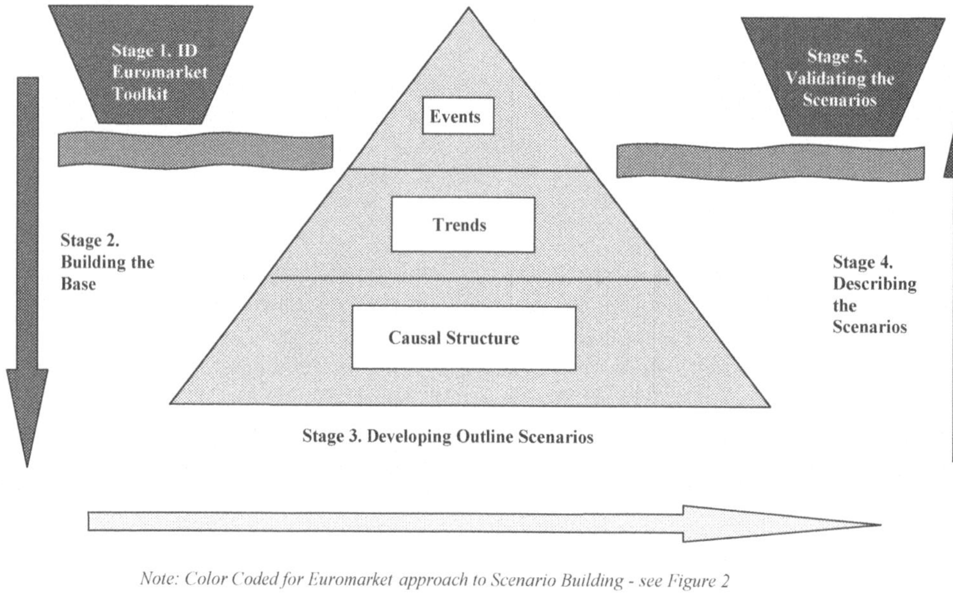

Figure 1: Iceberg Analysis

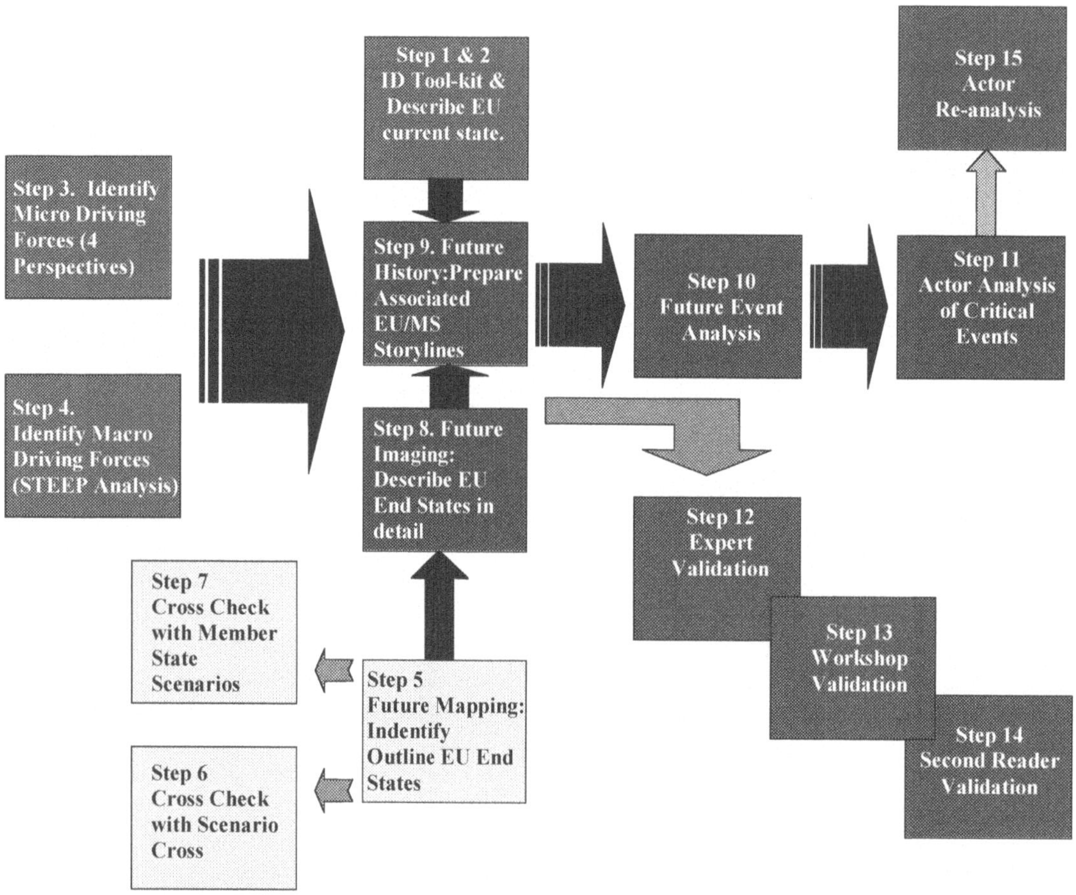

Figure 2: Euromarket Approach to Scenario Building (IDD)

The Euromarket Scenario Building Methodology

The Euromarket Project difficulty

Before moving on to discuss our approach to scenario building it is important to recognise a major unforeseen difficulty with the scenario building exercise in the Euromarket project. That is, the European water supply and sanitation services market does not consist of one single uniform current state - from which alternative (future) scenarios will develop. As is evident from individual Member State profiles we are dealing with multiple current states (which vary enormously across the EU). This makes the scenario building exercise (which normally embraces storylines from one existing current state) particularly difficult - as the national Member State institutional context is extremely important when considering potential liberalisation scenarios. By focusing solely on the EU level these national contextual issues would be completely lost.

We solved this problem by accepting that there were multiple current states and that multiple storylines should therefore be developed to a restricted set of future EU end states.

We assumed there would be two major forms of "current state".

I. The first current state would be at the *European level*. Here the "current state" is simply the 'average' position across the EU at the current time (2004). The EU is really a mixed structure (but with the majority of services based on delegated/direct public management – mainly relying on competition in supplier markets). We would simply use this average position as the current EU state and develop EU storylines away from this mixed structure toward majority positions using the other competition models. These *EU storylines* would be largely structured around generic actors and driving forces.

II. The second current state would be at the *Member State level*. These Member State storylines would be more tightly framed around specific actors and driving forces. The Member States selected were those eight originally identified in the Euromarket proposal – Belgium, France, Germany, Italy, Netherlands, Spain, Switzerland[6] and the UK. Each Member State was also represented by an organisation in the Euromarket team and this "local knowledge" was deemed vital for developing plausible *Member State storylines*.

The Euromarket Tool-Kit and Approach

Taking "off the shelf" approaches to scenario building is not generally appropriate. As noted above the scenario building approach needs to be tailored to solve the focal question being addressed – in this case the development of plausible liberalisation scenarios for the European water supply and sanitation sector. We therefore selected a number of scenario building tools that best suited this focal question. We believed they were particularly relevant to the scenario building objectives of the Euromarket project.

The selected scenario building tools (that form the basis of the dedicated Euromarket scenario building toolkit) are:

- Stage 2 – Building the Base:[7] *Historical/Trend Analyses* (eg to represent the results of earlier chapters) - possibly illustrated by *Causal Loop Diagrams* - and *STEEP analysis* to consider trends in the broader macro-environment.
- Stage 3 – Developing Outline Scenarios: *Trend extrapolation* (to develop initial outline Member State scenarios), *Future Imaging* (to map and develop outline European liberalisation end states) and the *Scenario Cross/Probability Effects Matrix* (to develop alternative outline European scenarios).
- Stage 4 – Describing the Scenarios: *Future History* via story telling (to develop plausible pathways from the current to the future end state), *Future Event Analysis* (to assess the detail of the storylines formed) and *Actor Analyses*[8] around selected critical events (to ensure internal consistency and provide additional, more animated, detail on the selected scenarios).
- Stage 5 – Validating the Scenarios: Two validation *workshops* (the first one focused on driving forces and the possible EU end states, whilst the second one focused on the full first draft scenarios), informal *external consultations* within 8 Member States and the introduction of *second readers* to contest the logic behind the proposed draft final scenarios.

---

[6] Not a Member of the European Union.

[7] Causal loop diagrams consist of arrows connecting variables in a way that shows how one variable affects another.

[8] Actor analyses have to some extent been used to frame the first four work packages. These have been primarily framed around specific actors – operators, regulators, and the EU (with markets being a catch all work package with more of a focus on customers). These work packages have mainly focused on assessing historic/current trends (ie timeline methods)

These twelve tools (three in each stage) are drawn from across the five main methods. Our approach relies heavily on six of these twelve scenario building tools[9] – *historical trend analyses* to identify the key driving forces, *future imaging* to develop the future end states, *future history* to develop the associated storylines (at both a European and Member State scale) to these end states, *future event analyses* to deconstruct the resulting storylines, *actor analyses* to validate and further describe the critical events and two *workshops* to validate the driving forces and the resulting scenarios.

Our approach involved integrating the above tools into 15 specific steps (see figure 2). Steps 5, 8, 9 and 13 are particularly important in the scenario building approach adopted. In essence we have adopted a modified future mapping approach to develop our liberalisation scenarios. Future mapping exercise occurs around a set of constrained futures that can be predicted at an early stage of the scenario building exercise (rather than a more traditional open ended scenario building exercise where more time is spent on developing the outline scenarios). With future mapping more time is spent on developing the storylines. However, that is not to say that mapping the future possibility space is unimportant. It is critical. And this is the subject of the next section of this chapter.

The Euromarket Scenarios

We have developed a set of contrasted scenarios.

The Euromarket liberalisation scenarios represent what is *plausible* - not necessarily what is either desirable[10] (ie normative) or probable (trend based)[11]. However, plausibility does not preclude the introduction of surprises. The inclusion of *surprises* within our range of *plausible* scenarios is important as history shows us that historical trends are characterised by strong fluctuations rather than smooth curves. These are often triggered by unexpected changes (ie surprises – whether they be improbable or probable).

We have therefore introduced both *wild cards* (high impact and low probability events) into the storylines and have included one *scenario XXX* (as a peripheral scenario) to illustrate this point.

IDENTIFICATION OF EUROMARKET SCENARIO OUTLINES

Mapping the "possibility space"

The real art of scenario development is to *reduce* a large range of possibilities to a handful of *plausible* directions that together contain the most relevant "uncertainty dimensions". The end states selected should, ideally, represent the complete future "possibility space" for the liberalisation of water supply and sanitation (WSS) services in the European Union.

We have defined the term *liberalisation* as "a *process* by which *competition* is introduced in situations or sectors hitherto characterised by exclusive or special rights, or monopoly granted to historical operators". And there are a limited number of potential *competition processes* in the WSS services sector. If all of these competition processes can be identified[12] and clustered then we can develop a set of EU end states that will conceptually map the whole future liberalisation "possibility space".

Transaction Framework

We have already developed a simple transaction framework to analyse the market dynamics. This transaction framework can be adapted to map the possibility space for future liberalisation (ie competition). The transaction framework identifies three main sub markets:

i. The market for *Consumer Transactions* (ie consumers who receive various water and sanitation services[13] from the operator in return for payment).

ii. The market for *Supplier Transactions* (ie where private/public companies provide various inputs – products, services, finance and possibly even bulk water – to the operator in return for payment).

---

[9] The nine tools in stages 2 to 4 formed the basis of 9 individual work activities completed during WP5.

[10] The Euromarket proposal states that we will not take a pro or an anti-liberalisation position.

[11] We can call the most probable scenario the *reference* scenario – whether it is trend based or not.

[12] Unlike some other broader scenario building projects the adopted modified future mapping approach is possible within Euromarket because the possibility space is relatively constrained.

[13] The consumer market involves the supply of two main services (water supply and sanitation) that really consist of a myriad of specific sub services (eg drainage, foul sewage, trade effluent, potable water supply, non-potable water supply, interruptible or standby water supply, peak or off peak water supply, connection to network).

iii. The market for *Water (Resource) Transactions* (ie where the water environment provides/receives raw water/wastewater discharges in return for appropriate controls, with payment possibly being made to environment agency to administer such controls).

Each distinct market comprises a group of buyers, a group of providers/sellers and a group of regulators. But it is important to recognise that these three markets are also intrinsically linked together – primarily through the operator. In addition, where there are competitive market segments within the consumer market, these can either be regulated by national competition authorities (who also focus on the supplier market) or by a sector specific economic regulator (within the consumer market).

Within this transaction framework there are four main actor groups (and each has 3 main sub-types):

1) Regulators (one set for each main market - consumer, supplier and water resource).
2) Operators (including public, private and social enterprises).
3) Suppliers (of products, services and finance).
4) Consumers (including households, non households and other).

The transaction framework can be modified with the addition of two more actor groups:

1) The *responsible bodies* (who are legally obliged to ensure the provision of such public services) and the *policy makers/legislators*[14] (who ultimately determine the rules that define individual actor relationships and shape the overall structure of the transaction framework).
2) The *lobbyists* and *opinion formers* (who seek to influence overall policy development).

The modified transaction framework - with a list of all the main potential actors (both at the final EU end state and in the supporting EU storyline) - is summarised in figure 3. The top part of the framework is based on transaction cost *economics* (re key actor relationships in the end state), whereas the foundation is based on transaction cost *politics* (re key actors influencing the storyline development). It is the responsible entities/policy makers and legislators that will ultimately influence (through action or non action) the shape of the final scenarios developed. These actors will, in turn, be influenced (to a greater or lesser degree) by various lobbyists/opinion formers (through formal and informal consultations). Indeed one could envisage their interaction as being the underpinning foundations of the modified transaction framework. The evolution of these foundations will ultimately influence the final shape of the top part of the transaction framework for each scenario.

There is an apparent fundamental difference between the supplier and consumer markets. In the supplier market the buyer (ie the operator) is involved directly in the transaction – whether this be *for* or *in* the supplier market. In the consumer market (except for large non household customers) responsibility for managing competition *for* the market (eg pricing and service quality) is wholly devolved from the consumer to the responsible entity, who may in turn then devolve some (or most) of this responsibility to another independent, but accountable, "regulatory" body. The relationship between the *responsible entity* and the individual consumer, and between the *responsible entity* and the regulatory body is then governed by transaction cost politics (rather than transaction cost economics)[15].

The responsible entities are usually elected bodies. Across Europe this is typically the municipal council. These entities usually have a democratic mandate to ensure that a universal water and sanitation service is provided to their constituency at a fair price. In reality these powers (especially over service quality and pricing) are usually (but not always) devolved to separate, accountable but quasi-independent, "regulators" – including municipal boards/committee.

Competition Processes

Within each of the three markets there is potential for two types of competition – either *in the market* or *for the market*[16]. Hence there are six possible generic competition processes within the three distinct transaction markets. In addition there is the possibility of regulated competition. Here some public authority (regulatory body or municipality), or some other body, attempts to mimic competition in/for the market through appropriate regulatory instruments (eg benchmarking, price/profit control).

A quick summary of the potential competition processes that could be used to structure the liberalisation scenarios are summarised in Box 1. This liberalisation "possibility space" (with the associated Box 1 acronyms) is shown

[14] More often than not these framework designers tend to operate at the national level.

[15] Note that the consumer market in figure 3 may be opened to form the butterfly structure discussed in chapter 2. However, the closed lower wing structure is more useful here as it highlights the critical actor relationships (between consumers, operators, responsible entities and "regulators") within the consumer market.

[16] Competition for the market is generally organised by the appropriate responsible/regulatory body.

graphically in figure 3. It is possible to narrow down the number of liberalisation end states by excluding competition processes within the water transaction market. Hence there are six competition processes that can form the basis of the EU end states. These different competition processes are associated with specific forms/models of regulation.

We recognise that, in reality, one or more of the competition processes (identified in Box 1 and positioned in figure 3) could be present within any given plausible end state. However, within any given end state one form of competition will predominate. The *dominant competition process* can therefore be used to determine the fundamental nature of each end state.

Traditionally competition processes *within* the two main transaction markets (consumer and supplier markets) have been seen as mutually exclusive (or at least problematic to operate together). For example, competition *for* the consumer market is not normally associated with either competition *in* the consumer market (ie granting exclusive rights is normally part of the contractual arrangement under delegation), or more importantly, intensive price cap regulation by an independent regulator. Competition *for* the consumer market may also preclude extensive competition *for/in the supplier market.* Here the winning bidder is normally allowed to use associate companies to provide various supplier inputs. This is in contrast to the (economic) regulation of the consumer market where competition for/in supplier markets is often seen as a vital complementary competitive process (that has to be protected from potential monopoly abuses). There are also notable difficulties[17] in regulating a system that involves both formal economic regulation (by a third party) and competition *in* the consumer market.

It is possible to combine the individual competition *processes* and construct a series of competition *modalities*. However, whilst these different competition modalities may represent more plausible futures they are also likely to have more similarities, with the increased danger of greater scenario overlaps.

Market Failure

In terms of market failure, water supply may be better characterised as a quasi-private good and sanitation as a quasi-public service. They are neither wholly public nor wholly private goods (the true nature of which will vary according to circumstance). But both services (as a result of high levels of asset specificity and sunkness) are subject to potential market failure.

In terms of third party access (via common carriage), the potential for competition *in* the consumer market appears to be restricted to a relatively small part of the infrastructure system. It appears that the majority of transportation/storage network (both before and after treatment) is monopolistic in character. Competitors in the potentially competitive parts of the market will therefore require access to these essential facilities via common carriage arrangements. Essentially the direct physical and local "piped" link between the consumer's premises and the water resource (for both water abstraction and wastewater discharge) undermines the potential for any large-scale vertical unbundling.

The potential for common carriage is also hindered by the need for institutional change (e.g. vertical unbundling) and the high regulatory costs associated with the need to minimise the risk of the other market failures (such as information asymmetry and the need to maintain coordination following market opening). Adequate regulatory procedures also need to be introduced to deal with the management of the two main externalities (health and environment) – namely how to monitor the responsibility for drinking water quality breaches (a difficult technical problem), and how to allocate/manage abstraction licences and discharge consents in a more fluid market place. Robust access pricing methodologies will also need to be developed to deal with the various subsidies that are inherent in the WSS market.

These potential market failures are not insurmountable but they do involve substantial transaction costs. In England and Wales there is an attempt to introduce competition *in* the market through the introduction of retail competition and water resource/treatment competition (via common carriage) for large water supply customers whose consumption exceeds 50 Ml per year. This initiative, launched in December 2005, will yield important evidence on the true economic viability of this form of competition.

Management Modalities

The six competition processes identified in box 1 have been used to structure the liberalisation end states that will dominate the EU in 2020. But we do not propose to base the EU end states solely on different competition processes. Different management modalities such as ownership structures (public or private), and/or the degree of separation between the responsible and management entity (direct or delegated) are also relevant.

---

[17] The boundary between regulation and competition becomes increasingly complex.

Box 1. Potential Competition Processes

Supplier Transactions

1. Competition ***in*** the (supplier) market (CiSM). Rivalrous competition across the board of supplier inputs. However, long term and complex relationships with private sector are relatively limited and traditional piecemeal public procurement is generally favoured.
2. Competition ***for*** the (supplier) market (CfSM). Various forms of Public Private Partnerships, PPPs (eg BOOTs, BOTs, DBFO, Management contracts), with various degrees of risk transfer to the private sector. This includes contracting out various service aspects (eg billing) and the possible use of secondary bodies to arrange financing. This can be thought of as a complex form of public contracting. It excludes concessioning/leasing arrangements.

Regulated Market Transactions

1. Benchmarking Procedures (BP) possibly in the consumer transaction market (and possibly where associates of the operator are used as suppliers).
2. Price/Profit Control Mechanisms (PCM) such as price/revenue caps (eg RPI-X) or rate of return targets. These mechanisms can be applied to both privately owned, community based and publicly owned operators.

Consumer Transactions

1. Competition ***in*** the (consumer) market (CiCM) that can include a range of service activities in a range of market segments including: self supply, geographic inset appointments (reflecting non exclusive rights of monopoly provider), retail competition and third party access regimes (eg common carriage). At present, unlike in other utilities, this form of competition is usually restricted to a limited number of market segments eg large industrial consumers.
2. Competition ***for*** the (consumer) market (CfCM) that can include concessions and possibly affermage contracts. This type of contracting arrangement with full risk transfer characteristics is sometimes defined as a PPP (see recent CEC Green paper). However, unlike competition in the market, competition for the market is typically organised on behalf of customers by the responsible entity (possibly through a contracting body).

Water (Resource) Transactions

1. Competition in the water (resource) market (CiWM) – competition to buy specific water rights.
2. Competition for the water (resource) market (CfWM) – competition for the right to abstract water/discharge wastewater, possibly through tradable licences

Competition processes I-VI can operate with public, community and private operators. However, it is noted that some of the above competition processes have traditionally been associated with specific ownership structures (ie competition in/for the *consumer market* and *regulated competition* have tended to occur where private operators predominate, whereas heavy reliance on competition in/for the *supplier market* have tended to occur where the public operator predominates).

The competition for asset ownership (and hence Competition for Strategic Control – CfSC in figure 3) could be seen as an important complementary (and so far omitted) competitive process within the overall transaction framework. Competition for asset ownership and strategic control of future revenues between the public sector, the community and the private sector is already evident in some Member States. This could have been explicitly recognised and be dealt with by using a peripheral scenario (ie a scenario XXX).

A scenario XXX could then have been framed around full "privatisation[18]" (in those Member States, the majority, with public ownership), or "nationalisation" (in the sole Member State, ie England, with private ownership). Under these two asset ownership scenarios it is likely that competition will be strongly associated with competition process III and IV (regulated market transactions) and I (competitive supplier transactions) respectively. Hence there is a strong overlap between asset ownership and the competition process, and hence the model of regulation. These ownership issues can therefore be wholly subsumed into these two competition processes. We did not therefore develop an end state dedicated to public-private asset[19] ownership arrangements.

However, this left one potential ownership structure – community based management – as a potential additional EU end state. This formed the basis of our peripheral (management/ownership orientated) scenario XXX.

[18] If privatised there may (depending on national merger and acquisition policies) also be further competition between the various private interests for ownership and control. However, this can be dealt with under competition in the supplier market (re finance).

[19] Privatisation of assets across the EU by 2020 is also seen as not being a particularly plausible scenario.

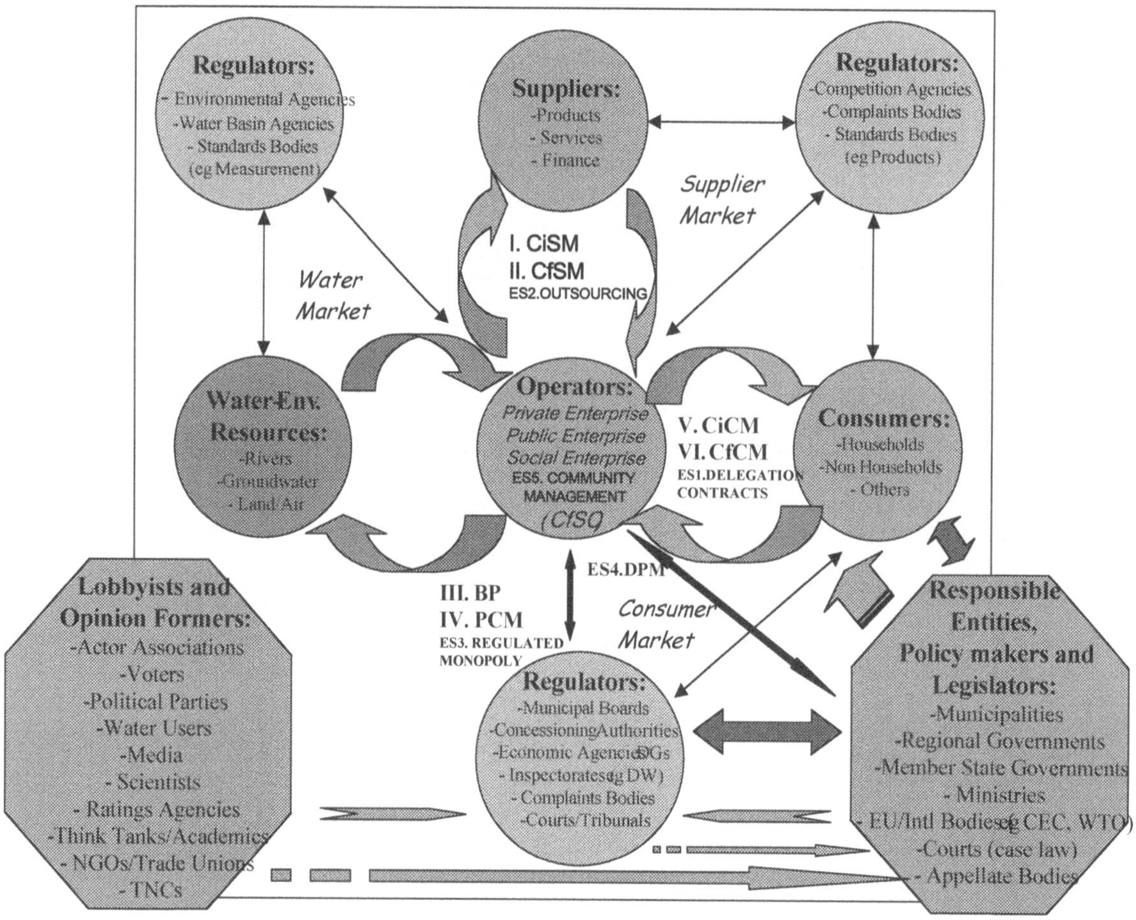

Figure 3: Framework for Water Supply and Sanitation Services - Competition Processes

Main Key to Institutional Frameworks
See Figures 3-9

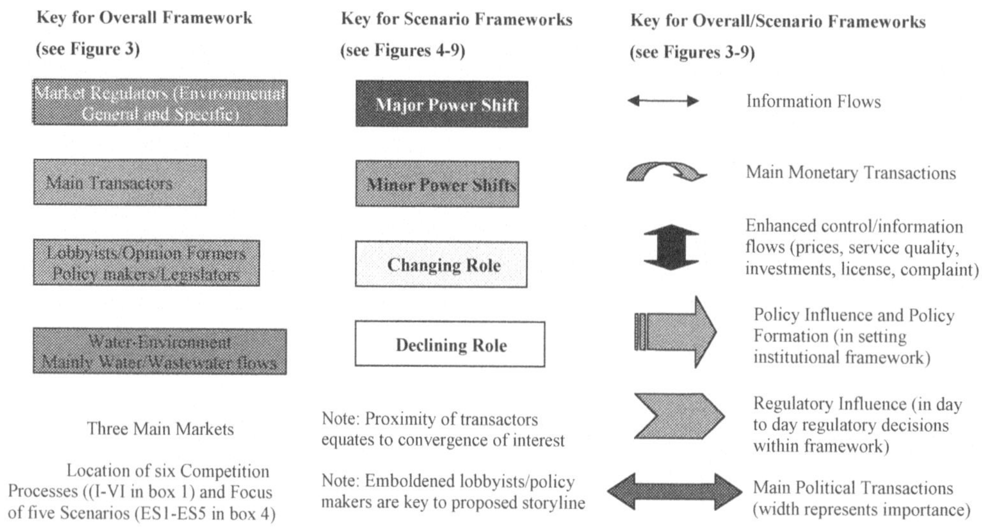

Selecting the Final EU End States

Including one scenario XXX option, there would then have been seven EU end states (six plausible end states based on mapping competition processes as outlined above, and one peripheral end state based on a specific management modality). This was too many to consider. Hence there was a need to reduce the number of end states to a more manageable number.

The Selected Outline EU End States

Based on the above discussion of market failure, competition in the consumer market (eg third party access via common carriage) did not therefore form a separate distinct scenario for the liberalisation of water and sanitation services.

To reduce the number of plausible end states to a more manageable number we merged competition process III (benchmarking), IV (profit/price control) and V (competition in the consumer market) together. These competition processes were viewed as being complementary. This merged end state is therefore based around a core monopoly network that is heavily regulated (via price/profit control and benchmarking processes) and a set of lightly regulated niche markets[20] that have been opened up to competition (via common carriage for a limited set of consumers etc).

We therefore proposed five initial outline "liberalisation" EU end states (see Box 2). The four most plausible end states (1-4) are based around contrasted competition processes. They can be conceptualised along two key axes – the nature of competition ("in" or "for") within the *supplier transaction market* (end state 2 on outsourcing versus end state 4 on direct public management) and the nature of competition ("regulated" or "for") within the *consumer transaction market* (end state 1 on delegation contracts versus end state 3 on regulated monopoly).

The competition focus of these proposed end states is illustrated in figure 3.

Box 2: Initial Outline EU End States

End State (ES) 1: Delegation Contracts (Symbolised by a gift wrapped box) – as developed by ENGREF and Paris VIII. A long-term (10-15 years) contract specifying very large objectives on integrated service is awarded to a private operator or a public enterprise. Criterion of choice: "Best bid". May include pilot regulation of the service by regional/national regulatory bodies.
End State (ES) 2: Outsourcing (Symbolised by a pair of dice and ingots of gold representing risk/reward) – as developed by IHE and TU Delft. Several short-term (1-5 years) well-specified contracts for different activities of the WSS service (e.g. network repairing, billing, management) possibly combined with longer-term contracts (eg DBFO, BOTs, Partnering) for infrastructure specific maintenance/construction. However, revenue risk and responsibility for service provision remains with public body – which is typically responsible for the integrated service. Criterion of choice: "Best price". No formal regulation.
End State (ES) 3: Regulated Monopoly (Symbolised by an inspector) – as developed by IDD and Ecologic. An unlimited monopoly (or a license for 50 years) that is totally privatised or totally public (but are typically autonomous). There is an independent authority of regulation with great power to set the price and/or control profits and investments, and instigate comparative competition. Niche (non monopolistic) markets are deregulated but are closely monitored by the well-resourced regulatory authority.
End State (ES) 4: Direct Public Management (Symbolised by a hierarchy of desks) – as developed by UCL and EPFL. Competition is largely restricted to public procurement for infrastructure development. No competition on service aspects. These are wholly managed by the public body. These bodies are typically non-autonomous local public water services under local 'municipal' control. Ultimately regulated by the local electorate.
End State (ES) 5: Community Management (Symbolised by a group of people holding hands) – as developed by IEFE and UNIZAR. Self-management by the users (shareholders).

Supporting Evidence and Implications

Two alternative scenario building tools - eight trend based Member State (MS) scenarios and the Scenario Cross (developed by the Euromarket team during the first validation workshop) – were trialled during this phase of the scenario building exercise. The five outline end states selected via future mapping were generally supported by the results of these two complementary scenario-building tools.

The majority of the Member State scenarios agreed on the need to cover outsourcing (4 out of 8 trend based MS scenarios), delegation contracts (6 out of 8), direct public management (6 out of 8) and regulated monopolies (7 out of 8). Competition in the consumer market was only identified in a limited number of Member State scenarios. In contrast,

[20] Monitored by the regulatory authority

community management (the peripheral end state) was not highlighted in any of MS scenarios presented. The results of the Scenario Cross supported the majority of the above end states (although regulated monopoly was not directly mentioned).

Whilst these initial end state descriptions were modified[21] in the subsequent stages of scenario building, the focus of the titles remained unchanged.

The responsibility for developing the detail of each scenario (EU end state and associated EU storyline) was delegated to individual Euromarket partners (see Box 2 for the scenario development responsibility).

DESCRIPTION OF EUROMARKET SCENARIOS

A summary of each scenario (the EU end state, the EU storyline and the associated actor and critical event analyses, and the 8 associated Member State storylines) is presented in this section. The full scenarios are available in the work package deliverable.

EU End States and EU Storylines.

A tabular profile of each EU end state is provided in table 1. Each EU storyline is summarised in table 2.

EU Actor and Critical Event Analyses

Actor Analysis

The transaction frameworks for each scenario are profiled in figure 4 to figure 9. Each individual scenario transaction framework illustrates three aspects:

1) The individual actors who are **central** to each scenario end state. Whilst all of the actors highlighted may play a role – it is their prominence and relative power that will help to differentiate each scenario. The key actors are identified (by being emboldened) within each actor group heading.
2) The **relative positioning** of each actor group – suppliers, operators, consumers and regulators. The closer the actor group the more common are their objectives. Indeed where actor groups overlap they may be seen as single actor, working to common objectives.
3) The change in **relative power** of each actor group within the transaction framework as shown by the colour of the circle (red for a major power shift, rose for a minor power shift, yellow for a changing role and white for a declining role).

Major policy developments and the associated changes in the "shape" of the transaction framework can be explained by: the force of new ideas from various opinion formers, the pressure of the interests of various lobbyists (especially key actor associations), and possible changes in the external/internal environment. The additional actor groups (policy makers/legislators and lobbyists/opinion formers) are shown in each scenario transaction framework. The lobbyists/opinion formers who win the storyline battle are highlighted in bold for each scenario end state. To summarise the key general lessons from the transaction analyses:

- Operators are the central actor in the transaction framework.
- Actors involved in major power shifts are generally also the lobbyists that win the ear of the key policy makers/legislators.
- Actors involved in major power shifts are generally involved in major conflicts as relative positioning changes.
- Scenarios can overlap. However, some scenarios (outsourcing and benchmarking) are easier to overlap than others (outsourcing and delegated contracts).

Critical Events and Trends

The key issues (around which different actors position themselves) relate to ideas, topics, problems or other matters (eg critical events or trends) that are open to discussion or dispute. For Euromarket the important issues are those where the

[21] The main changes were in End State 1 and 2 (via the development of two separate end states – one regulated and one not), End State 4 (via the development of two separate poles – high powered and medium powered benchmarking regimes) and End State 5 (via the adoption of New Public Management principles, including the possibility of more operator autonomy) – see next section for more detail of the final scenarios.

outcome may affect the future evolution of the liberalisation scenarios and, possibly more importantly, where the actors have diverging positions and are able to influence the outcome of the issue. The critical events and trends (and the associated dates, where provided in the EU storylines) identified during the development of the scenarios are listed in Box 3. However, one critical event, the publication/instigation of any EU liberalisation strategy/initiatives for WSS will strongly influence the plausibility of all six of the scenarios we have developed. The EU was identified as a key player in each scenario EU storyline.

- In Scenario 1 and 2 (Delegated Contracts) there was a EU Liberalisation Directive from DG Competition.
- In Scenario 3 (Outsourcing) there were DG Internal Market actions on the Procurement Directive[22] (widening and improving its application)
- In Scenario 4 (Regulated Monopoly) there was no EU pro liberalisation agenda but the EU launched benchmarking initiatives and established a European regulatory network.
- In Scenario 5 (Direct Public management) there was no EU pro liberalisation agenda but there were EU sponsored experience exchange initiatives.
- In Scenario 6 (Community Management) there was the EU implementation of Aarhus convention and the introduction of various EU Social Enterprise statutes and experience exchange initiatives.

Box 3: Critical Events

Scenarios 1 and 2. Delegated Contracts.
- Debt Crisis of TNCs (2005).
- Protest movement for more regulation of operators (sc. 1. 2013).
- Liberalisation directive amendment (sc. 2. 2015).

Scenario 3. Outsourcing.
- Public procurement becomes fashionable.
- Public procurement contracts become a standard, easy to apply, risk avoiding tool.

Scenario 4. Regulated Monopoly.
- EU publishes liberalisation strategy for the water sector (2006).
- Consumer views reflected in Eurobarometer report – public unenthusiastic about PSP (2008).
- Some high profile municipalities on the brink of bankruptcy (2010).

Scenario 5. Direct Public Management.
- Crisis management (2010).
- Resignation of private operator (2012).
- Official turn to DPM (2017).

Scenario 6. Community Management.
- Water service financial crisis (2009).
- Crisis of centralised water service (2012).
- Emerging new attitude toward collective way of living (2015).

Member State Storylines

Each Member State storyline is summarised in the Annex.

[22] Adopting the Services Directive would potentially enhance the plausibility of this scenario.

Table 1 and Table 2: EU End State and Storyline Summary

	End State 1. Delegation Contracts & Strong Regulation	***End State 2. Delegation Contracts & Extreme Competition***	***End State 3. Outsourcing***	***End State 4. Regulated Monopoly (Pole A & B)***	***End State 5. Direct Public Management***	***End State 6. Community Management***
Competition Modality	An EU liberalisation Directive. Competition for the market every 10-15 years. Not obligatory. Concession or lease contracts dominate.	An EU liberalisation Directive. Competition for the market every 5 years. Obligatory and must retain least costly bid. Delegation contracts (especially lease) dominate.	Operators outsource part of their task to external subcontractors. But revenue risks are not transferred. Depth and breath of outsourcing not specified. Depends on local needs of operator. Core service of "investment planning /regulation" always maintained by operator.	Pole A: Regulatory Body holds key levers of power. Heavily reliant on yardstick competition or price/revenue capping. Pole B: Muncipal influence in different regulatory areas.	No competition in/for customer market or for various service inputs. Outsourcing restricted to infrastructure provision and some finance (eg municipal bonds).	Competition modality. No competition in/for customer market. Outsourcing is possible, including for operation and maintenance.
Markets	No unbundling. Service provision fully integrated. But possible split between water and sanitation.	No unbundling. Service provision fully integrated. But possible split between water and sanitation	High degree of variation in the consumer market – including possibility of common carriage.	Integrated	Service provision fully integrated – water supply and sanitation.	Service provision generally not integrated – water supply and sanitation separated.
Operators	Major TNCs and private operators dominate. But size of operator varies. Urban areas served by both small and large private enterprises. But public management still present in some rural areas.	Major European TNCs dominate consumer (and supplier) markets. Limited SMEs. No public enterprises (squeezed out by TNCs).	Any operator (public /private/social) can benefit from outsourcing. High degree of variation in operators – direct/delegated public management and direct/delegated private management. Fragmented and integrated arrangements. Individual operators essentially decide the form and nature of market.	Pole A: Private companies (or in some MS public companies). Pole B: Autonomous municipal undertakings (private law). Multi-utilities important.	Local public authority and operator are within the same entity. Operator is under public law but is able to adopt some innovative business practices. Mainly local. Municipal multi-utilities are present. Move to new "modern" direct/ delegated public management under NPM principles.	Voluntary organisations (eg customer cooperatives, multi-actor associations and possibly mutuals). Outsourcing probably varies from a little (in smaller rural municipalities) to a lot (in larger urban ones).
Suppliers	Now typically either part of Operator holding companies or team up with	Now usually part of Operator holding company. Some niche product/service	Both small local niche players (eg leakage management, technical operations) and	There will be changes under pole A where private operators will move back up	No major move to outsourcing. Little change from current state.	Private sector increasingly important as supplier of new decentralised technologies

	End State 1. Delegation Contracts & Strong Regulation	*End State 2. Delegation Contracts & Extreme Competition*	*End State 3. Outsourcing*	*End State 4. Regulated Monopoly (Pole A & B)*	*End State 5. Direct Public Management*	*End State 6. Community Management*
	them on regular basis for major concession contract bidding.	areas remain independent,	large international companies (eg IT, laboratory services, vehicle management, customer billing centres) are present. Europeanisation of supply base is slow and function of EU policy maker actions (re making the internal market work better).	the supply chain and acquire strategic businesses. There may be more outsourcing under pole B.		(membrane and other separation technologies, more robust/simpler biological treatment facilities, automated features via ICT – enclosed systems) – enabling more "decentralised" community management.
Institutional	A public body (local or regional) is the responsible entity. Independent regulatory authorities exist. Has power to sanction operators but is independent from responsible authorities and public/private operators. New Fund to train up personnel in municipalities to manage bidding process. EU requires information for SGEI evaluation reports.	A public body (local or regional) is the responsible entity. But limited powers (as are obliged to delegate by EU Directive) and has limited expertise. Competition Authorities also play an increased role. But still arms length. No need for specific regulators as regular bidding (every 5 years) for the market.	Competition authorities play an important role. Government framework that clearly define legal responsibilities towards service provision and outsourcing contracts. Supply side concentration. European ramifications. European and national standards bodies facilitate functioning of European market by developing appropriate service/finance standards/protocols etc.	Pole A: Regulatory Authority - High powered benchmarking used. Pole B: Benchmarking Authority - Medium powered benchmarking (only publication & recommendation). Abuse control authorities for violation of competition law and excessive pricing.	Local authorities are responsible for both regulation and operation. No independent regulatory authority on economic/social issues. Assets are publicly owned and managed. Municipality (council and boards) play central role in regulatory decision making (price control, investment planning, tariff/budget approvals, staffing arrangements etc).	Responsibility transferred from public authority to communities. Coexists with other competition modalities. No independent regulatory authority on economic/social issues (exceptions such as in the British model are allowed). Community (or members of cooperative/ mutual) are responsible for economic *regulation*. But public regulators are responsible for public health and environmental aspects.
Economic	Ex ante regulation by competitive bidding. Ex post regulation by independent regulators – "controls" or "supervises" price and quality. Benchmarking. EBRD imposes	Operators very profitable. Manage to avoid competition authority investigations.	Full cost recovery pricing. High transaction costs. Pressure on public budgets. Considerable investments in new Member States. Benchmarking.	Full cost recovery. Cost causality and tariffs. Focus on renewals. Benchmarking.	Full cost recovery pricing (except in some rural areas). Benchmarking possible.	Move to Social Enterprise Operators. Different Social Enterprises favoured in different circumstances – cooperatives/ associations in rural areas, mutuals in urban areas.

	End State 1. Delegation Contracts & Strong Regulation	***End State 2. Delegation Contracts & Extreme Competition***	***End State 3. Outsourcing***	***End State 4. Regulated Monopoly (Pole A & B)***	***End State 5. Direct Public Management***	***End State 6. Community Management***
	conditionality on loans.					
Social	Regulatory authorities monitor quality of service (re public service obligations). Social tariffs can be implemented. Disconnection prohibited. Limited public participation. Funds for social protection and support to municipality.	Social provisions can be included in the contract.	Social provisions provided for by political instructions to operators and concessionaires . Consumer organisations important lobby group.	Pole A: customer representation bodies (also deal with complaints). Customer payments for poor service. Pole B: Democratic manadate of municipal control. Consumer representation on boards.	Disconnections prohibited. Social tariffs can be implemented. Increased public participation.	Public participation is a distinctive element.
Environment	National regulator with a focus on pollution control.	National regulator with a focus on pollution control.	Broad range of environmental problems.	Environmental quality issues are regulated by river basin authorities.	Environmental quality issues are regulated by river basin authorities.	Water basin authorities responsible for water quality/quantity issues.
Scenario Author	Paris VIII & ENGREF	Paris VIII & ENGREF	TU Delft & IHE	Ecologic & IDD	UCL & EPFL	IEFE & UNIZAR

	EU Storyline 2005-2010	*EU Storyline 2010-2015*	*EU Storyline 2015-2020*
Delegation Contracts & Strong Regulation	TNC change strategy – prefer lease in developed countries (eg EU) compared to concessions in developing ones. TNCs lobby for an EU Liberalisation Directive. New Member States use concessions more widely and GAT negotiations move to greater openness in services. EU Liberalisation Directive voted and agreed on in 2009 – obligation to tender every 10-15 years.	Protest movement (SOS) launched in 2013 – press for greater regulation (both to control high prices and ensure service quality). EU launches evaluation of performance and supports independent regulators.	EU Directive amended in 2018 to oblige establishment of independent regulatory authorities. A significant part of regulation mission dealt with by local authority.
Delegation Contracts & Extreme Competition	TNC change strategy – prefer lease in developed countries (eg EU) compared to concessions in developing ones. TNCs lobby for an EU Liberalisation Directive. New Member States use concessions more widely and GAT negotiations move to greater openness in services. EU Liberalisation Directive voted and agreed on in 2009 – obligation to tender every 10-15 years.	New pro liberalisation EC and Parliament. Further widening from GATs. Supported by Member States EU Directive amended to reduce contract period to 5 years – obliging responsible entity to accept "least costly bid". No regulatory authority required.	Transposition of Directive problematic in some MS.
Outsourcing	Drive for efficiency as a result of need for investment, poor state of municipal finances and demand for increased transparency. Outsourcing increasingly adopted. In response EU lowers threshold values beyond which contracts have to be awarded through public tendering. Subcontracts increasingly subject to European procurement law.	Outsourcing extended to areas where a high level of know how and specialist equipment required or highly labour intensive. Internationalisation of supplier market (especially CEE) and involvement of multi utility contracts and TNCs.	Regulatory failures/disappointments with private operators leads to greater reliance on outsourcing as favoured means of delivering efficiency. Evident need for (international) cooperation between operators to evaluate performance of contractors and knowledge centres – eg development and application of indicators and benchmarks. European support body established to support public procurement.
Regulated Monopoly	During this period six contrasting driving forces (already existing and emerging) created the ground conditions for the greater use of high-medium powered benchmarking and introduction of independent regulators/benchmarkers across the EU. Social: Public Attitudes against liberalisation (without adequate regulation) became harder. Economic: Municipal Financial (and Managerial) Resources came under extreme pressures and led to the increased the need for "divestment" of water supply and sanitation services. Environmental-Health: Higher Bills (as a result of increasing technical complexities and greater investments and moves toward full cost recovery) reinforced the need for institutional change. European: EU policy review on WSS liberalisation remained neutral but supported the need for modernisation and	During this period five complementary driving forces – two arising from European policy developments and three emerging from the increasing importance of European consumer rights – continued to drive the process of institutional reform in the European WSS sector. European: EU reviews of cost recovery and drinking water quality promoted the need for further increased transparency and encouraged further benchmarking activity. European: Implementation of Water Framework Directive (WFD) underpinned major structural changes in scale and integration of water supply and sanitation operations. Consumer: Abuses of monopoly positions stimulated the demand for greater economic regulation. Consumer: Consumer power rapidly became an important factor	During this period positive experiences of Member States that have adopted independent benchmarking/regulation and ongoing changes in the global macro-environment reinforced and widened the pace of institutional reform. European: Positive international experiences of benchmarking/regulation become well known across EU. Environmental: Climate Change impacted both directly and indirectly on WSS. Socio-Economic: Poor macro-economic performance and greater social engagement encouraged further public sector restructuring. The second driver proved to be particularly important in framing the future direction of the sector.

	independent/compulsory benchmarking. Operators: As part of their long term business strategies the TNCs gradually withdrew from the Western Europe (and focused on opportunities in Central and Eastern Europe). But the same TNCs were aggressively regulated in their remaining home EU markets. Stakeholder: Power of Trade Unions remained relatively strong across the EU and acted as important countervailing power to the (declining) influence of the TNCs. The first four driving forces proved to be particularly important in framing the changes that have occurred since 2005. Indeed they continued to be important in subsequent periods up until the present day - in 2020.	in pushing for water reform. Consumer: Limited direct competition emerged as result of industrial consumer lobbying, but common carriage remained controversial. The first four drivers proved to be decisive in the ongoing reform process.	
Direct Public Management	Public operators introduce NPM tools into organisation to improve efficiency. Local authorities get more autonomy and responsibility. Private operators mismanage major external events (accidents, sabotage, natural disaster) that result in the loss of lives. Resulting political/public crisis results in a return to DPM.	In line with WFD public operators modify governance structures to include greater public participation. Anti-globalisation spread to national/local level. Local movement demand water reintegrates the public domain. Call for disobedience – focus on water and is heard by general population (who have had poor quality service/high prices from private operators. They destroy their meter. Private operators resign and return to DPM. Private operators become insolvent highlighting corruption. Resulting political/public crisis results in return to DPM. Left wing political parties/thinking dominates Member States, EC and Parliament. Against profits from services of public interest.	Further improvement of management methods within public operators. As a result of previous crises there is a general turn toward DPM.
Community Management	Spreads in rural areas first (as public/private operators uninterested). Peri-urban areas follow with interesting involvement of farmers (who have experience of these cooperative arrangements re irrigation) and who have water supply infrastructure. New technologies enable self-treatment by customers and promote decentralisation. Existing technologies for centralised provision increasingly costly. Industrial customers also collaborate re water supply/trade effluent treatment.	Mistrust of public institutions and private companies. Cost breakthrough re decentralised technologies. Also threat of terrorism promotes fragmentation of system. Financial bottlenecks. Local authorities starved of cash. Private companies unable to make profits (required tariff increases opposed). Community management more important.	Mistrust continues to support community management and participation in decision making in other forms of organisation. This is supported by Aarhus convention and Water Framework Directive. Also supported by a crisis of individualism.

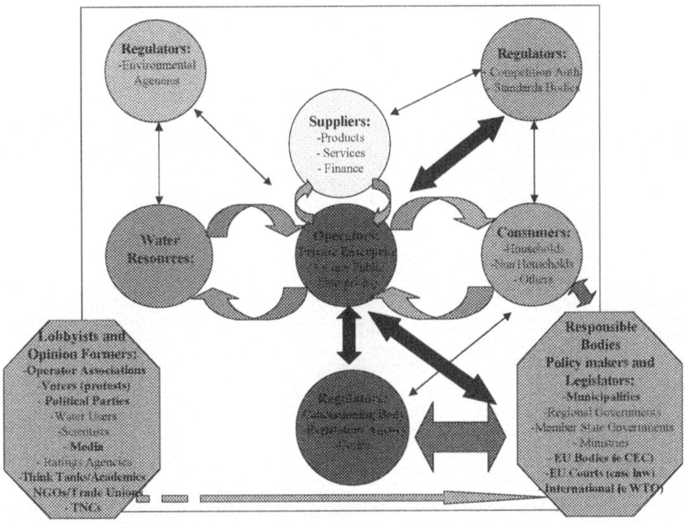

Figure 4: Framework for Delegated Contracts (and strong regulation) Scenario

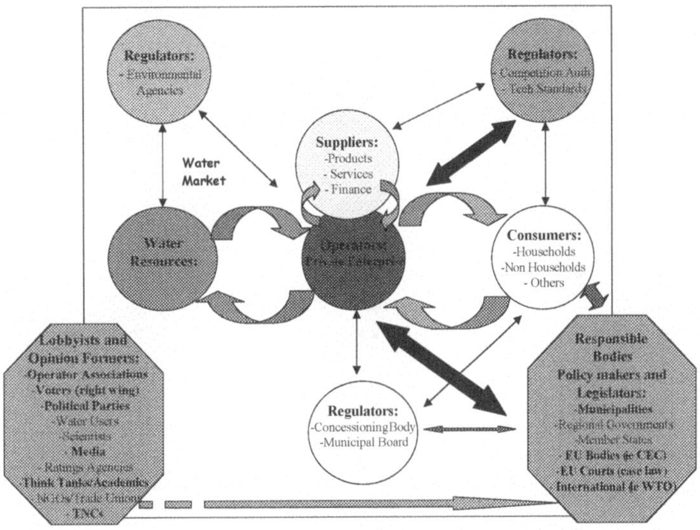

Figure 5: Framework for Delegated Contracts (and extreme competition) Scenario

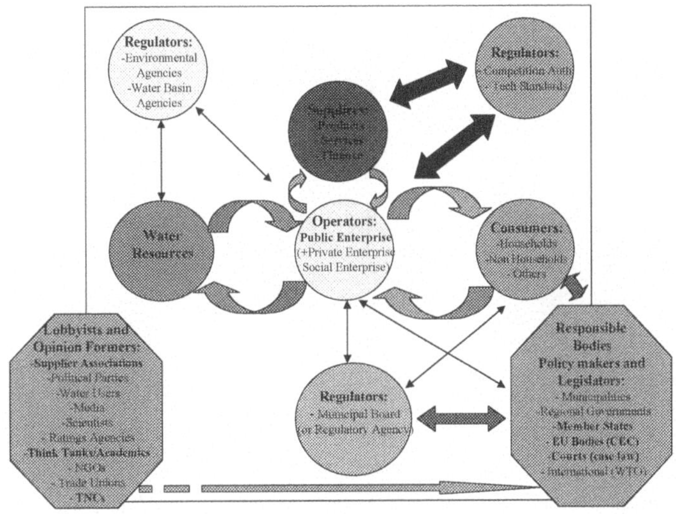

Figure 6: Framework for Outsourcing Scenario

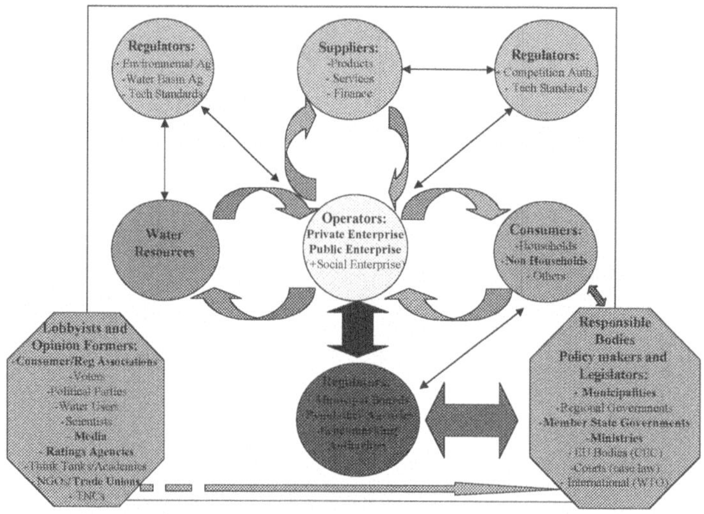

Figure 7: Framework for Regulated Monopoly Scenario

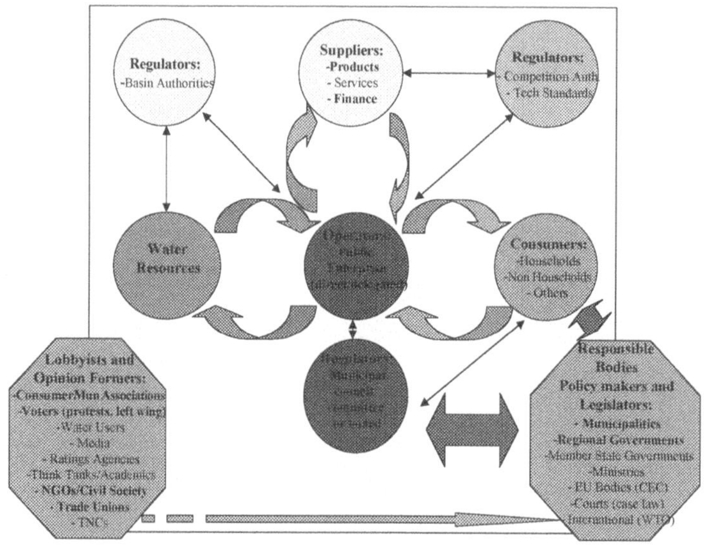

Figure 8: Framework for Direct Public Management Scenario

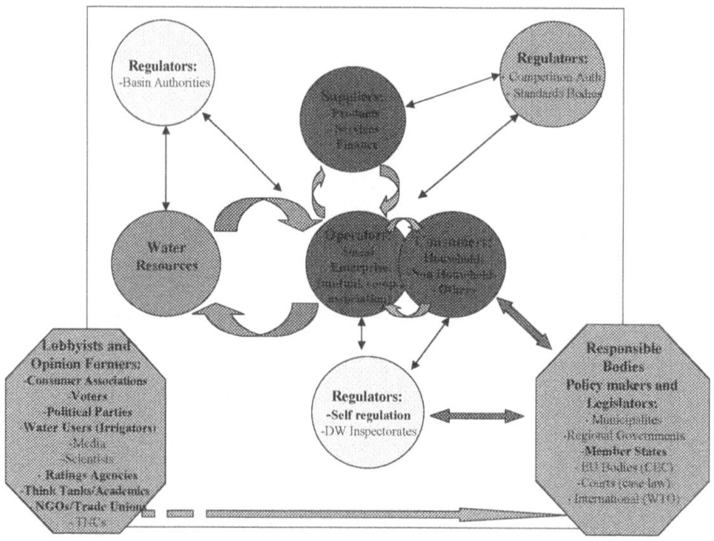

Figure 9: Framework for Community Management Scenario

CONCLUSIONS

End State Relationships

The relationship between the proposed EU end states is shown graphically in figure 10.

The outsourcing end state (3) is based on competition *for* the supplier market and this form of competition may, in reality, coexist with all the other end states selected. However, for the purpose of building the scenarios this overlap has been somewhat artificially constrained. In recognition of this potential for overlap, end state 3 (on outsourcing) is located in the centre of the triangular relationship between end states 1, 2, 4 and 5 (covering delegation contracts, regulated monopoly and direct public management respectively).

The sixth peripheral end state (6) forms the basis of our scenario XXX and is structured around a specific management modality – community management. Indeed this peripheral end state may coexist along side the others (and could also represent an alternative unexplored mode of competition – that of competition between the alternative competition modalities). To reflect its peripheral nature this end state has been located outside of the triangle.

As there will be complex overlaps in the competition processes[23], we need some other secondary means of end state differentiation – especially where the end states are quite close together (re competition processes) and can easily converge. This need for further EU end state differentiation occurs particularly with the three points of the triangle. Namely between:

I. The two EU end states that incorporate a more central role for the *public sector* ie Regulated Monopoly (ES4, pole B) and Direct Public Management (ES5) – especially where there is reliance on benchmarking/New Public Management[24] (NPM) tools and competition "in" some consumer/supplier markets.
II. The two EU end states that involve the establishment of an *independent regulatory authority* ie Regulated Monopoly (ES4, pole A) and Delegation Contracts (ES1) – where incentive regulation mechanisms such as i) price cap regulation/fixed price contracts, and ii) cost/profit-sharing regulatory mechanisms and contracts become important.
III. The two EU end states that embrace greater "Private Sector Participation (PSP)" ie Delegation Contracts (ES2) and Outsourcing (ES3)) - where lightly regulated competition "for" the consumer and supplier market predominates.

These points of overlap are identified via the text boxes in figure 10.

The means of further end state differentiation is provided for in table 3. However, further more detailed discussion of the mode of regulation (particularly in the three end states forming the three points of the triangle) is also required.

Regulatory Relationships

The nature of any given regulatory model will be largely determined by the nature of the main competition processes adopted. The Euromarket scenarios reflect the main different competition processes and hence they also reflect the main different regulatory models available (contract regulation, conduct regulation, incentive regulation, cost regulation, and self regulation). The generic end state relationship profiled in figure 10 therefore also reflects the different regulatory models that will be adopted within each end state (see figure 11).

For the community management end state (6), self-regulation predominates (normally via the members – typically the consumers - of the social enterprise), whereas for the outsourcing end state (3) the national competition authorities will typically regulate competition (ie conduct) in/for the different supplier markets. These regulatory models are quite distinct.

For end states 1 and 2 (delegation contracts), 4 (regulated monopoly) and 5 (direct public management) – the three points of the triangle that are focused exclusively on the consumer market - more traditional regulatory approaches are adopted (eg contract regulation[25] in scenario 1 and 2, incentive regulation under scenario 4 (pole A) and cost regulation in scenario 5). These three "alternative" regulatory models have important similarities. The nature of the link (ie the degree of separation and the split of individual responsibilities) between the responsible entity, the regulator(s) and the

[23] The predominant form of the competition process is still the primary means of differentiating the 6 proposed end states.

[24] NPM principles can include the use of market mechanisms, separation of policy from execution, an entrepreneurial culture based on managerial incentives and accountability based on direct relationships between manager and customer (rather than legal and political forms of accountability).

[25] Noting that depending on the nature of the delegation contract either cost plus, fixed price or incentive based regulatory instruments can be included.

operator is the key distinguishing feature of the consumer market. The regulation of the operator (public, private or social) then becomes a central distinguishing feature of the consumer market, and hence the four end states that focus exclusively on this part of the transaction framework (see figure 3).

It is evident that under certain circumstances (see figure 11) competition *for* the market (via delegation contracts, 1) can tend toward incentive monopoly regulation (with similar incentive structures to pole A of the regulated monopoly end state, 4). In addition there are also distinct similarities between the financial/investment/service controls (eg cost based) exerted under the direct public management (5) end state and rate of return regulation under pole B of the regulated monopoly (4) end state.

Extent of Liberalisation

Some experts might argue that the extent of liberalisation generally decreases as we move from end state 1 to end state 6. However this will ultimately depend on the detail of each scenario. For example, depending on the extent of outsourcing and operator governance arrangements, end state 6 (community management) may be more or less liberalised than end state 1, 2, 3, 4 or 5. And depending on the nature of incentive regulation (and the extent to which this process can substitute for real competition) and the reality of strong competition under delegation/outsourcing contracts then end state 4 may actually be more "liberalised" (ie competitive) than end states 1, 2 and 3.

For example, the current delegation contract market in France might not be currently classified as one of the most competitive/liberalised markets in Europe. In the author's opinion the delegation market in France is based more on a long-term trust-based partnership between the private operator and the municipality - with all the benefits and pitfalls (see Box 4 for more details) that one would expect with such a long term partnering arrangement. This is not an argument against this method of operating delegation contracts - it undoubtedly fits French culture and traditions, and has done so since the 17th century. But to counter some of the potential pitfalls of this form of delegation contracts the first European end state allows for the introduction of an independent regulatory body. However, as witnessed by recent events in the UK (where water companies have given false information to the regulator) an independent regulator is not the panacea to the inherent difficulties of introducing greater competition into the sector. Indeed, attempts by the economic regulator for England and Wales to introduce enhanced competition *in* the market (through inset appointments and common carriage) have also faltered. So again what might look like an increasingly competitive market on paper may in fact prove to be an illusion in practice. The devil (in terms of degrees of competitiveness and hence liberalisation) will be in the detail of each scenario (particularly the market structure and the associated incentive instruments).

Our scenarios typically assume that, in reality, optimal conditions for each dominant competition process exist.

EU Storyline Relationships

Water industry institutional structures are relatively stable - once formed they are difficult to break down. This partly explains the radical nature of some of the EU storylines (EU Liberalisation Directives, major enterprise failures, public protests etc). Indeed the nature of the storylines may also reflect the plausibility of the respective End State – the more radical the storyline the more implausible the End State? The more radical storylines also tend to be slightly more simplistic.

The nature of the individual EU storylines is very different. And these differences are partly determined by the status of the operator.

- Outsourcing/Regulated Monopoly (where different enterprises can coexist) scenarios have *evolutionary* storylines - with changes that reflect the need for greater efficiency (technical and economic) and altering social opinions and the influence of EU supporting actions.
- Delegated Contracts (where private enterprises dominate in 2020) and Direct Public Management (where public enterprises dominate in 2020) scenarios have *revolutionary* storylines - with major political-legal changes, social protests, enterprise failures and enterprise changes.
- Community Management (where social enterprises dominate in 2020) scenario storyline is *transformational* - with major technological change and major shifts in social attitudes.

Box 4: The Delegation Contract Market in France

Whilst the French management system has been in place for centuries recently, following the political and financial scandals of the 1990s, there have been strong demands for better regulation. Indeed this pressure led to legal reforms and the introduction of:

- The Sapin law (1993) that placed time limits on the duration of the service delegation and made competition compulsory.
- The Water Act (1992) that required the annual publication of two reports (one from the delegate and one from the mayor) that are supposed to be endorsed by the council, passed onto the prefect, and then made public.

Whilst the French management system has been in place for centuries recently, following the political and financial scandals of the 1990s, there have been strong demands for better regulation. Indeed this pressure led to legal reforms and the introduction of:

- The Sapin law (1993) that placed time limits on the duration of the service delegation and made competition compulsory.
- The Water Act (1992) that required the annual publication of two reports (one from the delegate and one from the mayor) that are supposed to be endorsed by the council, passed onto the prefect, and then made public.

However, recent investigations in 2000 and 2002 by the French competition authorities (the Conseil de la Concurrence, CdC) have identified continuing operational problems in the delegation market:

1. The formation of common subsidiaries between the two main competitors.
2. The low switching rates – often the incumbent is the only company to rebid for the contract (and where other small companies do bid the impact on the price can be substantial).
3. The limited advanced notice to other competitors on when the rebids will happen.

And various remedies were proposed by the CdC (including the suggested instigation of a surveillance authority and a study on the need for structural measures) to combat these apparent market failures.

Similar academic criticisms (eg Roche and Johannes, 2001) of the French delegation system also remain. They include:

- the structural asymmetry (between the TNCs and the municipalities) that undermines the negotiation and execution of contracts;
- limited real competition (for example, in 1998 CGE regained 100 out of the 111 contracts that were reopened for competition);
- poor incentives for optimal maintenance (especially of the invisible underground networks, at the end of the contract period); and
- the complete lack of penetration by foreign and small companies into the supposedly open "competitive" French market.

More radical modifications to the French contracting system are also now being considered – including the introduction of performance indicators to account for the level of service and a system of financial incentives to promote the attainment of these indicators.

These improvements are now being tested in the market place. In addition consultants are being used to advise the municipality (to redress some of the information asymmetry), and comparative analyses are now being undertaken (by research organisations (eg ENGREF), consumer associations, the media, the water agencies and the States statistical services) to highlight market anomalies. However one proposal, to establish an independent regulatory body, was subsequently abandoned by the French Government.

The peak of contract renewal is expected in 2003-05 and the results will give an indication of whether the transparency and the level of real competition in the French delegation market have improved.

Table 3: End State Differentiation

	Primary Feature: Dominant Competition Process	**Feature I. Nature of Regulation**	**Feature I. Independent Economic Regulator**	**Feature II. Clear distinction: operator & responsible entity**	**Feature II. Ownership of Operator**
1. and 2. Delegation Contracts	For the consumer market	**Contract Regulation**	**1.Yes (uncertain remit)** *2. No*	Yes	Private
3. Outsourcing	In/For the Supplier Market	Conduct Regulation	No	Yes	Public/Private
4. Regulated Monopoly	Pole A: Price/profit/revenue regulation Pole B: Benchmarking	**Incentive Regulation (managed via public body)**	**Pole A. Yes (price setting)** *Pole B. No (but benchmarking authorities)*	**Yes**	Pole A: Public/Private **Pole B: Public (private law)**
5. Direct Public Management	In selected supplier markets	Cost Regulation	No (via Municipal Boards)	**Not always**	**Public (in house or public law)**
6. Community Management	Various	Self Regulation	Not usually	Not always	Community

	Primary Feature: Dominant Competition Process	**Feature II. Ownership of Operator**	**Feature II. Financial Autonomy**	**Feature II. Financing of Infrastructure**	**Feature III. Unbundling**	**Feature III. Time Limits**
1. and 2. Delegation Contracts	For the consumer market	Private	Yes	Private	**No**	1. 10-20 years **2. 5 years**
3. Outsourcing	In/For the Supplier Market	Public/Private	Yes	Public/Private	**Yes**	**1-5 years**
4. Regulated Monopoly	Pole A: Price/profit/revenue regulation Pole B: Benchmarking	Pole A: Public/Private **Pole B: Public (private law)**	**Yes**	Pole A: Public/Private **Pole B: Public/Private markets**	Limited	Unlimited
5. Direct Public Management	In selected supplier markets	**Public (in house or public law)**	**No/Limited**	**Public (or private via public intermediaries)**	No	Unlimited
6. Community Management	Various	Community	Yes	Private (via loan capital or shares)	Possibly	Unlimited

Note: Distinguishing features are described in text.

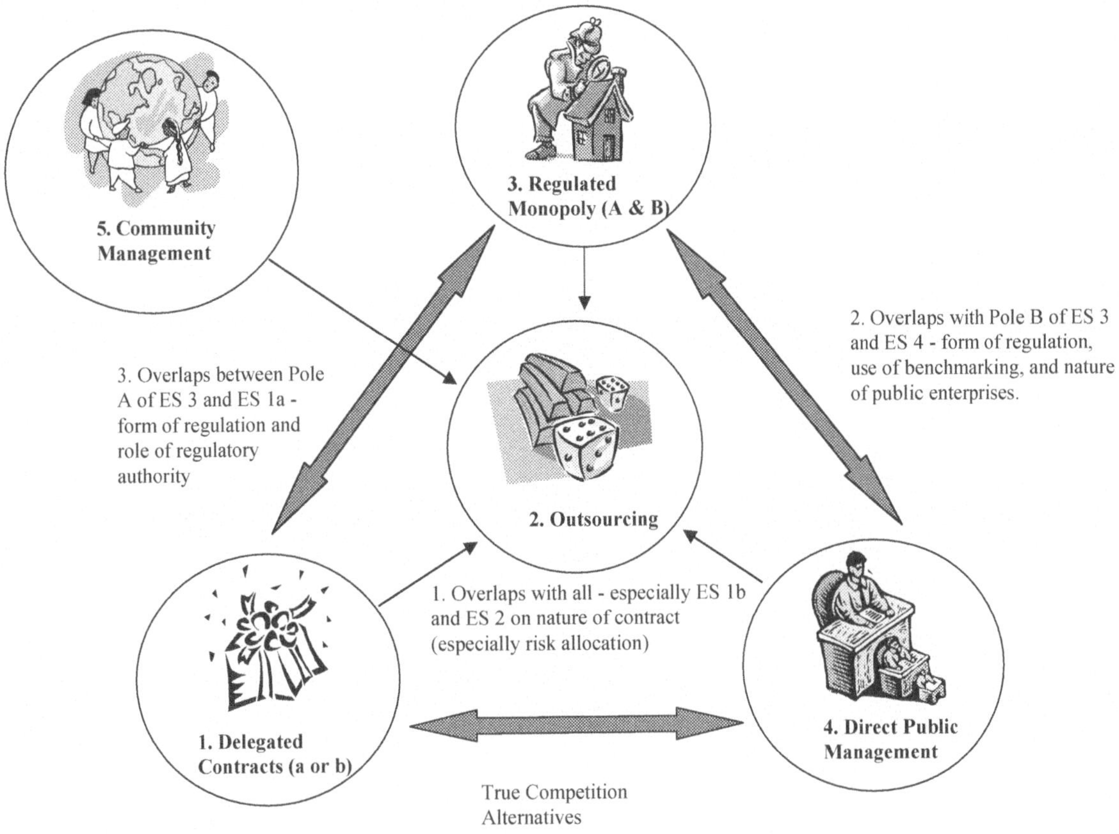

Figure 10. Relationship between End States - Competition Processes

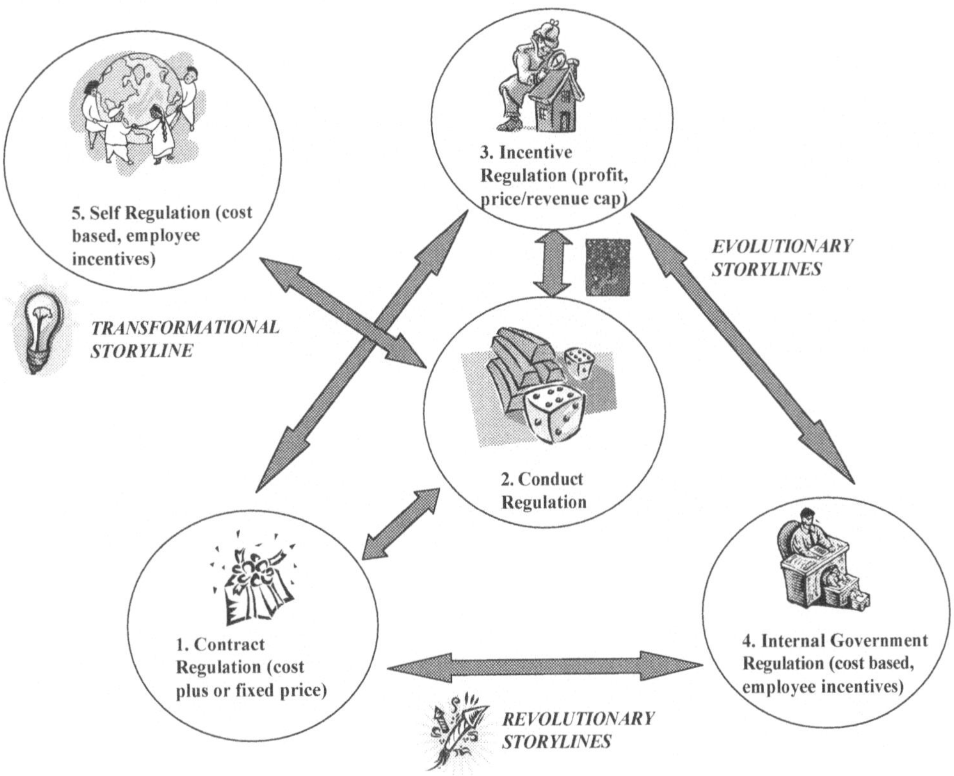

Figure 11. Relationship between End States - Models of Regulation and Storylines

There are strong similarities in EU storylines between the outsourcing and regulated monopoly scenarios and between the delegation contracts and direct public management scenarios (see figure 11). These storylines will therefore be discussed together.

Outsourcing and Regulated Monopoly

These two scenarios can occur with a mixture of operator types (private, public and social enterprises) across the EU. Hence there is no need for one operator type to triumph over the other. Here it is more an argument about what mechanism (benchmarking or outsourcing) provides more economic efficiency, noting that both can incur substantial transaction costs. Indeed it is the nature of these increased transaction costs and the way they influence the nature of the transaction system as a whole that largely characterises these two scenarios. These two scenarios have also countenanced some limited competition in the market (namely for industrial users). This has raised the need for change in the water resource market.

With outsourcing the nature of the operator-supplier relationship essentially changes. From just being a manager of the service the operator moves to also become a contract manager - involved in the design, negotiation, letting and monitoring of multiple outsourcing contracts. The role of the suppliers also changes – they essentially offer a broader range of services and are prepared to take on greater risks. To some extent they have become more powerful as non-core "expertise" moves away from the operator to the supplier.

With the regulated monopoly the nature of operator-regulator relationship will change. The operator will collate a variety of detailed data and information and provide this to the regulator/benchmarking authority at regular intervals. There may be new actors established – the regulatory/benchmarking authority - and there will be new relationships between these new bodies and the responsible entities under the old system. This may involve a loss of power by existing responsible entities.

These two scenarios are influenced by similar drivers (need for greater efficiency as a result of greater technical complexity and investments as determined by environmental factors such as diffuse pollution, climate change etc) and have been indirectly influenced by both changing social opinions (acceptance of outsourcing or the refusal to consider greater liberalisation without greater regulation) and specific EU supporting actions (projected changes to procurement Directive and possible approval of Services Directive[26] for outsourcing and EU initiatives on benchmarking for the regulated monopoly).

Both scenarios also assume that there is no specific EU liberalisation Directive. Hence the conclusions to the forthcoming EU water policy review are therefore quite important to the ultimate plausibility of these two scenarios.

In fact these two scenarios are not mutually exclusive. A regulated (benchmarked) monopoly will be able to benefit from the outsourcing of certain non-core activities. And outsourcing will often be complemented by associated benchmarking, possibly of both the consumer and supplier markets. This is reflected in the scenarios. So these two scenarios could be complementary (ie they could coexist) – one providing for additional competition in the supplier market with the other providing additional (but artificial) competition in the consumer market.

However, for the sake of contrast the two scenarios have been positioned as alternatives. We have played up the possibility of regulatory failure (in the consumer market) under the outsourcing scenario, and market failure (in the supplier market) under the regulated monopoly scenario. This is shown by the increasing power of the actors (suppliers and competition regulators) in the supplier market under the outsourcing scenario (see figure 6), as compared to the increasing power of the actors (consumers and economic regulators) in the consumer market under the regulated monopoly scenario (see figure 7).

Delegation Contracts and Direct Public Management

It is interesting to now look at the delegation contract scenarios[27] and direct public management scenario together. These two scenarios are contrasting especially in the nature of the operators that predominate – the direct public management scenario is centred on the predominance public enterprises, whereas the delegation contract scenarios are centred on the predominance of private enterprises. Both scenarios anticipate some limited outsourcing opportunities.

Both scenarios are ultimately underpinned by anticipated political support from key European policy making/legislative bodies (most notably the European Commission and the European Parliament), which then either translates into a water liberalisation Directive, or to the favouring of greater (delegated) public management. The media and public opinion informs this political positioning. Although the level of actual policy development differs – for

---

[26] Although this driver was excluded from the outsourcing scenario storyline.

[27] Note whilst these are two separate delegation contract scenarios for the purpose of this discussion they are treated as one.

delegation contracts it develops at the EU scale, whereas for direct public management it is generally at the regional/local scale.

In essence these critical changes are part of the policy development process (based in the interaction between policy makers/legislators and lobbyists/opinion formers) identified in the introduction. For example, EU political support for the proposed Liberalisation Directive is partly stimulated by lobbying from the TNCs and certain pro liberalisation Member States. Opposition to this (and implied support for direct public management) is provided for by the trade unions and a coalition of NGOs. The conclusions to the forthcoming EU water policy review are therefore particularly critical to the ultimate plausibility of these two scenarios.

Similar drivers also influence these two scenarios.

1. Broad *European political changes* are anticipated in both scenarios – a move toward (a rightwards shift in the political make-up of European/National Parliaments), or away (a leftward shift in the political make-up of European/National Parliaments) from liberalisation, depending on which scenario is being discussed.
2. *Social protests* (as animated by key anti liberalisation lobbyists) against liberalisation are also envisaged in both scenarios – in the delegation contracts scenario it results in a new variant of concessioning (with a regulator) and in the direct public it reinforces the need to move away from involving the private sector. However, in the more extreme version of delegation contracts this social protest fails to influence the key policy makers/legislators.
3. *Enterprise failures* are envisaged for both scenarios to support the suggested pathway to the future. For the delegated contracts scenario existing public enterprises fail to deal with the technical complexity and greater financial needs of the sector. Whereas for the direct public management scenario the private enterprises fail to deal with a crisis and there are important corruption scandals.
4. *Enterprise changes* are envisaged for both scenarios to support the suggested pathway to the future. For the delegated contracts scenario the private enterprises (reluctantly) accept regulatory oversight. Whereas for direct public management the public enterprise will embrace the use of new DPM tools. Hence both operator types are prepared to adapt to the prevailing policy climate.
5. *Improved consumer representation* is recognised as being important in both scenarios. Although in both scenarios the real extent of this representation remains uncertain.

There are important difference in the operator-regulator relationship – most notably the existence of two "regulators" under one delegated contract scenario and the importance of the municipality in controlling public enterprises.

Community Management

Community management is possibly the most complex scenario in that it embraces both changes to the nature of the operator (from public/private enterprise to social enterprise) and associated adjustments in both the operator-supplier and operator-regulator relationship. The nature of the social enterprise will also probably depend on the scale of operator – for smaller agglomerations in rural areas associations/cooperatives are likely to predominate whereas in larger more urban areas the mutual will probably be preferred.

The drivers are also distinct from the other scenarios. They emanate from anticipated technological change (based on advances in membranes, biotechnology, and Information & Communication Technologies) and a possible (if somewhat) surprising shift away from individual values to more community based ones.

In reality the alignment of operator and consumer objectives is likely to undermine the role of the responsible entity – although tasks pertaining to service quality regulation will probably remain. In addition it is likely to enhance the position of suppliers – both as the providers of the new technology and as likely operators (as the larger social enterprises will probably opt to outsource their management operations to the private sector).

Member State Storylines

The Member State (MS) storylines are extremely important. They have been prepared by "locals" and are founded in the reality of national cultures and traditions. They reflect the path dependency of water supply and sanitation institutional arrangements and the inherent barriers to change that exist in all Member States on this politically sensitive sector.

MS Storyline Plausibility

To illustrate this importance we have gauged the degree of plausibility of each scenario (for each Member State) by assessing the nature of the Member State storylines. For each Member State storyline the author has assessed it as being easy (E), difficult (D), or uncertain (U). We can then compare the ratios of E:D:U for both the individual scenarios and the individual Member States (see table 4). It is evident that outsourcing (3) and direct public management (5) appear to be easily achieved in a majority of Member States. These scenarios may be seen as slightly more plausible. The delegation contract (1) and the regulated monopoly (4) scenarios are more troublesome whereas the scenario storyline on community management (6) is difficult to achieve in most Member States.

Table 4: Plausibility Measure of Member State Storylines

	Delegation Contracts & Strong Regulation	***Outsourcing***	***Regulated Monopoly***	***Direct Public Management***	***Community Management***	***Summary E:D:U***
Belgium	E	E	E (Pole B)	E	U	**4:0:1**
France	E	D	U (?)	D	D	**1:3:2**
Germany	D	E	E (Pole B) [& D (A)]	E	D	**3:2:0**
Italy	E	E	U (Pole A)	E	D	**3:1:1**
Netherlands	U	E	D (Pole A)	E	D	**2:2:1**
Spain	D	E	U (?)	E	D	**2:2:1**
Switzerland	D	E	E (Pole A)	E	D	**3:2:0**
UK	D	E	E (Pole A)	D	U	**2:2:1**
Summary E:D:U	**3:4:1**	**7:1:0**	**4:1:2**	**6:2:0**	**0:6:2**	

E=Easy, D=Difficult, U=Uncertain.
Storyline plausibility based on author's judgement.

A number of Member States struggle to reach the respective end states without some quite radical storylines.

It is also evident that some Member States are particularly resistant to change. In France and Spain (where delegation contracts are important) moves to the majority of the end states is problematic. The Netherlands and England/Wales also show considerable resistance to change (in reaching all the proposed end states). The UK appears particularly entrenched in its existing institutional arrangement. In contrast Belgium and to a lesser extent Italy and Germany (all of whom already have quite mixed regulatory models) are possibly more flexible to potential change.

MS Storyline Relationships

There are interesting similarities between the Member State storylines.

For the delegation contracts scenario all the Member State storylines are predicated on the European liberalisation Directive. Indeed this probably the only way this scenario could happen and, even then, it is anticipated that there will be strong opposition from Germany and the UK. Furthermore it is anticipated that there will be a need for a subsequent Directive amendment to incorporate a powerful national regulator in each Member State. Indeed the need for a` regulator was anticipated earlier in a number of Member State storylines.

For the outsourcing scenario, in four Member States the EU move to reduce the thresholds for public procurement kick-started the drive toward greater outsourcing. In two Member States the drive for greater efficiency was the over-riding driver whereas in two other countries it was stimulated by national legal changes.

For the regulated monopoly scenario, in four Member States the need to increase the efficiency of incumbent operators (especially on price and service quality) was the critical driving force. In three Member States (who opted for pole A) municipal financial problems led to greater privatisation and the need to establish an independent regulator.

For the direct public management scenario, in four Member States major environmental disasters (most notably flooding) increased the importance of public management. Indeed flood management is a typical public good and integrated water management policies (as encouraged by the WFD) could lead to greater public involvement in the sector. Three of these Member States also mentioned the possibility of terrorist attacks – another rationale for involving the State more. Other interesting drivers identified in the Member State storylines included the potential failure of liberalisation in the energy sector and the need to avoid greater EU inspired competition by defining operations as "in-house".

For the community management scenario, only three Member State storylines led to major moves to this end state. In two of these Member States a change in social values and important legal judgements are key drivers. For the majority community management remains of relatively minor importance (only in rural and rich peri-urban areas) although it has important knock on consequences for other operators of more "centralised systems" re improved involvement of consumers. In some Member States there are national legal barriers to the wide-scale introduction of community management.

Overall Implications

It should be noted that liberalisation in water and sanitation services is either:

- very difficult – particularly in the case of competition *in* large parts of the market (as a result of the potential for severe market failure, and the associated large transaction costs); or
- particularly intensive (and possibly costly) from a regulatory perspective – particularly in the case of competition for the market, or comparative competition (to remedy the more minor market-regulatory failures).

The key problem to be considered appears to be the assessment of possible economic benefits of increased liberalisation/incentive regulation when compared to the transaction costs associated with the need to control various possible market/regulatory failures.

This assessment of costs and benefits can only really be undertaken on a case-by-case basis - as it will be determined by the local conditions (infrastructure, institutional, political etc) that exist prior to any proposed change.

Essentially the optimal liberalisation path (if there is one) will be different in each Member State, and possibly even, in each municipality. This path dependency is one of the reasons why we developed both EU and Member State storylines. The Member State storylines were essential to ground the scenarios in some form of reality – particularly the politics of the liberalisation process.

REFERENCES

Davis G (2003), Scenarios as a Tool for the 21st Century, Probing the Future Conference.

Fahey L and RM. Randall (1998), Learning From the Future – competitive foresight scenarios , John Wiley and Sons.

Henley Centre (2001), Benchmarking UK Strategic Futures Work, A report for the Performance and Innovation Unit.

Kahn A. and Wiener AJ (1968), Year Two Thousand, Collier-Mac.

Lindgren M and H Bandhold (2003), Scenario Planning – the link between future and strategy, Palgrave Macmillan.

Roche PA & Johannes B (2001), Regulation in the water and sanitation sector in France in Regulation of network Utilities: The European Experience, Oxford University Press.

ANNEX

Member State Storyline Summaries

Market Drivers	*Delegation Contracts & Strong Regulation*	*Outsourcing*	*Regulated Monopoly*	*Direct Public Management*	*Community Management*
Belgium (UCL)					
1. Regionalisation process. 2. Obligation from existing Directives 3. Strong financial pressure in context of economic recession 4. Political opposition to liberalisation 5. Water corporatisation is strong. 6. Industrial policy to create national champions. 7. Uncertain interest of water TNCs	In **Belgium** the move towards concessioned markets in Belgium was conditioned to the adoption of the Water Liberalisation Directive in 2009 that opened up the sector to competition. The Federal State pushed forward an integration and liberalisation of the WSS sector. The decision followed the regionalisation of the public debt (remaining above the 100% of GDP if aggregated). As a consequence, the Regions urgently needed cash to afford the service of the debt and reduce indebtedness. A regulator was established in response to social protests, thereby anticipating amendment to Water Liberalisation Directive.	In **Belgium** the move to outsourcing was largely driven by adoption of Directive on public procurement and state aids. At the early stages of outsourcing it became clear that there were disadvantages such as "fuzzy" contracts and it missed the implementation of important technological improvements. The sector drew lessons from its experiences and the model of outsourcing was generalised in 2010-2015. In the end state, water management is outsourced in most extent to CIBE, but CIBE does not hold a monopoly in exploitation (anymore). Bulk water trading has drastically increased. Three main providers share the market.	In **Belgium** price reductions for large industrial users (via the introduction of grey water networks in Flanders and as a result of industrial policy to attract new industry in Walloonia) induced social mobilisation to obtain similar price reductions for domestic customers. Benchmarking was adopted in 2013 to constrain domestic prices. A new regulator was established in Walloonia to run the benchmarking operation and Aquafin took over these same responsibilities in Flanders. The results of the benchmarking (re cost frontiers) were made public and the results were enforced by the regulator. Pole B.	In **Belgium**, there has been a radical shift towards the DPM model following large natural disasters (most notably floods as a consequence of climatic change) that created strong public discontent. This stimulated an investigation by the Auditors that revealed mismanagement and corruption scandals. Social protests forced the commune to react and to withdraw from private regional companies. Financial benefits and voter support at municipal elections encouraged direct public management.	In **Belgium**, following the inadequacy in WSS provision by centralized systems (especially for what concerns wastewater treatment and drinking water quality in rural areas), many community initiatives developed in order to manage water services. These mainly arose from affluent rurbans in peripheries of Brussels and Antwerp and the farmers of the polders. Farmers became important players in community management of water resources in rural areas.
France (Paris VIII)					
N/A	Delegation contracts developed further in **France** with the new Directive on competition for the market (2009), and this movement was enforced for one main reason: local/regional authorities responsible for water and waste water management	In order to make the Outsourcing scenario happen in **France,** significant changes are assumed in order to evolve towards greater unbundling. The reason is that in the current state, the operator is in charge of the whole service. Two driving forces enabled	In **France** EU policy (improving performance evaluation and transparency), local government lobbying (in recognition of poor control over existing operators) and NGOs (more transparency, more information to consumers, better control of operators, evaluation/sanction for breaches of contract) encouraged greater	In **France**, the *régie* model became more and more widespread following strong distrust of private operators by the population. There were social protests against claimed excessive water company profits. A coalition of actors pushed the Government to	In **France** community management remains residual. In 2020 only limited rural areas and specific communal organizations (sharing common goals) will experience this form of management. Corruption scandals create discontent with both private and public operators. This stimulated the above

Market Drivers	*Delegation Contracts & Strong Regulation*	*Outsourcing*	*Regulated Monopoly*	*Direct Public Management*	*Community Management*
	were less and less inclined to cope with drinking water and wastewater treatment as they lack sufficiently trained and numerous personnel. Therefore, they tended to rely more and more on private operators' expertise and technology. And competitive bidding was mostly won by private operators (mostly lease contracts as is the tradition in France). This was observed in both the drinking water and sanitation sectors (régies loose progressively the sanitation sector with the ongoing process of the implementation of the wastewater treatment directive).	outsourcing. First, at the EU level, there was an explicit policy towards promoting competition for the market through outsourcing. Secondly, there is a major scandal in Toulouse, leading to a national debate on the alarming state of distribution networks in France and the need to promote more competition in different pipe sub-services. In the end state, TNCs have the majority of the unbundled market, but smaller private operators, specialized in small niches within the outsourced market, are also present.	benchmarking/regulation in France. Indeed there was a national protest movement supporting regulation (2010). An independent national regulator was established in 2015. However, this did not undermine the predominance of the delegation model as the regulator assists the municipalities in negotiating their contacts with the private operators. Hence it does not exactly fit the end state envisaged.	promote public management. Subsidies provided to train up municipal staffs in order to enable more municipalities to take back responsibility for operations. This proved to be a success. New methods of public management have also showed that this model was successful.	groups to consider alternatives. A new community management model developed in a small village and this was copied elsewhere. Centralised provision remains dominant but is influenced by the trend towards CM. However, a greater attention could be devoted to this form of management, as a consequence of change in social value, and individuals seeking new social identity structures.
Germany (Ecologic)					
1. Modernisation (including benchmarking) is preferred to privatisation. 2. Equal fiscal treatment of water supply and wastewater undertakings. 3. Possible relaxation of the principle of locality which restrict geographic operation of municipal operators. 4. Unfavourable financial position of many municipalities. 5. Anticipate EU policies to increase competition in water sector.	In **Germany**, the EU liberalisation Directive of 2009 predetermined the future. This type of compulsory public tendering challenged the autonomy of municipal organisation and interfered with the municipal right to self-administration, such as the creation of inter-municipal associations (*Zweckverbände*), or other municipal co-operation and public-private partnerships. So it met with severe criticism from certain German actors. The compulsory competitive tender for concession	In **Germany** the municipalities control the provision of the water services. In connection with the national modernisation strategy, in order to lower their costs and to comply with EU demands, municipal enterprises increasingly awarded subcontracts to external firms for specific service segments of the WSS. In order to push competition through outsourcing further, the EU lowered the threshold values beyond which contracts have to	In **Germany** there was an increased level of private sector participation and concentration in the initial period. This was mainly driven by municipal financial pressures. The EU pushed for the introduction of competition wherever possible and regulation (with transparency in costs, prices, and tariffs) where not. For later years two alternative storylines have been presented for Germany: Pole A: "Privatisation" following municipal budget collapse in 2012. Equal fiscal treatment was introduced which led to the formation of merged water and sewerage service utilities and a renewed concentration process. This increased market power & price. To	In **Germany**, the pressure (mainly from EU) to introduce more competition at the EU level led municipalities to actually opt for greater DPM (eg Regiebetreieb, Eigenbetrieb and Eigengesellschaft), which was not considered under the EU mandatory tendering procedures. This reaction was determined by municipality wish to maintain the communal right to self- administration (ie to keeps its autonomy in local affairs). Social pressure for more transparency, negative cross border	In **Germany** there was a general shift in social values, towards solidarity and local communities. In addition, following the WFD implementation, the public participation led to a heightened public interest in water service management issue and on a better co-ordination of different social group at local level. As a consequence, German associations and co-operatives also experienced a revival. Municipal finances were in crisis but a high court judgment declared that municipalities did not actually own the assets – but merely held them in trust. They could not therefore be sold.

Market Drivers	***Delegation Contracts & Strong Regulation***	***Outsourcing***	***Regulated Monopoly***	***Direct Public Management***	***Community Management***
	agreements drove municipal enterprises, generally Small and Medium sized Enterprises (SME), partly out of the market as losing the bid meant they had to be phased out. This process was mainly prevalent between 2009-13. Consequently, and in the context of the national modernisation strategy, there was a general demand for a more lenient application of the locality principle (*Örtlichkeitsprinzip*), so that municipalities would be allowed to operate beyond their traditional municipal territory. In fragmented market there was an evident need to support municipalities in their negotiations with the large private companies. The Directive amendment for greater regulatory oversight was introduced by increasing the capability of the abuse supervision bodies	be awarded through public tendering by 2012, which complicated the tendering processes. In the end state, the Federal Ministry of Economics and Labour strengthened its efforts to provide guidance and offered concrete assistance such as example contracts. The technical scientific organisations (ATV-DVWK) covered technical aspects of the contracts and the associations (BGW, VKU) provided technical assistance to their members. By 2020, the degree of outsourcing of water service elements had significantly increased.	counter this dynamic regulatory powers were transferred from Länder to Federal Government and a central regulating body was established. Pole B: Capacities of abuse supervision bodies were increased and data collection intensified following EU pressure for increased transparency (2009). Municipalities & associations demanded introduction of voluntary & anonymous benchmarking system. But the EU favoured compulsory benchmarking with unified core indicators. Finally the Federal Government developed a binding public benchmarking system.	leasing experience and ecological incidents (ie increasing floods and droughts) reinforced this trend. To respond to competition agenda benchmarking procedures were introduced. Initially this was voluntary but then compulsory.	As a result assets and their management was transferred to connected users (instead of private firms) who had established associations or cooperatives at a local level.
Italy (IEFE)					
1. Legal constraints and opportunities provided by Galli Law. 2. Liberalisation of other utility sectors. 3. Need for technical and managerial expertise. 4. Municipalities under financial pressure. 5. Financial	In **Italy**, two main driving forces should be considered. The first refers to the relevance of the legislative framework (this driving force should be considered as a common event, as it influence also the other scenarios building). Recently some decrees concerning the	In **Italy**, an increase in technological complexity and financial requirements for new investments stimulated the growth of outsourcing. In the following years, the WSS outsourcing business continued to grow more complex and there was a great	In **Italy** regulation was originally (in 2005) maintained through contractual arrangements between the operators and the municipalities. Post 2007 municipal financial pressures (reduced general revenues and high water investments) encouraged a sell off by those municipalities who had maintained a local public owned utility. The market power of the	In **Italy**, despite Gali law many municipalities retained control of water/sanitation assets. To avoid competitive procedures they decide to form in house entities through merger of local operators. Still rely on outsourcing to some extent. Creation of separate accounting system	In **Italy** CM remained a residual form (eg in mountain and rural areas) of WSS management, regarding its traditional form. National laws ensured WSS was managed at the ATO level. In some cases of community management outsourcing was commonplace.

Market Drivers	*Delegation Contracts & Strong Regulation*	*Outsourcing*	*Regulated Monopoly*	*Direct Public Management*	*Community Management*
requirement to implement EU Directives and maintain existing assets.	evolution of the utility sector have been passed. These decrees influence the degree of competition in the Italian WSS. In particular, the General Utility Sector Legislation favoured the liberalisation of WSS, by establishing that ATO operator should be chosen through a competitive procedure. As an effect, it promotes the entrance of foreign WSS operators in Italian WSS market, and the expansion of Italian WSS operators outside their captive market. As a consequence, the legislative framework influences the dynamics in WSS Italian Market, by making the market more open. This is enforced with the EU directive of 2009.	necessity to develop a number of collateral activities. Subcontracting was considered as a useful way to take advantage of professional firms with high expertise. Range of contract types present including BOT, DBFO and service contracts. In the end state, there is still an increasing demand for outsourcing of high technological requirements activities, since the local incumbent did not have the capacity to deal with this complexity. Some problems with contract coordination arise. In 2020, outsourcing of the WSS activities is 80% of the total.	existing private law regional operators was further enhanced by merger and acquisition activity (to increase both geographic and utility scope). Increased regulation and benchmarking was then needed to counter this changing market dynamic. The Comitato di Vigilanza was made responsible for early benchmarking initiatives (the data being collected by regional authorities) with the results being published annually. Finally a regulatory body was established although its powers and modus operandi are still uncertain. However price cap regulation is considered to be the most likely outcome of ongoing discussions. Pole A.	improved accountability of system – especially as local public finance problems continued. Special purpose companies were established for infrastructure maintenance. The introduction of new public management measures made the sector more transparent. Moreover, financing problems related to water infrastructure also contributed to the development of DPM. Where the ATO operator route was selected often there was not the political will to delegate the service and the local authorities continued to have ultimate control. Flooding incidents also highlighted importance of municipality.	However, the success of community management in the above residual areas showed the potential benefits of the alternative forms of user involvement (for example through the selling of ATO operator' shares to customers or through their representation on boards).
Netherlands (TU Delft)					
N/A	In the **Netherlands**, the EU Liberalisation Directive has been adopted. This followed gold plating and other public sector excesses, and successes of delegation in neighbouring countries and Delfland. Public companies bid against private companies for the time-bounded contracts (encouraged with the EU directive of 2009). The bids	The pre-condition in **the Netherlands** is characterized by limited financial resourced and substantial financial demands. Therefore, the water sector was looking for efficiency gains. This resulted in the outsourcing of non-core activities, later leading to a diverse range from maintenance of the pipes, to BOT contracts for water treatment plants, to outsourcing of the cleaning of the	In the **Dutch** storyline EU policy stimulated transparency and further liberalisation in the WSS sector. National discussions highlighted a need/want for wide-scale liberalisation – municipalities/provinces (under financial pressure) wanted to sell for the cash, experiences in other sectors highlighted the potential benefits of liberalisation, and there were ongoing technical difficulties in dealing with new drinking water quality requirements. Following a national debate (2007) regional private law	In the **Netherlands**, WSS are already managed by multi municipal bodies and regional water boards. This is further enhanced by innovations in public management, the need for an integrated approach, and the perceived failure of liberalisation in other network industries (e.g., gas). As a result of climate change, the implementation of WFD and older dykes/polders sinking	In **Netherlands**, the CM model developed from the public management model. The main driving forces are the emerging of a new interest towards self-organized form of WSS provision and the development of new technologies, which make possible a decentralized purification possible. Following a terrorist attack on a large water company that could not then supply its customers via the network activists from a housing estate went to the municipality and the Government to claim

Market Drivers	*Delegation Contracts & Strong Regulation*	*Outsourcing*	*Regulated Monopoly*	*Direct Public Management*	*Community Management*
	result in lower prices to consumers. Transparent implementation is key and even before the amendment to EU Directive the Ministry of Environment was given the mandate to act as a strong environmental regulator and the Knowledge centre on PPP was converted to the national economic regulator. In several public limited water companies, the management has undertaken a management buy-out to be able to truly compete as a private company. Foreign companies, as Suez, Veolia and RWE have entered the market and have won a substantial portion of the contracts. Also some Dutch public limited companies were able to win contracts. Wastewater treatment has been joined to collection and single delegation contracts are being signed in 2020.	buildings. In 2020, there is an internationally blended mix of companies supporting the Dutch public limited water companies, the water boards and municipalities. We see TNCs, but also low price suppliers from former Eastern Europe. However, the effect of outsourcing in terms of pricing and service delivery is almost negligible: the substantial transaction costs in the back office tendering, contracting and contract monitoring of all these different suppliers is more or less balanced by the resulting efficiency gains.	companies were formed that covered both water supply and sanitation. Some of these companies also covered other utility areas such as wastes. Bulk supply and retail supply was also separated, (some) assets were sold to the private sector and a national regulator was established. Regulatory benchmarking highlighted the good performance of multi-utilities. As a result of the regulatory practice of rewarding the most efficient players the multi-utilities gradually became the dominant mode of operation. Pole A.	a fundamental rethink was required. Flooding became an ever- increasing risk. In the context of the WFD the river basin approach became the polder approach involving integrated land and water management. This complex and integrated approach could only be achieved in the "public realm". Some public/private partnerships were adopted to develop new "living with water" areas. This proved a great success	the right to self provide. Dutch law prohibited this so the activists formed a broader unit and following a major pollution incident (resulting from under-investment by operator) and a break through in membrane technology they took their case to the European Court. The court decided in their favour. So in 2012 communities had the right to organize their own drinking water, provided that regulatory bodies would oversee the quality of the management. Some communities take up these rights. Regarding wastewater treatment, the purification water boards (*zuiveringsschappen*) continued to have responsibility on wastewater treatment and water quality control.
Spain (UNIZAR)					
1. Diminishing financial resources of municipalities. 2. EU liberalisation policies. 3. Financial and technical capacity of main TNCs. 4. Discipline and equilibrium in public budgets. 5. Social and political reaction in defence of welfare	In **Spain**, with strong competition (enforced by the EU directive of 2009), and without the possibility of subsidies from public institution, public companies have tended to disappear due to the pressure from the TNCs (which are already settled	In **Spain**, poor service and higher prices stimulated a search for a new management strategy. A new law was passed which strengthened the responsibilities of the public sector. But the law also included an	Following EU policy initiatives **Spain** introduced regional regulatory institutions to oversee their growing concessioned WSS market. These institutions are able to negotiate concession contracts on behalf of the municipality. Depending on the powers and position of the regulatory institutions two	In **Spain**, public management was encouraged and strict regulation rules were applied over prices following the victory of left wing governments, which discouraged PSP. Public financing (through public institutions) was also encouraged because	In **Spain** the CM has been used traditionally in rural areas for the management of water and for irrigation. In these irrigation districts the extension of responsibilities to their villages has been relatively easy. In new housing estates especially in the peri-urban area of the big

Market Drivers	***Delegation Contracts & Strong Regulation***	***Outsourcing***	***Regulated Monopoly***	***Direct Public Management***	***Community Management***
state 6. Sustainability of aquatic ecosystems 7. Governance crisis requires greater transparency and participation 8. Tradition of public financing of water infrastructures.	in the Mediterranean Spanish Coast). These large firms would also control the high technology markets (of reuse and desalination) that would be very active under the perspective of strong competition. But the price increases, resulting from withdrawal of public subsidies, would be associated with the liberalisation process and would generate much euro-scepticism.	exhaustive list of activities in which the private sector can participate (but excluding bulk water supply) and provided limits on contract lengths and allowed for the possibility of reversibility clauses. The TNCs compete aggressively in the supplier market. But there is sufficient space for small national companies who provide highly specialised inputs.	alternative storylines have been presented: Case A: Financial position of public bodies strengthened (by continuing EU support) and Regulators unfavourable to private sector interests. Private operators abandoned core concessioning business and focus on supplier markets. Case B: Regulators favourable to private sector interests. Private operators remained (and expanded) their activities in the core concessioning business.	of the important need to comply with European objectives. And the historic predominance of public bodies in water management and bulk supply highlighted potential role of public sector in water. To gain efficiencies the scale of management would be increased – supra-municipal in rural areas and metropolitan in urban areas.	cities and in touristy areas CM has grown under the public health control of local and regional institutions. It has also been extended to large industrial users and industrial estates.
Switzerland (EPFL)					
1. Public opinion against liberalisation. 2. Legislation on consumable goods. Also. 3. Liberalisation of electricity and gas markets – highlighting specificities of water. 4. Financial availability.	In **Switzerland**, in 2011, a new referendum was organised where, surprisingly, 52% of the population and a majority of cantons voted in favour of the entry into the EU. It joined in 2012 – a speedy entry. Switzerland then has to implement the EU directive of 2009 obliging competition for the consumer market. This started in the big cities. But the small towns and rural areas failed to immediately open to competition. There were social protests in 2013 and politicians emphasised the need for more regulation and transparency and actively lobbied for a change in the	In **Switzerland** the 2002 federal law on consumable goods increased the responsibility of the operators and in this way, it pushed professionalization. This tendency, in turn, resulted in a new form of competition in the Swiss water supply market with short-term outsourcing contracts for small engineering companies in low populated areas. This enabled smaller communes to bear their new legal responsibilities. The larger urban areas remained in public hands, but pressure to outsource more is increasing. In sanitation, outsourcing became more important more	In **Switzerland** major reforms were triggered by a corruption scandal in one municipality in 2011. Investigations by media & civil society uncovered further irregularities and this pushed national politicians to introduce benchmarking at the federal level. But in the event (because of a lack of resources and the need to maintain local authority autonomy) by 2015 this initiative failed. However, encouraged by national and local politicians the Swiss Association of water and Gas (SSIGE) launched an alternative benchmarking system, but only for water supply. This system is compulsory and the results are made public. Indeed SSIGE may issue recommendations to the local authorities when the price is judged too expensive. The national association is now considering benchmarking in the	In **Switzerland**, new managerial measures were introduced for enhancing public management. This included accounting separation to enhance transparency. There were some pressure for more regionalisation and more liberalisation (especially following experience in the energy sector). However the communes were reluctant to relinquish control and mobilised public opinion. The impact of liberalisation of the energy sector actually highlighted perceived public good specificities of the water sector (as compared to other network industries). This actually led to the turn towards DPM.	In **Switzerland** CM developed along centuries in rural communities. In 2012 CM continued to operate in small rural areas, without expanding in new ones. Academics started to publish on the success of the community management model in Switzerland (eg Wittenbach) and elsewhere. This intellectual interest had an unexpected effect and community management started to expand into small towns. Indeed following financial troubles and bankruptcy in one large city it split into hundreds of community organizations that then took on water supply and sanitation responsibilities.

Market Drivers	*Delegation Contracts & Strong Regulation*	*Outsourcing*	*Regulated Monopoly*	*Direct Public Management*	*Community Management*
	liberalisation Directive. The amendment to introduce a Regulatory Authority in each Member State was welcomed.	quickly. The main events happened between when the increasing use of chemicals in the agricultural and industrial sector led to new demands in terms of technology for wastewater treatment. In the end state, outsourcing (using proprietary wastewater treatment technology) is common.	sanitation sector. Pole B		
UK (IDD)					
1. Public opinion. 2. Scale and timing of investment programme. 3. Search for cost savings. 4. Stakeholders pressing for greater liberalisation. 5. Affordability agenda. 6. Continuing negative cash flows. 7. Pro-competition legislation. 8. Financial market conditions. 9. Government concerns over stability 10. External environment. The current Government and the economic regulator have made it clear that they are seeking financial and regulatory stability for the companies over the next decade. The current policy aim to retain vertical integration and continue with statutory rights and duties for the whole supply chain.	In **England and Wales** this scenario needs to be set in a relatively constrained institutional context. However, in 2009 the industry had reached a plateau in terms of cost levels and service performance. A regulatory review was instigated which highlighted the potential of delegation contracts. At the same time the EU introduced a proposal for a liberalisation Directive focused on competition for the market. Despite opposition from the UK (and Germany) the Directive was adopted. In 2015 under the auspices of the EU Directive the Secretary of State issued notices to quit on all water companies. In one day the whole water and sanitation sector became a 25 year concession – whereby ownership of the assets would be taken back into public control and	In **England and Wales** high price rises, over 20%, played a major role in the development towards outsourcing. These price rises led to a lot of discontent, especially amongst household consumers. The Regulatory Authority was not successful in lowering the prices: the moves toward liberalisation (e.g. common carriage) failed, and new investments were needed as result of the WFD. Shareholders gradually began to exit the market, to be largely replaced by bondholders. In addition, the Regulator set tough targets and refused to countenance a greater role for mutuals. These tendencies contributed to the appetite for institutional change. There was an emerging consensus that outsourcing was the only way to	In the UK (or more accurately **England and Wales**) there has been little change between 2005-2020 – the original regulatory model (price cap combined with comparative competition) has continued in a similar fashion to 2004. There is still much diversity in institutional structures and strategies. Common carriage has been successful in niche large user markets. Some European benchmarking initiatives have also been adopted/used by the regulator. Pole A.	In **England & Wales**, the pressure to reduce costs led to the merger of companies. This was allowed by the regulator - despite some evidence against economy of scale arguments. By 2015 the industry was dominated by 3 large equity financed companies. And the consolidated industry appeared to be working well. The system was called into question following the failure of security systems, increasing financial pressures (following on from impacts of climate change) and corruption scandals involving two of the remaining mega-companies. There was an investment strike by the capital markets as a result of uncertainty. And companies were unable to fulfil some of their principal duties. The Secretary of State petitioned high court for a Special Administration Order and Government decided to fund investments created	In **England and Wales** the investments needed to comply with EU directives together with the strong regulatory opposition towards price increases were the main drivers. In 2009 the periodic review led to small price decreases despite continuing investments. As a consequence, shareholders gradually began to exit the market to be largely replaced by bondholders. A few companies became either wholly debt financed, or only with a very thin equity wedge. However, a few of the larger companies continued to favour the equity model. These large equity based companies began to offer specialist services to the largely debt-financed companies. In response to the tough price review of 2009 a large number of companies also put proposals to the Regulatory Authority to become mutualised - with ownership and some risks of ownership being transferred to the customers of the utility or selected members. The regulator accepted but insisted on major

Market Drivers	***Delegation Contracts & Strong Regulation***	***Outsourcing***	***Regulated Monopoly***	***Direct Public Management***	***Community Management***
	consequently the rights to use them openly rebid. In 2040 the Regulatory Authority would become the Governments concessioning authority. It was anticipated that under the terms of the new contract that the Regulatory Authority would review prices with the new private operators every five years.	achieve the tough targets set by the regulator, although the RA insisted on detailed procurement plans. Outsourcing focused on both infrastructure and non-infrastructure elements. The RA left it to the companies to determine the extent of their outsourcing. The resulting contracts were mainly won by the remaining equity financed English water companies		by climate change. Several buy-out initiatives at the local/regional level re-established DPM after being absent for more than 50 years.	changes to the associated licences. By 2020 outsourcing became a key feature. Again this was monitored by the Regulatory Authority.

Chapter VI

Economic Implications of Water Scenarios

Authors: Aad Correljé, Delphine Francois, Antonio Massarutto[*]

INTRODUCTION

Part 1 of this book analyses the current situation of the European water supply and sanitation sector and elaborates on the ongoing process of change. Part 2 sheds light on possible future characteristics of the sector, through the examination of the set of hypothetical scenarios presented in Chapter V. Within this context, this Chapter VI addresses the economic implications of these scenarios. Each of these scenarios is built upon more or less specific institutional arrangements. As a consequence, the scenarios fundamentally differ from each other, for example, in terms of the public/private *ownership* relationships involved and the operational coordination of the water systems, through the market, hierarchy and/or forms of regulation. These arrangements embrace 'schemes' of incentives and restrictions to the behaviour of the public and private actors involved, thus determining the economic implications of the scenarios.

The assessment of the *economic implications* in this chapter goes beyond the analysis of the traditional *indicators* for financial-economic performance, like investments, costs, profits, return on investment (ROI), productivity rates, etc. Our analysis of economic implications will address the degree of effectiveness and efficiency with which the water supply and sanitation systems, under the several scenarios, are able to respond and react to various, diverging, societal and environmental demands. These responses are a consequence of interacting public policy and private actors' strategies. *Economic implications* say something about the kind of trade-offs that are established in this interaction and the manner in which they are achieved, to be summarized as "who does what and how"?

We argue that the six scenarios, through their institutional setting, incorporate specific patterns of allocation for responsibilities, tasks, values, risks, resources and information. Together these elements constitute *incentive* structures, or regulatory models, which are believed to drive business strategies, investments, the industry's capability to attract resources (capital, skills, staff, and knowledge) which eventually make up the performance of the supply and sanitation systems.

Faced with a range of societal demands, these different patterns will yield outcomes that may vary in their emphasis on particular financial, economic, social, environmental or technical characteristics and objectives. The patterns,

---

[*] Aad Correljé is based at the Delft University of Technology (NL). Delphine Francois is based at the Delft University of Technology (NL) and The University of Gent (B.). Antonio Massaruto is based at the University of Bocconi (It.). This chapter is based on Deliverable 6 of the Euromarket project, made possible by the valuable contributions of Vania Paccagnan, Massimo Di Domenico and Elisabetta Linares of the University of Bocconi (It.), and Laetitia Guérin-Schneider, Serge Garcia and Fady Hamadé of ENGREF (Fr.) and Jan Jaap Bouma of the Erasmus University (NL). This publication has received support from the NGI Public Values Project at TU Delft.

 Water and Liberalisation: European Water Scenarios. Edited by M. Finger, J. Allouche & P. Luís-Manso. ISBN: 1843391139. Published by IWA Publishing, London UK

however, also have an impact in the manner and swiftness in which these outcomes are reached, as a consequence of the associated incentives schemes and the allocation of risk to the actors involved[1]. The degree of consistency between these elements in the light of the objectives set, by and large, will determine the broader effectiveness and efficiency of the systems.

In this chapter, we will sketch the main economic implications of the six scenarios, on the basis of institutional economic analysis and the examination of a number of empirical cases. This analysis will focus on the identification of specific sets of structural constraints and drivers in the several scenarios and on the interpretation of the consequences thereof for the particular ways in which inputs, like cash flows, investments, operative resources, knowledge and insights and labour, are employed to undertake the supply and sanitation of water.

The next section will describe the theoretical framework that is used to analyse and explain the outcomes, in the light of the characteristics of the scenarios. Drawing from a body of theoretical and empirical literature a roadmap will be built (Figure 1), linking perspectives and conditions for the alternative governance approaches (regulatory models), to the characteristics of the several segments of water systems and their organization.

Subsequently, the six scenarios will be framed in terms of these regulatory models. We will reduce the complexity in the number of potential variables, by distinguishing three main sets of *criteria,* within which the most important characteristics will emerge:

- *Rents and their distribution*, as a reflection of the direct operational and management characteristics and the role of the several stakeholders involved;
- *Investments*, as a reflection of the objectives of the system managers, as well as the potential to mobilize the resources required for investment, both for improving the system and for maintenance or replacement;
- *Supply characteristics*, as some measure of the output of the water systems in terms of shorter and longer term social and environmental utility and technical, environmental and market related characteristics along the value chain.

The main conclusions of this analysis will briefly be reported per scenario and a number of the main trade-offs between the systems and their consequences will be highlighted. For each scenario, the three sets of criteria will be evaluated from the perspective of the stakeholders. These involve: a) the consumers, b) next generations of citizens, c) taxpayers, d) the environment, e) the objectives of (public) entities politically responsible for the supply and sanitation of water, f) the private investors in (parts) of the systems and finally, g) the supply chain itself. Finally, the main conclusions will be presented. These are based on a combination of insights drawn from the theory and other scholars' empirical studies, plus a number of case studies, analysing examples of current water systems representative for the scenarios.

REGULATORY MODELS, THE MARKET AND PERFORMANCE

Guidelines for regulatory reform: a roadmap

Economic theory has developed a robust and consistent analytical framework for understanding the outcomes of alternative ways of organizing the provision of services of general interest (SGI). This framework is based on the concepts of market and non-market failures. Market failures are caused by those features of a given SGI that deviate from those guaranteeing an efficient functioning of the "invisible hand" and involve, for example, natural monopoly and market power, externalities, public goods, information asymmetries causing incompleteness of markets. Non–market failures, in turn, arise from inefficiencies and distortions caused by public intervention and political decision making. We refer to Annex 1 of Deliverable 6 of the Euromarket project (Euromarket, 2005) for a thorough discussion of the main theoretical arguments and empirical evidence, from which we derive the main dimensions to be checked, in order to evaluate expected features of alternative regulatory models and outcomes of regulatory reforms. The "regulatory roadmap" shown in Figure 9 below resumes the main steps. The analysis developed in this section will apply this structure to water supply and sanitation (WSS) services. It will:

- Discuss the *dimensions of general interest* (DGI) that characterize the WSS industry;
- Verify the possibility of introduce *competition in the market* without regulation or if free competition can be made *compatible with DGIs* through external regulation (quality regulation), imposition of general service obligations, cross-subsidies etc;

[1] Risk is defined as the likelihood of differences (positive and negative) between expected and actual costs or revenues. The major sources of risk concern: a) the need for maintenance and renewal of network assets; b) the risk of new regulations; c) unexpected pollution of existing resources; d) adverse climatic events; e) lack of commitment of regulators; f) variations in demand (Pricewaterhouse&Coopers *et al.*, 2004).

- Verify the possibility to introduce competition in *unbundled phases* (e.g. once essential facilities or monopoly segments are managed separately);
- Verify the suitability of mechanisms fostering *competition for the market* (e.g. competitive tenders), in the industry as a whole or in given segments;
- Verify the possibility to implement regulatory models that *simulate the market* and/or provide market-like incentives to monopoly operators;
- Verify the possibility to reduce non-market failures through the adoption of *market simulation mechanisms within the public sector*.

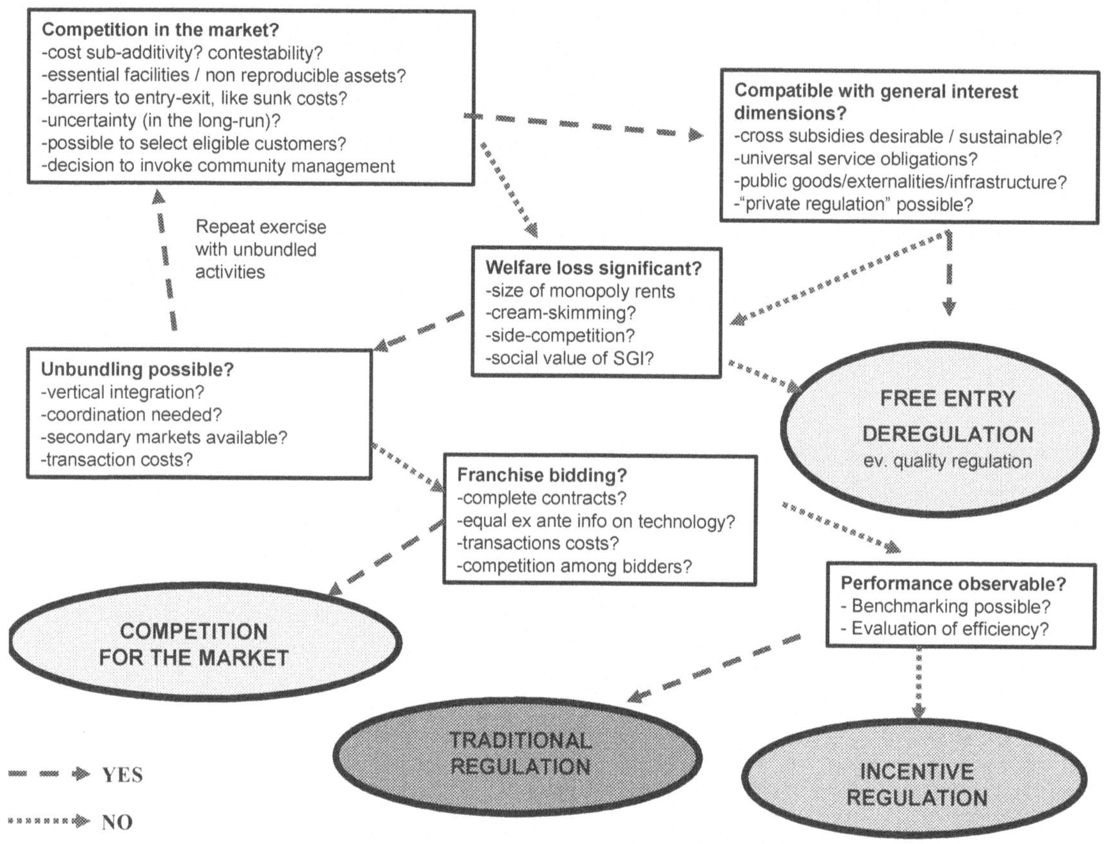

Figure 9: A policy roadmap for regulating services of general interest

Dimensions of general interest

The water industry is characterized by multiple market failures, from which potential dimensions of general interest arise (Cowan, 1993; Massarutto, 1993). First of all, water as a *scarce resource* will normally generate monopoly rents if its use is not regulated, or when it is provided by means large scale, specific artificial facilities, like reservoirs, transfer schemes. Secondly, there are *externalities* related with water use and the environment. These arise when human pressures begin to alter the quality of available resources (Barraqué, 1998; Merrett, 1997; Green, 2003). Thirdly, the water industry affects *public health and environmental quality*. It not only provides utility to individual consumers but also can be seen as a collective good to the community. Moreover, the supply systems allow controlled access to the resources and compliance with environmental rules (Kraemer, 1998; Noll *et al.*, 2000). Consequentially, all urbanized areas in Europe have centralized systems for supply and compulsory connection to sewerage and sanitation[2]. Fourthly, there is the requirement of *service resilience*, involving both the short-term supply of peak demand, as well as the long-term adjustment of the system to the evolution in the patterns of demand and/or resources availability. This dimension

[2] Nevertheless, technological development and the high costs of extending networks outside urban areas, may stimulate reliance on smaller scale local systems.

links levels of investment and the capacity of the system to generate the required financial resources with the choice for technology, with a lock-in potential and sizable sunk cost. A certain excess capacity of the infrastructure and flexible technologies allow lower adjustment costs (von Hirschhausen et al., 2004). Finally, as an essential service, water supply poses questions of *affordability and accessibility* (Gleick, 1998).

Some of the DGI can arguably made compatible with private sector participation (PSP) achieved through the imposition of effective external regulation (Brown, 2004; Clarke et al 2004; Galiani et al 2005; Summerton, 1998). Others stress the many pre-requirements for successful PSP, pointing to weak regulation, information asymmetry, weak contractual powers, public sector inefficiencies and high capital cost of PSP (Gleick et al., 2002; Hall, 2001, Massarutto, 2003a). Nevertheless, the more effective regulation is in transferring the responsibility for performance quality to the private operator, the higher is the operational risk that the operator faces and the higher the required remuneration (Pricewaterhouse&Coopers et al, 2004). Indeed, the high capital intensity and the very long depreciation schedule of physical assets make WSS a risky business, which is rare in an industry with a rather stable and predictable demand (OECD, 2000).

Competition in the market

As has already been indicated in Chapter I, the economic literature is rather unanimous in recognizing the limited scope for competition in the market for water as far as WSS are provided through physical networks. Several theoretical and practical reasons are mentioned (see Massarutto, 1993; Muraro and Valbonesi, 2003; Spulber and Sabbaghi, 1994; Cowan, 1997; Beecher, 2001; Rees, 1998).

Important is that the potential geographical size of the market is limited by the high cost of water transport and by the environmental unfriendliness of concentrated impacts in single points. Even if the most important segments, i.e. transport, storage, distribution, sewage collection, sewage treatment, are considered separately this remains true. Most segments of the value chain involve facilities with significant economies of scale. Moreover, a large portion of the costs of WSS are sunk. 80-90% is infrastructure investment and the costs of customer services are only modest. Local water production costs essentially are related with environmental and hydrological conditions. So, efficiency gains from competition in this segment are likely to be modest too (Noll *et al.*, 2000; Spulber and Sabbaghi, 1994). Although, exception may exist under particular circumstances, there is limited scope for the provision of water services through third party access to the system or via different substitute goods and services.

These arguments support the conclusion that wherever a network system is required, a unique infrastructure is the most efficient solution. However, there is no consensus about the optimal territorial scale and size of a water network. While some economies of scale have been demonstrated for certain segments, like drinking water treatment, sewage treatment, storage and long-distance transport, networks in urban areas remain a local monopoly. In principle there are no major obstacles from the economic point of view, to serve small communities via single networks. Management of distribution networks on a smaller territorial scale can be envisaged for pipeline maintenance and operation, with a single operator for water resources management, supplying bulk water to these systems. This structure was quite common in many countries in the Mediterranean region, while England and Wales have experimented with a comparable model; the so-called "inset appointments" (Cowan, 1997; OFWAT, 2000).

Unbundling

As compared to the energy and telecoms sectors, unbundling has not received much attention in respect of the WSS. Nevertheless, an analysis of the potential for unbundling of the water value chain may provide interesting insights. Four main options are relevant:

- Unbundling along the functional components of the value chain;
- Unbundling of the different phases of the production system, involving respectively, the extraction of raw water, storage, transport, bulk supply, treatment, distribution, wastewater collection, wastewater treatment, sludge disposal;
- The separation of regulated water services from other activities;
- Horizontal and territorial or geographical unbundling of local systems.

Functional Unbundling

Figure 10 provides a representation of the value chain of WSS. We can distinguish a *management core*, in which we can identify three functions that might or might not be combined within the same entity, namely: 1) planning; 2) legal responsibility; and 3) the operation of the system. Combinations of these three functional spheres constitute the scenarios used in the Euromarket project.

The management core interacts with two other spheres, namely, firstly, the *regulatory sphere*, involving requirements of service quality and dimensions of general interest and the competition rules and, secondly, the *market sphere*, from which inputs are obtained. The boundary between the management core and the regulatory and market sphere is quite elastic and permeable. There is interdependence between arrangements occurring in the management core (operators market) and those occurring both with the supply chain (suppliers market) and water resources and environmental policy (water market). The more the management core is based on market-type relations like tenders and contracts, the more the industry will tend towards vertical integration in the suppliers market. Vertical integration helps private operators to reduce risk arising from tendering and accepting the concession contract, while raising information barriers against potential new entrants and/or regulators. In contrast, a more integrated hierarchical structure in the operators market, either dominated by responsible entities (Direct Public Management) or by the privatized operator (regulated monopoly) seems to have a greater tendency to purchase inputs from the market and via outsourcing (Kraemer, 1998). This can be explained by the fact that local monopolies are not necessarily reaching the optimal scale for all segments. If historical, institutional and political factors impose small-scale local monopolies, they probably will acquire inputs from the market.

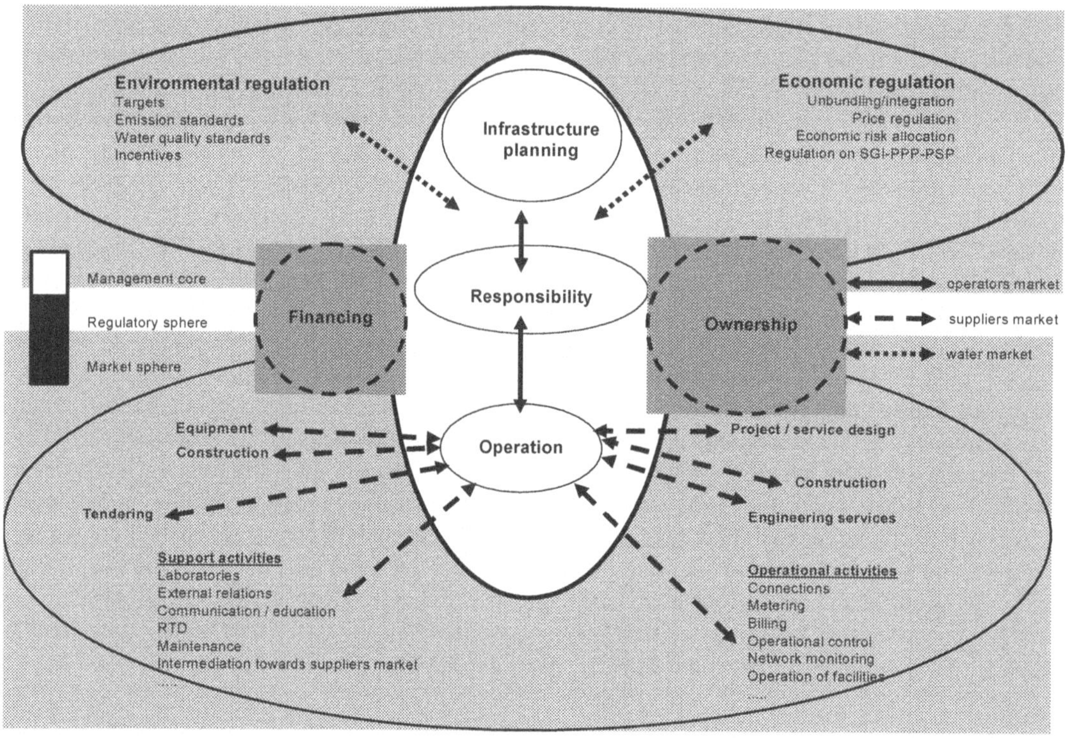

Figure 10: The value chain of water services

The most critical element and the major source of economic risk are asset ownership, management and development (AMD) (World Bank, 2005; Pricewaterhouse&Coopers *et al*, 2004). The more the private sector is facing market and operational risk, the more it tends to refuse responsibility for AMD. Nevertheless, this responsibility is more easily accepted when operational and market risks are reduced, by offering longer-term contracts and by allowing extra costs to be passed on in tariffs. AMD contracts are exposed to significant transaction costs, arising from information asymmetries and opportunistic behaviour. Indeed, as the investment is sunk, responsible public entities might be tempted to impede price increases over the marginal cost, thereby appropriating the "quasi-rents" and making it impossible for private companies to recover their fixed cost (Noll *et al*., 2000). Alternatively, the company may capture regulatory decision-making and "invite" additional technological and environmental requirements, thus justifying over-investment in assets associated therewith (Lobina and Hall, 2003).

In fact, these activities are never clearly allocated in the management core or in the external spheres. Rather, they are positioned somewhere in-between, as a continuum of solutions, based on shared competences, responsibilities and risk. The more they are in the private sphere; the higher is the time horizon and the correlated risk premium. When these competences are separated from operation and remain in the public sector, contracts are more easily released for operation only. The public sector can later on adopt other arrangements for PSP in that particular phase (such as Design, Build, Finance and Operate (DBFO)), while pursuing autonomous strategies for operation.

Unbundling of phases

Concerning the second option, involving unbundling of phases, it does not seem that there are relevant economies of vertical integration between water supply and sanitation, with perhaps the exception of smallest utilities, where economies are arguably generated by the opportunity to share some administrative and fixed costs (Garcia *et al.*, 2003). Both activities are only occasionally managed together and there are no clear signs that joint operation would be of any economic benefit. Eventually, at a larger scale patterns of wastewater reuse may have an impact, because the supply-side relies on good quality surface water (Massarutto, 1993). When system design, investment decision and possibly financing of infrastructure are managed centrally by environmental regulation and planning authorities, the scope for a greater segmentation of phases seems higher. In contrast, when these functions are delegated to the operator, external effects and costs are more likely to arise, imposing significant transaction costs upon the system.

Separation of non regulated activities

The separation of non-regulated activities, has received attention with special regard to multi-utilities. Some empirical studies suggest a better performance of multi-utility companies relative to WSS (Vaccà, 2003; Drusiani *et al.*, 2004). At the same time, there is evidence that most private companies are abandoning multi-utility strategies and seem to prefer to focus on only one domain, with exception of the natural gas and electricity sector which increasingly converge towards integrated management (Finger and Allouche, 2001; Ninni, 2005). It has been suggested (Cozzi and Massarutto, 2002; Massarutto, 2003b) that this apparent paradox can be explained by the fact that economies of scope arise on the demand side to a greater extent than on the supply side. Local multi-utilities can more easily generate positive externalities, like a more efficient network maintenance programme, a better allocation of their fixed costs, and the reduction of the risk associated with sunk investments, via an integral financial and operational planning of infrastructure and inter-service cross subsidies. The operating company can only capture these economies if it is a local monopolist, integrating the different services in the same market. A regulatory strategy aimed at separating WSS from other services may reduce information barriers and allow a more credible benchmarking in one segment. Yet, at the same time, it could eliminate positive externalities. A compromise solution would maintain horizontal integration for AMD only, while horizontal unbundling would be adopted for operation (Drusiani *et al.*, 2004).

Territorial unbundling

The benefits from territorial unbundling are considered controversial. Some authors provide evidence of the existence of economies of scale that justify large management units (Mizutani and Urakami, 2001). Many, however, seem to reject the hypothesis that costs are significantly lower on larger territorial scale (Antonioli and Filippini, 2001; Saal and Parker, 2004). This conflicting evidence might be due to misunderstandings in the specification of production functions and to a systematic overlap of different concepts: scale economies in the *single facilities*, in the *territorial size* of the market and in the *water companies* (Massarutto, 2002). As stated above, economies of scale are arguably present in most production *facilities*, leading to natural monopoly conditions. Yet, the possibility to exploit these scale economies is obviously linked to the size of (urban) settlements and available water resources. The optimal *territorial* size of WSS utilities therefore depends on economies of density, as much as on economies of scale, since transport costs are high (Garcia and Thomas, 2003). A different issue is the optimal size of *water companies*. Empirical evidence shows that larger companies are better equipped for competing in an open market, as they are able to diversify risk and to internalize key (input) segments of the value chain, such as engineering services. This, however, is not necessarily so in each and every single local market. Indeed, a local Small and Medium-sized Enterprise (SME) may have similar operation costs, particularly if it outsources the production phases entailing economies of scale or replication.

Competition for the market

Competitive tendering is the solution that has received most attention (Muraro e Valbonesi, 2003) and is a rather popular model in practical PSP experiences (Noll *et al.*, 2000), also because of the support given by international institutions like the World Bank (Lobina and Hall, 2003). Water concessions e.g. are expected to create value by boosting service coverage and quality, and by improving the efficiency of utility operations (Berg, 2000). However, a more detailed analysis reveals that this is true only under some conditions. Critical factors involve contract completeness, transaction costs, information asymmetries among participants and the number of suitable candidate bidders. For a description of the concept of competition for the market we refer to Chapter I.

The economic literature (Massarutto, 1993 and 2002; Cowan, 1997; Noll *et al.*, 2000; Rees, 1998; Beecher, 2001) as well as practitioners' experiences (WrC and Ecologic, 2002; OECD, 2004; Pricewaterhouse&Coopers *et al*, 2004; Lobina and Hall, 2003) underlines the critical importance of infrastructure life and related risk. Private firms have limited incentives to invest without appropriate guarantees for compensation at the end of the contract. Thus the investments would either be postponed indefinitely, with the risk of collapse of the system and the need for new investment by the state. Alternatively, guarantees for investment recovery may stimulate operators to over-invest to avoid the risk of service failures. Operators with a significant information monopoly could thus be stimulated in justifying unnecessary new investment; so-called "gold plating" (Kraemer 1998; Rees 1998).

These effects arise from the inherent incompleteness of contracts intended for the very long run and specified in generic terms. As a consequence, contract renegotiation rules, penalties for incompliance and the sharing of risk are crucial elements (Beecher, 2001). The more concrete the task, the less complex is the tender. The role of public entities in assuming at least part of the risk in respect of the long-term asset management is also important. Contracts should provide a clear definition of the categories of risk that the private firm is assuming. Otherwise the tender is unlikely to be interesting to bidders. This can be achieved, for example, by maintaining asset ownership – and related responsibility in the long run – in public hands, and formulating appropriate solutions for the transfer of sunk investments from the former to the new operator (Boitani and Petretto, 2002). Such arrangements are particularly important in phases when the asset base is expanded, of course (Kraemer, 1998).

A second critical pre-requirement for efficient competitive tendering is equal ex-ante information. The cost of water services, while being easily standardized in respect of each individual system components, can not be foreseen for the system as a whole, because of the many site-specific variables. Therefore, incumbent operators accumulate a significant information monopoly, discriminating against potential new entrants, especially if the tender is open-ended and subject to bargaining. International experience provides clear evidence that incumbent replacement only occurs in exceptional cases and only when the contract covers simple activities for short periods (Massarutto, 2002; Lobina and Hall, 2003).

Incumbents may reinforce their position by vertically integrating parts of the value chain where profits are made, or by introducing their own patented technology during their contract (Prost *et al.*, 1999). Another way of reducing risk is by establishing a mixed venture company as a Public-Private Partnership (PPP). The PPP has advantages both for the public authorities and the private company, since it allows the public partner to gain from the exploitation of the service while maintaining a hold over the system. The private partner is assured of the commitment of the public authorities to maintain the company viable and, thus, acquires advantages in the next round.

The third critical factor involves the number of competing candidates. No systematic studies are available that analyze the concentration of the WSS industry and the likelihood of effective competition. National markets are quite differentiated with this respect. Yet, despite the large number of local water companies, there is not much competition for tenders. Most water companies are not interested in bidding outside their home-area. If they engage as bidders, they generally participate with a minority share, supplying technical services and know-how. Examples include the French multinationals and a few companies from other EU countries (see Castro, 2004; Lobina and Hall, 2003; Drusiani *et al.*, 2004; Euromarket, 2004). More lively is the market for specialized activities along the value chain, from DBFO for assets and facilities to outsourcing of operational functions (WRc and Ecologic, 2002).

Monopoly with incentive regulation

Following the roadmap, the next step is to verify whether the introduction of incentive regulation is possible in those segments that remain unsuitable for competition. We thus assume that total or partial privatization of the operators will not lead to abuse of market power. If regulation provides the appropriate incentives, this model is expected to generate efficiency gains (Rees, 1998; Beecher, 2001; Cowan, 1993; Littlechild, 1983). These may involve a permanent incentive to reduce costs in the regulated segment and the encouragement of private investment, attracted by a relatively low market risk. Yet, if the regulator suffers from information asymmetries and the regulatory regime is not optimal, these advantages are weighed down by regulatory capture. Costs are not reduced and simply transferred to consumers, while in case of revenue cuts, investment and service quality will be reduced. Quality regulation is required, since incentive regulation will foster a trade-off between cost reduction and service quality (Noll *et al.*, 2000).

A crucial issue is the long-term viability of this approach, particularly if it is considered together with private capital market expectations (Bakker, 2003). The full privatization of water assets is not necessarily welcome to the companies, if it is accompanied by a permanent responsibility to maintain them in order to fulfill service obligations. To ensure an adequate cash flow for asset renewal and maintenance, the corporate cash flow should be equal to the depreciation and capital cost. But private capital markets have a time horizon that is usually shorter than the depreciation schedule and require, either, depreciation over a shorter period, or a higher risk premium[3].

This problem has surfaced in the British water industry, where the need for asset replacement appeared. Some companies did try to step back from total asset ownership (Saal and Parker, 2004). Thereafter, the Office of Water Services (OFWAT) became slightly more generous in allowing capital cost recovery. Massarutto (2005a) has provided estimates for two Italian Regions, showing the substantial increase in water prices, if the long-run full cost would be taken into account, including the true depreciation of assets, together with a market rate of return.

Non-market failures and limits to state intervention

It is not surprising, considering the above, that WSS are seen as a sector with limited scope for PSP. Adam Smith himself admitted water as one of the few exceptions to the domain of the "invisible hand". The "golden age" of public works in the water sector is to be placed in the last century, following the progress made in the construction industry and the big boost to water demand, caused by urbanized growth and the use of hydropower. In this model, "hydraulic structuralism" (Massarutto, 1997; Bressers *et al.*, 1995), the public goods approach fostered public responsibility in the promotion and financing of water development projects. There was an emphasis on engineering and infrastructure. Policy was dominated by the "iron triangle" of a) the large and concentrated users with good access to the political arena; b) the construction industry, getting the construction contracts funded on the public budget and c) political actors, whose influence on public spending on water projects was the source of their power. "Water needs", as expressed by various (potential) uses were considered independent variables and "issues of general interest" to be governed at the political level.

More recently, it became evident that public involvement in the WSS is affected by distortions in the decision-making process, due to voting procedures and information asymmetries, capture of regulatory decisions by individual interests, reduced incentives to efficiency and productivity in public organizations, and so on. Given the structure of WSS management under public operation, this may steer towards excessive investment and x-inefficiencies. Moreover, to develop the necessary competence and skills in the public sector, specific investment and sunk costs have to be incurred, while private companies could more easily spread these costs over a larger number of operations (Drusiani *et al.*, 2004). Apart from efficiency considerations, there are other reasons that provide limits to direct state involvement in WSS, calling for private sector participation. The first of these is the so-called "fiscal crisis of the state", implying a general drive towards a reduction of public budgets and debt. A second reason for private sector participation has to do with the shift that water services are experiencing in respect of their management paradigm, towards a more intensive model (Barraqué, 2003) of water management, with a focus on rehabilitation and treatment, more complex technology and qualitative, rather than quantitative solutions. More emphasis is put on information, quality control and pollution prevention (Kraemer, 1998; Drusiani *et al.*, 2004). While posing new and complex regulatory problems, the new paradigm emphasizes the need to increase the autonomy of WSS operators from the public administration via innovative private-public partnerships.

The performance of private and public water management systems

The arguments above suggest that the focus on contrasts between "private" and "public" water systems is not a so much a relevant way of framing the discussion, as all systems entail a mixture of public and private elements. There are many different ways to involve the private sector. These can be modeled on the theoretical archetypes as discussed so far, but in reality they will deviate a lot. Studies on these issues can be divided in two categories: those comparing panels of private and publicly owned water systems, and those analysing the impact of regulatory reforms in particular cases. It is no surprising that the empirical evidence on relative performance of private and public water systems has not reached any clear conclusion so far. In most cases, it ends in a "no contest" judgment, especially if comparison is not limited to costs and profitability, but also includes service quality, investment and general interest dimensions (Renzetti and Dupont, 2003).

We argue that in order to derive relevant implications, the focus for attention should be on the regulatory environment, the way risks and rents are shared between actors, the system of incentives and penalties, the capacity of regulators to extract information from operators, etc. This environment also includes public health, safety and environmental protection standards, which may be determining factors in the choice of technology and capital investment. The size of the utilities, their geographical location (rural versus urban), hydrological characteristics

[3] Ofwat assumes a risk premium differential cost of capital of 4.75% over risk-free activities; Cooper and Currie (1999) argue that this value is underestimated and the correct risk premium should be 6.6%.

(surface versus groundwater) and other site-specific features also seem influential factors, as well as horizontal integration and the opportunity to share fixed costs with other services.

Although efficiency gains are often put forward as a main reason for PSP, most studies report an increase in water prices and in the overall amount of economic resources devoted to the water sector. It is quite difficult to assess whether the price increases are a consequence of either the elimination of subsidies from the public budget and increasing cost recovery records, or improved service performance and maintenance, or new investment driven by environmental regulation, or an increase of overall costs, particularly capital and transaction costs. Another factor that most empirical analyses seem to ignore is the allocation of risk and, thus, perspectives for costs and price increases in the long run and responsibilities on replacement investment. It can be reasonably concluded that private sector participation per se is not a welfare-improving solution, unless appropriate regulatory tools are put in place. Yet, the effectiveness of regulation is inversely proportional to the attractiveness of the industry to the private sector, in particular in situations characterized by higher risks.

In respect of transaction costs, Lobina and Hall (2003) argue that some components of economic risk are present only under PSP, involving for example the risk of varying political commitment and market and regulatory risks. This constitutes a first source of transaction costs, reflected in a higher risk premium required by private investors. Secondly, transaction costs arise from legal, administrative and design requirements necessary to establish competitive tendering procedures. As efficiency gains can only be expected when tenders are repeated regularly, short-term contracts are laid down, in a rather specific and detailed way. A third category of costs flows from the enforcement of the contracts, requiring costly regulatory systems and the costs of legal issues, like litigation, renegotiation and termination of contracts etc. Finally, corruption is seen as a potential source of transaction costs, allowing inefficient choices to be made by the operator.

It has been shown in the sections above, that from an economic perspective, the WSS sector involves many dimensions of market failure, including: the monopolistic structure of the industry causing the exploitation of market power and correlated welfare losses; the presence of externalities and public good dimension, requiring a collective specification of demand and the willingness to pay; the vulnerability of operators to risk, especially in the long run, with the correlated tendency of divesture, loss of asset value, under-investment, insufficient quality. To overcome these problems of coordination, the WSS sector can be organized under a range of alternative *regulatory regimes*, including *competition in the market*, *competition for the market*, *private monopoly with regulation* and *public management*. These regimes involve different combinations of coordinative *tools*, like the discipline of new entry, tariff and quality regulation, competitive tendering, unbundling, incentive regulation, yardstick competition, etc. To a certain extent, these tools are specific for the type of regime involved, but real-world variants of these regimes apply a wide range of specific mixes of these tools. Assessing the economic outcomes of these alternative regimes requires an understanding of the consequences of the regimes for the level and structure of their output.

SCENARIOS AND ECONOMIC IMPLICATIONS

From Regulatory Model to scenario

The relation between the alternative regimes and the outcome in terms of welfare is never straightforward. Three main issues are at stake. Firstly, regulatory regimes, as well as ownership patterns, do not have an absolute impact on the systems' performance. No regime emerges as absolutely superior in theory, while empirical analysis yields both positive and negative examples of all regimes. Secondly, what appears to make a real difference is the quality of regulatory approach and the capacity of policy-makers to implement and maintain appropriate incentive schemes, under each of the regimes. Finally, each regulatory regime has its own critical features and institutional pre-requirements. This is a consequence of the fact that regimes are built upon regulatory tools that address specific market failures but ignore others. Moreover, the several options allocate responsibilities, risks and decision powers to actors in different ways, with the consequent variation in incentive schemes.

The framework presented above is based on a broad range of theoretical and empirical insights in the management of water supply infrastructures, organized around the analytical roadmap in Figure 9. To fit this framework to the scenarios we project manifestations of the four abstract models of governance of this framework (*competition in the market, competition for the market, incentive regulation and traditional regulation)* onto segments of the value chains that constitute the six Euromarket scenarios (See Chapter V). Hence, for each scenario it is established what specific transactions are being governed under which alternative model of governance. The next step involves the definition of a set of relevant economic indicators to characterize the scenarios and to cover the empirical elements of the case studies undertaken. To this end, based on the analysis above, a "check list" has been developed of critical *features*, *issues* and *categories of risk*, specific to each of the four models (see Annex I). This allows us to connect these issues to the several scenario-specific market segments[4].

Table 7: From Regulatory models to scenarios

Scenarios \ Regulatory models	Free competition	Competition for the market	Monopoly with incentive regulation	Traditional monopoly
Delegation A		Long term contracts (I)		Large bulk water systems
Delegation B		Short term contracts (I)		Asset management and development (II)
Outsourcing	Procurement (II)	Design, build, finance and operate of assets and infrastructure (II)	Procurement (II)	
Regulated monopoly	Firm ownership (II)	Procurement (II)	Public Private Partnerships (I) Fully privatised	Asset management and development (II)
Direct public management		Procurement (II) Private partnership (I) Finance (II)		Publicly-owned water companies (I)
Community management	Direct competition (IV)			

I = Operators market: transactions between the WSS operator and entities holding the responsibility for service provision
II= Suppliers market: transactions between the WSS operator and suppliers of inputs along the value chain
III= Water market: transactions between WSS operator and entities holding the property rights on natural resources
IV = Customer market: transactions between WSS operators and final consumers.

The critical *features*, *issues* and *categories of risk* identified vary in appearance and importance per model of governance and per scenario and they are not connected unambiguously to specific stakeholders involved. This makes it difficult to employ these elements directly as a meaningful indicator for performance characteristics of the scenarios. Therefore, these elements have been condensed into three broad dimensions, *Rents and their distribution*, *Investments* and *Supply Characteristics*, involving a limited number of key indicators relevant to specific groups of stakeholders, which form the basis of the scenario assessments below (See Table 8).

The first category, *Rents and their distribution*, deals with the "net dividend" created by improved productivity and the way this dividend is shared among the several stakeholders. The empirical literature is inconclusive in respect to the existence of this net dividend and links it to the effect of the regulatory regimes for specific transactions, typical to the scenario. While believing in a certain potential of competitive regimes to foster some improvement in productivity, these gains can be annulled by overinvestment in capital and transactions costs. Moreover, all empirical studies show quite clearly that regulatory reforms have major distributive impacts, depending on regulatory decisions concerning cost recovery and risk sharing.

The second category, *Investments*, concerns the capacity of the water industry under different scenarios to mobilize the resources required for investment for improving the system and for maintenance or replacement. A pre-requirement for a viable supply system is a continuous accumulation of capital, compensating for technical and financial depreciation of assets. Here, again, we do not expect that scenarios exhibit inherent differences, or that ownership and models of governance matter as such. Instead, we do expect that the economic incentives to invest will be different in the scenarios and that regulation should address specific issues, in order to ensure a similar investment performance.

[4] Because the constructed End States as well as the observed real world regulatory approaches are actually defined and made up as mixes of the regulatory models, it is impossible to make a short cut. Indeed, theoretical insights, as well as empirical analysis, are grounded in these models and not in the compound structures constituting the End States.

The third dimension, *supply characteristics*, involves more or less technical, environmental and market related characteristics along the value chain, as well as the dimensions that may require the development of explicit regulatory tools, like standards and performance indicators.

In the sections below, we will evaluate the economic implications of each of the scenarios, on the basis of the framework outlined and the findings of a number of case studies. For each of the scenarios, a summary will be provided, focusing on the main elements of the particular institutional-economic environment of the scenario and the three dimensions[5].

Table 8: Indicators for assessing economic outcomes of scenarios

Stakeholders \ ***Criteria***	***Rents and their distribution***	***Investments***	***Supply characteristics***
Consumers	Price/ unit Fixed fee Connection Cost of Information		Indicators for water quality, service and security of supply
Next generations	Implicit costs transferred to future generations	Value of assets over time Investments Real depreciation of assets	System resilience Accumulation of knowledge in the WSS
Taxpayers	Net subsidies Public debt reduction	Risks and guarantees still on the public budget	
Environment	Cost of environmental regulation enforcement	Continuous improvement Technological innovation	Indicators of compliance and capacity to improve WSS sustainability over time
Responsible entities	Subsidies Revenues Fees Transaction cost	Size of investments Type of investment	Indicators for water quality, service and security of supply
Private Investors	ROI Dividends Asset value Transaction cost	Size of investments Value / cost of shares Risk charge on investments	
Supply chain	Opex/unit Capex/unit Revenue/unit Wages Transaction cost	Expansion Innovation Productivity	Performance indicators for growth, technology use , environment, labour productivity Cost-recovery

Scenario 1: Delegated contracts and strong regulation

Scenario 1 is characterised by competition *for* the market via compulsory invitations to tender every 10-15 years. Infrastructural assets remain in public hands, whereas their operation is financed and controlled by the private operator. The existence of potential economies of scale, the specificity of technologies employed and potential advantages of scope in overhead activities may drive operators to expand their size, via horizontal and vertical integration of water production, treatment, storage and distribution and sanitation and other services, like waste collection. This is illustrated by the historical dominance of large multi-utility companies, like Suez and Veolia and RWE.

As a consequence there is a tendency towards the development of an oligopoly market structure, also because of the contract length of 10 to 15 years. This context is conducive to information asymmetries and problems of adverse selection. Examples are a lax attitude to reduce costs, high investments and claims to cross-subsidize environmental and other projects. The incumbent operator will know much more about the network than its competitors in tenders. This

[5] The full assessments plus relevant excerpts from the case studies undertaken are presented in Euromarket, 2005.

context requires a powerful regulator independent from both the authorities as well as the operators. The regulator also monitors and controls qualitative aspects of the service, by checking a set of performance indicators.

Fundamental factors are: a) reciprocal trust and commitment of both parties, as conditions change often under a long term relationship; b) a correct information base, shared between municipalities and candidates; c) operators have to secure economic viability vis a vis the regulator, while the authorities have avoid the risk of cost inflation, excessive investment and/or the transfer of rents to the parent company, via internal transactions.

Rents and their distribution

In principle, delegated contracts plus strong regulation will reduce pressures upon the public budget, through private funding and the principle of cost recovery. Public authorities negotiate the contents of contracts, forcing the competitors to reveal their information and striking a trade-off between the rent extraction and efficiency or quality of service. Sometimes subsidies may be required from environmental agencies to facilitate investments in sanitation services.

The operators charge a tariff, negotiated at the beginning of the contract, with the aim to cover all costs of the provision of water, plus reasonable revenue on investments. RPI-X prices are set for relatively long periods and include adjustment rules and indexation formulas, allowing operators to retain a certain amount of the extra revenues flowing from reductions in their costs. This stimulates early and drastic cost reductions, to be harvested by the authorities in the subsequent round of tariff setting, or tendering.

As regards the consumers, normally, the tariff would involve a fixed fee to cover fixed costs and a variable element to account for the volume of water consumed. Direct subsidies by the responsible authority are forbidden, but social tariffs are possible. Disconnection and "self-disconnection" are also prohibited.

This scenario involves fairly high transaction costs for the tendering procedure, including the expenses of local authorities and tendering parties for the design of the terms of references, the bids, the evaluation, the negotiation and the enforcement of contracts between the principal and the agent. The system of regulation also commands significant transaction costs, related to data gathering, monitoring and regulation.

A final element is that collusion between competitors for the contract is socially costly. Indeed, an increasingly larger scale plus a reduced number of competitors commanding the required technologies, provides an environment in which competition among bidders may be reduced considerably.

Investments

The responsibility for investment lies with the operator and depends on contractual arrangements with the authorities. Contracts normally limit investment levels to a certain maximum, leaving responsibilities for unforeseen events in the public sphere or providing for renegotiation. Also responsibilities for maintenance and modernisation are addressed. There length of contracts, on the one hand, may provide an incentive to private investment as it allows for recovery of the investment, while on the other it increases the long-term uncertainty to investors. Provisions on renegotiation and cost pass-through are decisive in ensuring investments, since conditions foreseen in initial contracts are very likely to be changing over time.

Supply characteristics

Generally, it is argued that a good quality of WSS services is assured by the professionalism of operators. Public authorities delegate the operation of services to specialized firms, able to deal with stringent drinking water standards, with the increase in technical complexity of water supply and with waste water treatment. In essence, there is no contradiction between environmental and other quality requirements in respect of supply and the scenario as such, as long as costs can be passed on in adequate tariffs, striking a balance between improvements in quality falling to the operators and the tariffs to consumers. Generally, it is expected that the firm operates more efficiently because of incentives to generate profits and the provision of services with fewer staff.

Final Remarks

The essence of this scenario is that the performance of the firms involved and the trade off between objectives, interests and outcomes are determined very strongly by the quality of the contractual framework and the role of the regulator. Given a sufficient degree of competition, the main challenge is in striking an acceptable balance between several feasible objectives and interest. Whereas, depending on the type of contract, the public authorities may have succeeded in reducing their direct financial and operational involvement, now there is the need to formulate regulatory adequate guidelines that succinctly prescribe the targets of the private firms involved. This happens, firstly, in setting and negotiating the delegation agreement, and later in the process of monitoring and (re)negotiation of terms. This context is extremely vulnerable to strategic behaviour by means of regulatory capture, asymmetric access to information, etc. A prerequisite for this approach is that substantial resources have to be committed to maintaining or developing an adequate knowledge base, to master the development of the sector. Also the often-associated public ownership of assets requires a development of a longer-term perspective for the infrastructure and for the (bulk) supply system. It is obvious that the more complex and demanding the socio/political and natural environment is, the more complicated it will be to formulate adequate and efficient marching orders.

Nevertheless, economies of scale and scope may arise. Both the comparative element in evaluating the individual firms, as well as the development of broad, longer-term oversight over the sector as a whole will tend to be reinforced, to the advantage of the water policy. However, another main requirement is that there has to be enough competition between sufficient amounts of serious contestants.

Scenario 2: Delegated contracts and extreme competition

The theoretical scenario 2 is more extreme than 1, as it involves a tendering procedure every 5 years. Public authorities lose much of their responsibilities, as EU Directives will govern the market. They also lose much of their technological and economic expertise. Upon the basis of a set of minimum requirements, they are obliged to select the least costly contract offered. The fact that contracts are of short duration suggests that markets are self-regulating, as competition arises at the regular pace. There are no real examples of such systems available in the EU context, however.

The consequences of this approach may be twofold. One the one hand, it can be envisaged that more competition provides an easier entry to the market for new firms. Particularly, the support for European operators to engage in these tenders and the opening-up of national markets to foreign firms imply easier entry. The reduced length of the contract, moreover, downplays the informational advantage of the incumbent operator. On the other hand, the enhanced competition and the emphasis on economic efficiency may force firms to submit too ambitious bids, thus enhancing the *ex-post* uncertainty in the transaction. This may either reduce the interest of firms in submitting a bid after a number of unfortunate experiences, or it may drive firms to horizontal and vertical integration of services, to reduce their risk and enhance their resilience. An integration of the several markets in the water sector, i.e. extraction, treatment, transport, distribution and customer services, will reduce competition among bidders (for delegated contracts) and among suppliers (in the context of outsourcing).

Rents and their distribution

In theory, a drastic strategy in promoting competition would lead to a reduction of the public financial involvement in the WSS sector, while private involvement would increase. It is a question, however, to what extent the private initiative would be prepared to engage in actual investments, given the short length of contracts. Nevertheless, competition leads to lower prices, as operation costs are reduced progressively and municipalities are obliged to select the lowest cost bidder. Through their wide involvement in the market, a selected set of operators may acquire superior technological expertise and knowledge on costs, allowing them to out-compete contesters.

In respect of the consumers, the binomial tariff and the principle of cost recovery remain the general rule. A fixed price contract gives strong incentives to reduce the costs of operation and can benefit to users. Yet, it is not only competition through which costs are low, the level of quality of service is moderate, as the lowest cost bidders have been given preference in the tender.

The fact that contracts are of short duration may justify the perspective that markets regulate themselves, via the competition regularly brought about by the tenders. This would imply that there is no *ex post* regulation of the market and there is no independent regulatory authority, apart from the regular environmental checks and balances. So costs of regulation are only limited. The repeated bidding processes for WSS services, however, imply high costs to the authorities, as well as the bidding operators.

Investments

The responsibility for investment, for expansion and renewal, depends on delegation arrangements between the authority and the operator. However, as stated above, the short duration of contracts does not incite private operators to invest. As a consequence, a large part of the investments required are to be taken care of by public entities. Therewith, in a context of reduced public budgets, it may become difficult to undertake expansions and modernization of systems, or to enhance their environmental performance. Within a context of excess capacity and 'gold-plated' facilities, however, the resulting cost cutting may yield an adequate performance.

Supply characteristics

Obviously, the primary objective of this mechanism of delegation is to lower the costs of the service and to reduce prices charged to the users. It is expected that firms operate more efficiently, driven by incentives to profit-making and providing services with fewer staff. A likely implication is that the general quality of service is set at the minimum level. Generally, the EU directives pose a minimum level on quality of water, which is well enforced. The knowledge and experience of operators will allow achieving a minimum quality at low cost. Yet, it remains a fact that the overall quality of service is a concept difficult to define in contracts and difficult to control.

Final Remarks

Under this scenario, the (potential) performance of the firms involved and the trade off between objectives, interests and outcomes tends to short term, low cost solutions either, or an abandonment of the sector all together, when the private risk is considered too high. The implication is that this approach may not be the most effective way of coordinating a developing water sector, or a sector in which large (environmental) problems have to be solved. Basically, for a well kept, stagnating or declining system, it may be the appropriate way of managing it, at low cost for public authorities. Also the tendering process can be rather straightforward and standardized, as low cost is the main criterion. Yet, also here competition has to be a credible threat to bidders. What may get lost, probably, is the longer-term perspective over asset and resource management. Moreover complex socio/political and environmental issues may be suffering. Lack of a real regulatory agency, most likely implies a non-existent institutional memory on the public side, which downplays the possibility of evaluating behaviour.

Scenario 3: Outsourcing

Outsourcing of operations and maintenance tasks to external sub-contractors is becoming a major strategy to WSS firms. Outsourcing may take on a variety of forms and is used by public firms, as well as by private operators and concessionaires. It may involve either competitive public procurement for larger lumps of activity, or direct procurement of specific tasks of a limited dimension. The latter form of outsourcing has become common practice, while outsourcing of larger lumps of activities is gradually becoming more important.

It seems likely that the most interested operators will be those which are not able to achieve economies of scale and scope on their own, as they operate within relatively strict constraints, in respect of their geographical width, the components of the water value chain covered, limitations in their discretion to apply specific techniques or specialist staff, and the segments of consumers serviced.

Most common, throughout public as well as privately operated systems, are service contracts for short periods (from six months to two years) with precisely defined tasks, covering services ranging from the design and building of facilities, installing and/or reading meters, monitoring losses, repairing pipes, and operating and/or maintaining specific stations, and administrative tasks, to the provision of staff cars. But sometimes all encompassing management contracts were sub-contracted too, particularly in the form of development, operation and maintenance contracts. This involved the transfer, or sharing, of the responsibility for the design, construction, operation and maintenance (of a part) of a service or system to a third party, at a fee for performing the tasks agreed for a longer period.

Professional consultancies with a high expertise in these fields - and often with experience in other sectors like energy and telecommunications – are offering their services, in competition with each other. Competitive markets develop for the different types - and bundles - of services. Strong rivalry in the several segments may force sub-contractors to ferociously compete on price and quality. Even larger (multi-utility) operators participated in this competition, by offering their experience, services and unused capacities to colleague operators (see Charles, 2004).

Rents and their distribution

Outsourcing can be an effective way to reduce operational costs, while maintaining ownership and/or control over the water services provision. Operators, or other public authorities, remain in charge of the risk of exploitation, but pay the sub-contractors a fee for their execution of the different activities of the whole service. Main elements of efficiency are: firstly, the recourse to the advantages of scale, scope and specialization of large subcontractors concentrating on their core competencies; secondly, the opportunity of competitive procurement; thirdly, the reduction of staffing costs of the operator; fourthly, the higher efficiency of external workers; finally, it is also argued that outsourcing may enhance cost control, as the firms know the exact price of the services they outsource.

Traditionally, the costs involved with transacting and coordination, including the costs of gathering information on transactions, and for the monitoring and the enforcement of contracts, had been internal to the overall organization of the water sector. In the newly emerging context, however, these costs are explicitly quantified and allocated to the public authorities responsible for coordination, the operators and the tendering sub-contractors. The unbundling of services, according to different activities and tasks, allows for a definition of targets, standardisation of costs for in- and outputs and the establishment of relatively clear technical parameters for comparison.

In respect of the distribution of the rents, outsourcing allows the operators to profit from enhanced efficiencies and innovation, if these advantages can be internalized by sharing the rents with the sub-contractors. Obviously, the degree to which rents are appropriated by either the operators or the contractors depends on the degree of competition among the subcontractors for specific services. This is a function, on the one hand, of competition among sub-contractors and, on the other, the way of tendering. Consultants and sub-contractors show a tendency to concentrate, combining the different knowledge aspects in the field and reducing the number of players available. Rompala and Whitely (1998) signal the development of more advanced forms of contracting, which include incentives related with specific performance standards or the establishment of real joint operational units between the principal and the agent, to share insights and creativity, while reducing opportunism on the side of the subcontractor.

An important cost element, however, are the transaction costs related to the coordination over a number of actors with conflicting interests and to the growth in the number of contracts, each of which with different design. Information asymmetry exists between the tendering operators and the specialized sub-contractors and consultancies. Municipalities or regional enterprises, thus, have a growing need for assistance in relation with the outsourcing process and the further development of related management processes, in respect of technological, economic, legal and other issues.

Investments

In essence, there are no direct relations between outsourcing and the level and type of investments undertaken in the several types of scenarios involved. Nevertheless, an important development may be that large private equity investors may shift their investments to the specialist sub-contracting firms.

Supply characteristics

Water supply and sanitation systems have to adjust their operations, via additional investment in the replacement of groundwater by surface water, in storage facilities and in the management of wastewater. Also the sanitation systems in quite a number or Europe's larger cities require a fundamental overhaul and modernization. Particularly, the technological development and the scale economics of standardization of technologies associated with outsourcing, provides for a reduction of the costs arising from these requirements.

Final Remarks

Outsourcing can take place under many regulatory models. The focus of this scenario is on enhancing efficiency and rationalizing the operational process of the firm. The trade-off between objectives, costs and outcomes are determined by the capability of the management to decide, firstly, what to outsource and what to retain and, secondly, to evaluate and monitor the tenders and the work in a realistic manner. Both requirements are not automatically present in firms and need to be developed. This often catalyzes other processes of changes as well, including shifts in the composition of the staff. Outsourcing, thus, may induce investments in improving the management and quality of the system, by the water companies as well as their subcontractors. This enhances the industry's ability to meet standards and other demands at relatively low cost via specialist sub-contractors. It is obvious that the collection of rents, either by the firm or its consumers, is a function of the further policy setting and the regulatory framework. In respect of the subcontractors, the degree of competition is important. A balance has to be found between too loose relationships, which are intensive in knowledge exchange and transaction cost, and too rigid relations, that lead to a disappearance of the competitive pressure.

Scenario 4: Regulated Monopoly

Regulated Monopoly is centred on operators ranging from private companies to highly autonomous public (municipal or supra-municipal) entities. They are generally organised under private law funded with private, public or mixed capital. Assets are owned by the operators. This scenario assumes the absence of EU liberalisation directives in the water sector. Whereas a range of policy instruments is used, there is a heavy reliance on incentives and informative instruments, including benchmarking.

Benchmarking may take two polar forms. Under pole a), involving high-powered benchmarking applied to monopolies by an independent regulator; operators have management autonomy with respect to appointments, investment decisions and financing. But consumer service levels and budgets, prices and tariffs are largely controlled by the regulator. Under Pole b), medium-powered benchmarking is applied to municipal entities, by decentralised regulation via benchmarking authorities, state agencies or customers' councils. Private involvement exists in the form of share-holding or larger scale outsourcing operations. The autonomy of these operators tends to increase in parallel with their growth in scale, from primarily city-based operators towards larger geographic, multi-municipal or catchment-based areas. The basic drivers behind this process of horizontal and vertical concentration involve economies of scale and scope in operation and management and the specificity of the technology. Nevertheless, smaller operators may be able to efficiently run the local services. So, the firms operate integrated services (e.g. production/treatment, stocking/pressurisation and distribution in the case of water supply) to reduce costs of transactions and coordination. The involvement in other utility sectors is frequent and multi-utilities have become a prominent feature of the water sector of many Member States.

Rents and their distribution

In respect of the evolution of the rents and their distribution, the existence of semi-autonomous private and public monopolies and their regulation has a number of consequences. Firstly, such operation may imply a reduction of the burden on the public budget, as they operate on a cost-recovery basis. Secondly, the fact that they are autonomous, regional monopolies implies that the threat exist of monopolistic rent-seeking behaviour by these firms.

The operators bear the exploitation risks, instead of the responsible public authorities. The contribution of taxpayers is thus reduced to the minimum and, even under a public system; the financing of services is totally private. Operators are allowed to design and charge tariffs that facilitate the recovery of full costs from the consumers. This may include external environmental and resource costs, when it has been agreed politically how these costs can be accounted for. Volumetric charges are the main component of customer bills, which reflect long run marginal costs (LRMC). To ensure that consumers retain control over their consumption and bills, they may be supported to rationalize their use. Disconnection is banned and there are guidelines dealing with payment arrears. Social tariffs can be in place but these have to be mandated by the central governments and incorporated in the general tariff structure. Only poor families, the elderly and people with health problems that require higher water use are eligible for such exemptions. There are no direct subsidies, but indirect (cross) subsidies remain possible, for example for agricultural use and from urban to rural customers.

The establishment of an effective system of regulation implies significant costs. These costs emerge from the duties of the firms under the benchmarking exercises, the monitoring, processing and publication of the collected information and the budgets of the regulatory authorities. Yet, in contrast to systems in which there is a recurrent process of tendering, the transaction costs associated with benchmarking are diminishing over time.

Investments

Long-term monopolies, in principle, may provide operators adequate incentives for asset management and investments. An operator with a longer-term engagement while owning the assets will have an incentive to manage his assets correctly. His interest is to maintain and repair the network in such a way as to avoid higher operating and investment costs at a later stage. Depending on the state of the local systems and the need for specific investments to deal with environmental and hydrological problems, the responsible authorities and regulators may impose new investments upon the operators – with an adequate allowance for cost recovery. Otherwise, only investments in maintenance will be required. If the objective is to cut the costs and a drastic reduction of the water price, operators may be forced to focus on short-term results and neglect the need for maintenance of their installations.

In respect of the sources of funding, both the public and the private operators are becoming interesting objects for risk-averse investors, looking for a longer time commitment, like pension funds. If the economic life of the assets is sufficiently guaranteed, by stabile and adequate regulation, these operators can be financed at relatively low cost, via equity or bond financing. Even private companies heavily rely on bonds.

Supply characteristics

The primary objective of operators is to efficiently supply a vital public service, meeting consumers' needs, the managerial organization, the scale effects and the enhanced professionalism tend to increase the quality of service. Service quality levels are generally good, also as a result of continuous peer-reviewed performance benchmarking. These collective benchmarking procedures induce technology development and implementation in respect of water production and treatment, enabling a continuous efficient improvement of the water quality and a reduction of adverse environmental and hydrological consequences.

An important driver in this respect is the fact that public as well as private operators are able to develop technologies and approaches to improve their services in many aspects. Subsequently, they are able to advocate the newly achievable performance as the sector standard, to be recognized by the regulator and thus included in the permissible level of cost.

This allows them to invest new amounts of capital in sector-wide activities which are profitable, by regulatory definition. Thus they expand their business at a relatively low risk within their regional monopolies. The main question, of course, is when these quality improvements will move beyond the frontier of being reasonable, where they just generate profits to the operators while contributing very little in terms of additional social well-being.

Final Remarks

This scenario is characterized by three sets of drivers that determine its economic performance. Firstly, long-term supply monopoly provides a high degree of certainty and adequate incentives for the management of assets and investments. Secondly, it also implies that potential monopolistic rent-seeking behaviour of operators has to be controlled, either via stringent regulation or by softer benchmarking, seeking a balance between adequate revenues and consumer protection. Thirdly, given the more or less stringent cost-plus style of controlling revenues, more income can be generated only by adding activities to regulatory asset base. This, either, drives horizontal expansion via the addition of (adjacent) supply territories, or vertical expansion, via the development and application of new technologies, recognized as being useful by the regulatory authorities. This may provide economies of scale and scope and incentives to improve the quality performance of the sector, possibly via the internalization of (environmental) external effects. The trade-off between objectives, costs and outcomes, however, will be determined by the capability of the regulatory authorities to decide what minimum service and environmental requirements should apply and at what cost these should be remunerated. The increase in scale of the firms and the associated decline in the number of operators in the market make it increasingly difficult for the regulator, though, to acquire insight in the cost functions of these activities, to set reasonable levels of remuneration.

Scenario 5: Direct Public Management

Under the Direct Public Management (DPM) scenario, the responsible entity also carries out the management, the operation and the financing of water services, including the required investments in the infrastructure. So, public authorities have complete responsibility over the operation of the water system and the relationships with the consumers. There is no independent regulatory authority.

There is no competition in, or for, the customer market, neither is there competition for the water resources. Competition only emerges in respect of public procurement, or outsourcing arrangements for specific infrastructure developments and supplies of goods and services. As municipalities are not specialised in such activities, they tend to outsource them "in" the market for required supplier inputs, or "for" the market by way of contracting out of larger turnkey (design and build) infrastructure projects, specialized high technology, like Geographic Information Systems (GIS) and control of the pipes, Information and Communication Technology (ICT) and finance (see scenario 3 above).

This scenario explicitly acknowledges that consumers express their concern in respect of water services via the political agenda. Indeed, the political clout is seen as the ultimate responsible entity for whatever concerns citizens have about water provision, no matter the organization of the sector. Nevertheless, increasingly a balance is struck with private involvement, particularly as a consequence of budgetary pressures. There is an increasing role for private investors in bonds, while outsourcing of supplies and tasks to private sub-contractors also gains importance.

Rents and their distribution

Direct management implemented by the municipalities is not characterized by a wide segmentation of the domestic users. Most customers are treated in the same way and all have access to the service. The tariff can still be based on a volumetric system, often driven by the objective to rationalize water use, and tariffs for social considerations are perfectly feasible. The general consumers' satisfaction may not be homogenous in all the situations and depends on the quality of service and the tariffs charged.

Operators act as local monopolists with captive consumers. Depending on political decision-making, budgets and principal-agent relationships, these systems may provide a high quality water supply to the entire community, or just a mediocre, sub-standard service.

An important role – especially in terms of principal-agent relationships - may be played by employees, either organized in unions and syndicates, as municipal agents, or as specialists with a very strong local knowledge and a high status.

Many operators are municipal entities but sometimes they are organised at a higher scale, i.e. inter-municipal. The legal and political responsibility will nevertheless often lie with the municipalities involved. Overall, a larger scale of operation has increased economic efficiency. Under pressure of municipal budgetary constraints, together with the objectives of good accountability and transparency, activity based-costing and cost recovery, the municipal systems are being overhauled and modernized, in management and exploitation. This may imply that, in addition to the exploitation, long-term environmental costs are also included in the tariffs. Moreover, most of the direct subsidies will be skipped, except for infrastructure development in isolated and less developed regions.

Outsourcing and tendering of inputs are becoming increasingly important as a means of enhancing the municipal systems' access to technology and operational strategies. Important in this respect is the fact that, as a consequence of labour relations and cost accounting, in-house operations by public suppliers may be substantially more costly. Therewith, private suppliers and consultants appropriate part of the rent, depending on their specific expertise and the structure of the suppliers market. Tendering and monitoring will imply a slight increase in the costs of transaction, as compared with the formerly full public provision of water services.

Investments

A continued role of full public management implies an increase of public investment in the water infrastructure. Particularly as a consequence of the Water Framework Directive (WFD) and delayed investments in expansion and overhaul of the pre-existing systems – both in water supply and sanitation – considerable investments are made. Nevertheless, whereas these investments are undertaken by public operators, these entities rely increasingly on private sources of funding via bonds etc.

An important principal-agent struggle arises between the operators' inclinations to motivate investments associated with the stringency of the environmental and technical standards to be applied, and public bodies' attempts to control cost. It is obvious that when operators are subsidized from the general budget, there will be considerable pressure to reduce costs. When the operators' costs are already recovered on their revenues, the discussion on such investments will take on a broader political dimension, involving elements of sustainability, security and health issues.

Supply characteristics

As stated, most politicians stress security of supply, high coverage of territories and a responsible approach towards the several environmental and social external effects, as main arguments to retain public management. It is generally expected that public operators support these characteristics in aiming for high supply standards. The extent by which they actually succeed in these ambitions is dependent on the financial and policy space they manage to acquire from the political realm. Another important factor is the application of modern (asset) management strategies allowing for specialization, quality monitoring, public participation, and the coordination with other public services and tasks in multi-utility provision.

Final Remarks

Under Direct Public Management, regional bodies keep direct control over the management, the operation and the financing of water services, including the required investments in the infrastructure. Depending on political decision-making, budgets and principal-agent relationships, these systems may provide a high quality water supply to the entire community, or just a mediocre, sub-standard service.

A number of drivers seem important. Firstly, as a consequence of the WFD and in response to lagging maintenance and expansion, considerable investments may be required. As public entities, instead of on their budget, rely increasingly on semi-private sources of funding like pension funds and bonds, part of the risk is transferred to the investors. This requires considerable efforts to account for their performance and the (cost) effectiveness of management practices. Secondly, larger local scales of operation may increase economic efficiency, supported by good accountability and transparency, activity based-costing, full cost recovery and light benchmarking and oversight. Outsourcing and tendering of inputs may contribute to enhancing the municipal systems' access to technology and operational strategies. Another important factor is the application of modern (asset) management strategies allowing for specialization, quality monitoring and the coordination with other public services. Thirdly, generally, politicians and public operators stress security of supply, high coverage of territory and a responsible approach towards the several environmental and social external effects, as main arguments to retain public management.

Scenario 6: Community Management

In the past, forms of Community Management developed where the centralised provision of water and sewerage services to (rural) areas with a low population density was hampered by a lack of potential profitability to private investors, or weak political pressure to develop public services. The technical simplicity of services did induce the local community to engage in the provision of such services, organised by voluntary user co-operatives. In a more advanced stage, water management entities were formed by landowners, private enterprise and/or public corporations, in new peri-urban areas near cities, or tourist resorts. Customers may be directly involved in the decision making process (e.g. rural communities, industrial users), retaining the ownership of the infrastructure and directly controlling investments and the operation of the system, or, alternatively, they are involved only indirectly in the governance of centralised water services, through representation in the boards of the delegated operators.

A wide range of variants can be observed, with all kinds of peculiarities and specific local characteristics. This large variety makes it difficult to draw general conclusions. Community Management, like outsourcing, does not constitute a scenario alternative as such and will not substitute other forms of WSS management. The economic consequences of Community Management, therefore, must be assessed in relationship with other models of WSS management.

Rents and their distribution

Depending on the specific context, an analysis of the rents and their distribution under a Community Management approach will have to consider the position of the consumers, as shareholders/owners, or as stakeholders, operating entities, public bodies involved and private investors and sub-contractors. As a rule, under Community Management the consumers or stakeholders are the owners of the infrastructure. They bear the full exploitation risk of the systems and, unless they can rely upon subsidies, consumers must cover the costs of operation and maintenance (O&M) and investment capital through tariffs. The funding of the system depends upon the communities arrangements for allocating expenses to its members. An efficiency driven system will rely on a volumetric approach with a capacity charge, either based on metering or related to the size of households. More socially driven systems may relate tariffs to property ownership, for example.

The O&M tasks may involve voluntary work, but often operation and maintenance activities are contracted out. Therewith the same problem of collective decision making emerges that can be observed in DPM, namely the principal-agent discrepancy between the stakeholders, as principals, and the agents, that carry out the real work and implementation of decisions made, at an 'adequate' remuneration, of course. Also the discussion on operation and investment cost versus quality standards, etc., is important. Depending on the preferences (*willingness to pay/accept*) and character of the stakeholders, participants and owners, local circumstances, traditions, etc., the balance will be established between acceptable costs and tariffs and quality levels.

Moreover, in daily practice, third-party operators are extensively involved in the O&M tasks. Thus, the contracts and arrangements under which these are carried out are of crucial importance in analysing the consequences for the generation and sharing of rents. Indeed, whereas outsourcing contracts may enhance efficiency by specialization and economies of scale, asymmetric information between the principal and the agent will affect the appropriation of these rents and their distribution. The transaction costs can be lowered through institutional arrangements, like reporting and good accountability, towards members, capital suppliers and public environmental authorities. Trust is essential for effective community management.

Investments

The above structure of interests and involvement, of course, very strongly dominates the objectives and the nature of the investments. Are the investments needed to deliver the quality of the service made timely? What kinds of decisions are made on behalf of asset sustainability, i.e. the capacity of the infrastructure to provide the requested water services in the future? Of course, public authorities influence the investment requirements, through standard setting on health and environmental aspects.

Supply Characteristics

In essence, the community involvement plus national standards of supply may yield a system, the characteristics of which are fully geared towards the needs of the local community. This may guarantee good levels of service, adequate coverage of the area and attention for the longer-term needs and sustainability of the system. This, however, requires an effective institutional design to overcome principal-agent conflicts, plus the ability to efficiently contract out services and supply purchasing.

Final Remarks

A wide range of variants can be observed in Community Management, with all kinds of peculiarities and specific local characteristics. This large variety makes it difficult to draw general conclusions. Consumers or stakeholders are the owners of the infrastructure and bear the full exploitation risk of the systems and, unless they can rely upon subsidies. Consumers must cover the costs of O&M and investment capital through tariffs. The funding of the system depends upon the communities arrangements for allocating expenses to its members. Depending on community decision-making, budgets and principal-agent relationships, these systems may provide a high quality water supply, or just a mediocre, sub-standard service.

CONCLUSION

This chapter has examined the scenarios in respect of their economic implications. An assessment of the economic outcomes of the forms of organization implied in the scenarios shows that these may involve pre-conditions and regulatory solutions that can result in very different outcomes, while local peculiarities and path dependent factors play an important role. A large number of critical factors influence the resulting individual and social welfare, and many of these factors can be accommodated in the several scenarios. Nevertheless, it also has been shown that the scenarios generate specific concerns and regulatory issues that require specific solutions.

The six scenarios encapsulate a range of trade-offs between a five sets of main characteristics, namely: 1) the degree of private involvement; 2) the role of competition; 3) the duration of relationships between authorities and operators; 4) the way in which societal objectives and preferences can be formulated and made operational, and 5) the access to and management of knowledge, insights and capabilities in respect of the supply of drinking water and sanitation services.

Generally, a reduction of operational cost and an increase in productivity are likely to be stronger if there are competitive incentives. Private sector participation is not the primary driver here, the incentive essentially depends on operator's leeway in managing the (use of the) assets and the right to collect the extra revenues generated. In this study, it is suggested that the incentives for an efficient operation are larger in the *Regulated Monopoly* model (especially with high powered regulation via price caps) and lower under *Direct Public Management*. A critical issue regards the number of firms effectively competing on the market. The models involving *Delegated Contracts* have a potential for cost reduction, but are strongly dependent on awarding criteria and contract (re)negotiation patterns. The leeway referred to above, is also a main variable in the grip the public authorities have on the operators performance in terms of environmental and social characteristics of supply.

Another main element is the duration of relationships. The relative short-term perspective of the *Delegated Contracts* models forces public authorities to a greater or lesser extent into asset-ownership, as investors refuse to invest large amounts of capital in fixed assets. The longer term orientation of *regulated monopoly* and *public management* models yields a relatively stable situation for investments, which however demands active monitoring and evaluation to avoid over-investment in useless assets, in turn.

The role of public authorities and the consequent requirements in terms of insight and capacities at the various hierarchical public strata also varies per scenario. The *Delegated Contracts* models force local public authorities into the *ex ante* formulation of tight contract specifications to bring about the envisaged service levels via tender procedures. The tendency of the industry to increase market power and minimize market risks through vertical integration, the shifting of risks and collusion may be compensated by a strong authority balancing the contractual power of responsible entities and/or by the simplification of the objects put out for tender, via short term contracts, individual activities and facilities.

Regarding scenarios dominated by *Regulated Monopoly*, the boundary between private and public is becoming increasingly vague by the appearance of mixed companies and the partial privatization of public firms and the involvement of public firms as bidders in tendering procedures. This development may require a shift in emphasis from local, municipal, public influence towards nationally oriented regulation, either via the light of more stringent variants. Critical issues here concern the capacity of regulators to provide adequate incentives, by balancing between the collection of monopoly rents and securing industry viability.

Quality and environmental objectives are fundamental, but difficult to express in sensible parameters. This requires regulatory mechanisms that provide permanent incentives to quality improvement and an immediate responsibility of the operators. Beyond direct economic incentives, instruments focusing on reputation, voluntary engagement and loyalty may allow significant steps in this direction. Involvement of users through participation, community ownership and/or shared governance of water systems could as well represent an important innovation, reducing the risk of regulatory capture as well as of information asymmetries in the correct specification of demand and quality objectives.

As stated, *Direct Public Management* is facing a challenge in the need to both politically justify its continued existence towards the voters and consumers, and in securing the economic and operational circumstances that allow for enhanced levels of investment, via semi-private sources. This may require considerable improvements in management practices. Larger scales of operation may increase economic efficiency, supported by good accountability and transparency, full cost recovery, light benchmarking and oversight. Outsourcing and tendering of inputs may contribute to enhancing the municipal systems' access to technology and operational strategies. Obviously, if public entities fail to achieve these objectives, there will be a growing pressure towards either a concessionary system or arms' length operation under regulatory overview.

By and large, the fundamentals of the emerging model of the EU water sector seem to involve two sets of key features. The first of these has to do with the financial economic performance, involving financial solidity and credibility of operators' vis-à-vis private capital markets, emphasising rational and regular maintenance and financing of assets and risk management. The second basic element involves the management and acquisition of specialized professional skills, technological capabilities and access to innovation in the international markets.

Whereas these characteristics may be applicable to the future EU water systems in general terms, this chapter has shown that there are fundamental issues at stake in determining the precise role and localization of the main components of system operation and asset and risk management, as a function of local peculiarities. In addition to the physical and spatial features of national systems, these seem to be in the manner in which the different scenarios deal with the availability of knowledge and insights in the several private, public administrative, political and regulatory strata of the system and the use thereof in the several decision-making processes. This latter aspect, certainly, deserves more attention in the discussion as to the appropriateness of either of the scenarios, or regulatory models.

REFERENCES

Antonioli B., M. Filippini (2001), "The use of a variable cost function for the regulation of the Italian water industry", Utilities Policy, 10, 181-187

Bakker, K.J. (2003), "From public to private to... mutual? Restructuring water supply governance in England and Wales", Geoforum, 34, 359-374

Barraqué, B. (1998), Ownership or usership: property rights regimes on water resources in Europe, in Correia, F.N., Eurowater. Institutions of water management in Europe, Balkema, Amsterdam

Barraqué, B. (2003), "Past and future sustainability of water policies in Europe", Natural Resources Forum, 27, 200-211

Bayliss, K., Hall, D. (2001), A response to the World Bank's "Private sector development strategy: issues and options", PSIRU, University of Greenwich, London, October 2001, http://www.servicesforall.org/html/tools/PSIRU_PSD_Bayliss&Hall.pdf

Bayliss, K., Hall, D. (2002), Can risks really be transferred to the private sector? A review of experiences with utility privatization, PSIRU, University of Greenwich, http://www.psiru.org/reports/2002-11-U-risk.doc.

Beecher, J. (2001), Privatization, monopoly and structured competition in the water industry: is there a role for regulation? In Holwarth e Kraemer, eds, Environmental consequences of privatising water services in Germany, Ecologic, Berlin

Berg, Caroline van den (2000), Water Concessions. Who wins, who loses and what to do about it, in Public Policy for the Private Sector, October 2000, note number 217. The Worldbank. Accessible: www.worldbank/Org.html/fdp/notes

Boitani, A., Petretto, A. (2002), Il servizi pubblici locali tra governance locale e regolazione economica, in Robotti L., ed., Competizione e regole nel mercato dei servizi pubblici locali, il Mulino, Bologna

Bressers, H., O'toole, L.J.JR., Richardson, J., eds. (1995), Networks of Water Policy, Frank Cass, London

Brown, R.L. (2004), "Water supply and public-private partnerships", Water Resources Journal, Dec, 12-55

Castro, J.E.(2004), Barriers to and conditions for the involvement of private capital and enterprise in water supply and sanitation in Latin America and Africa: seeking economic, social, and environmental sustainability *(PRINWASS)*, Final Report, (European Commission - 5th Framework Programme, INCO-DEV, Contract ICA4-CT-2001-10041), 2004

Charles, R. (2004), "Economics driving growing outsourcing trend in utilities", Pipeline & Gas Journal, October 2004

Clarke, G., Kosec, K., Wallsten, S. (2004), Has private participation in water and sewerage improved coverage? Empirical evidence from Latin America, World Bank Policy Research Working Paper n. 3445, The World Bank, Washington DC

Cooper, I., Currie, D. (1999), The cost of capital for the UK water sector, Regulation initiative discussion paper series n.28, London Business School

Cowan, S. (1993), "Regulating several market failures; the water industry in England and Wales", Oxford Review of Economic Policy, vol.9 n.4

Cowan, S. (1997), "Competition in the water industry", Oxford review of economic policy, vol.13 n.1, 83-92

Cozzi, G., Massarutto, A. (2002), "Dalla municipalizzata all'impresa pubblica locale", Economia delle fonti di energia e dell'ambiente, 1

Drusiani, R., Fazioli, R., Massarutto, A., Matino, P. (2004), Regolazione e ristrutturazione del servizio idrico italiano, CLUEB, Bologna

Euromarket (2004). Analysis of the Strategies of the Water Supply and Sanitation Operators in Europe. Deliverable 3. Euromarket Project.

Euromarket (2005). Analysis of the Economic Implications of the Scenarios. Deliverable 6. Euromarket Project.

Finger, M., Allouche J. (2001), Water privatisation. Transnational corporations and the re-regulation of the water industry, London & New York: Spon Press

Galiani, S., Gertler, P., Schargrodsky, E. (2005), "Water for Life: The Impact of the Privatization of Water Services on Child Mortality", Journal of Political Economy, 113, 83-120

Garcia, S., Thomas, A. (2003), "Regulation of public utilities under asymmetric information", Environmental and Resource Economics, 26, 145-162

Garcia, S., Moreaux, M., Reynaud, A. (2004), Measuring economies of vertical integration in network industries: an application to the water sector, IDEI Working Paper

Gleick, P. (1998), "The human right to water", Water Policy, 1, 487-503

Gleick, P., Wolff, G., Chalecki, E., Reyes, R. (2002), The new economy of water. Benefits and risks of globalization and privatization of freshwater, Pacific Institute

Green, C. (2003), A Handbook of the Economics of Water, Wiley

Hall, D. (2001), Water in public hands, PSIRU, London

Kraemer, R.A. (1998), Public and private water management in Europe, in Correia (1998), Selected issues in water resources management in Europe, Balkema, Amsterdam

Littlechild, S.C. (1983), Regulation of British Telecommunications Profitability, HMSO, London

Lobina, E., Hall, D. (2003), Problems with private water concessions: a review of experience. On www.psiru.org

Massarutto, A. (1993), Economia del ciclo dell'acqua, FrancoAngeli, Milano

Massarutto, A. (1997), Politiques de l'eau: scenarii pour le XXI siécle, Pôle Sud, 6

Massarutto, A. (2002), La regolazione economica e finanziaria dei servizi ambientali: acqua e igiene urbana, in Robotti, L., a cura di, Competizione e regole nel mercato dei servizi pubblici locali, il Mulino

Massarutto, A. (2003a), "Politica ambientale ed energetica e servizi pubblici locali: dalla pianificazione alla responsabilizzazione", Economia delle fonti di energia e dell'ambiente, n. 1

Massarutto, A. (2003b), I servizi pubblici tra sviluppo locale, regionalismo e liberalizzazione, in Grandinetti, R., Massarutto, A., eds., Servizi pubblici e politiche territoriali, FrancoAngeli, Milano

Massarutto, A. (2005a), An ecological economic approach for evaluating carrying capacity of river basins, presented at Inside the black box – Second international workshop for implementing economic analysis in the WFD, Paris

Massarutto, A. (2005b), Stakeholder-based institutions, subsidiarity and community-based management of water systems: an alternative approach to public participation, HarmoniCOP final conference, Osnabrueck, 4 October

Merrett, S. (1997), Introduction to the economics of water resources, Rowman and Littlefield, Lanham, USA

Mizutani, F., Urakami, T. (2001), "Identifying Network Density and Scale Economies for Japanese Water Supply Organizations", Papers in Regional Science, April 2001, v. 80, iss. 2, 211-30

Muraro, G., Valbonesi, P. (2003), I servizi idrici tra mercato e regole, Carocci, Roma

Ninni, A. (2005), "Mergers and acquisitions in the European public utility sectors: an overview", Economia delle fonti di energia e dell'ambiente, forthcoming

Noll, R., Shirley, M., Cowan, S. (2000), Reforming urban water systems in developing countries, SIEPR Discussion paper 99-32, Stanford Institute for economic policy research, University of Stanford, Stanford Ca.

OECD (2000), Global trends in urban water supply and wastewater financing and management: changing roles for the public and the private sector, CCNM/ENV(2000)36/Final, OECD, Paris (www.oecd.org)

OECD (2004), Competition and regulation in the water sector, DAFFE/COMP/WP2(2004)1, OECD, Paris

OFWAT (2000), Competition Act 1998: application to the water services, http://www.ofwat.gov.uk

Pricewaterhouse&Coopers, Office Internationale de L'Eau, BIPE (2004), MEIF – Evaluation methods for investing in the water sector: forward-looking financial strategies and water pricing, European Commission, 5th Framework Programme, Final report, http://www.meif.org

Prost, T., ed. (1999), R&D and Water Management Systems in a perspective of sustainable development, European Commission, DgResearch, 4th Framework programme, Final report, Bruxelles

Rees, J.A. (1998), "Regulation and private participation in the water and sanitation sector", Natural Resources Forum, Vol. 22 (N. 2), 95-105

Renzetti, S., Dupont, D. (2003), "Ownership and Performance of Water Utilities", Greener Management International, 42 (Summer 2003), 9-19

Rompala, J., Whiteley, D. (1998), "Better Outsourcing: It's All In The Contract", Watermarque, 1.2, November/December 1998 http://www.ib-net.org/asp/resources_benchmarking_publications/ 1_2/rompala.asp

Saal, D.S., Parker, D. (2004). "The comparative impact of privatization and regulation on productivity growth in the English and Welsh water and sewerage industry, 1985-99", International Journal of Regulation and Governance, 4 (2), 139-170

Spulber, N., Sabbaghi, A. (1994), Economics of Water Resources: from regulation to privatization, Kluwer, Amsterdam, NL

Summerton, N. (1998), "The British way in water", Water policy, 1, 45-65

Vaccà, S., ed. (2003), Problemi e prospettive dei servizi locali di pubblica utilità in Italia, FrancoAngeli, Milano

von Hirschhausen, C., Beckers, T., Brenck, A. (2004), "Infrastructure regulation and investment for the long term: an introduction", Utilities Policy, 12, 203-210

The World Bank (1997), Private Sector in Infrastructures, The World Bank, Washington DC, USA

The World Bank (1999), Water. Competition and Regulation. The World Bank Group, Washington D.C.

The World Bank (2005), Approaches to Private Participation in Water Services: A Toolkit, 504

WRc and ECOLOGIC (2002), Study on the application of the competition rules to the water sector in the European Community.

ANNEX

Table 3a: Regulatory regimes: main features

Regulatory model / Main features	*Free competition*	*Competition for the market*	*Monopoly with incentive regulation*	*Direct management*
Main Responsibility	User	Responsible entity	Operator	Responsible entity
Ownership of assets	Companies / users	Responsible entity	Operator	Responsible entity
Risk allocation	Companies / users	Mixed/shared between company and public budget	Operator/customers (depending on pricing rules and cost pass-through)	Responsible entity/public budget
Main coordinative tools	Market	Contracts Tenders	Regulation Legal obligations	Planning Hierarchy
Asset management	Companies	Shared (companies propose and manage, municipalities accept, own and take risk)	Companies	Responsible entity
Criteria for selecting operator	Market	"Beauty contests" Best economic offer Lowest price	Market for corporate ownership Tenders for partners in PPP	Political choice of managers Political definition of management units
Planning horizon	Short	Depends on contract	Long	Long
Price regulation	Market and antitrust	Tender and contract	Incentive (price caps)	Cost-based
Size of market	Individual / community	Municipal	Regional	Regional
Likelihood of WS+S	Very little	Little	Medium	High
Likelihood of horizontal integration and multi-utilities	Little	Little-medium	Medium-high	High

Table 3b: Regulatory regimes and main issues

Regulatory model / Issues	*Free competition*	*Competition for the market*	*Monopoly with incentive regulation*	*Direct management*
Who cares for Pubic Goods and externalities	Environmental regulator Health regulator	Responsible entity	Environmental regulator Health regulator	Responsible entity
Who cares for quality?	Quality regulator	Quality regulator	Quality regulator	Responsible entity
Vertical integration in the suppliers market	Some	Notable	Little	No
Main sources of transactions costs	Coordination of the system	Tender Contract enforcement	Regulatory authorities Benchmarking	Outsourcing Procurement
Scope for outsourcing	High	Limited to some activities to be provided on-site (like meter reading)	Medium-high, depending on the size of the system	Medium-high, depending on the size of the system and budget constraints
Likelihood of productivity gains	High	Medium-high (depending on dimensions in the tender, awarding criteria, discretion of responsible entity)	Medium-high (with price caps and yardstick competition) Low (with RoR regulation)	Medium-low
Incentives for investment	Low	Medium-high (depending on risk sharing, contract length and exit clauses)	Medium-low (depending on degree of cost pass-through	Medium-high (but constrained by public budget availability)
Cost of capital	Medium	High/Very high (depending on risk-sharing)	High	Low (depending on the financial ranking of the country and budget constraints)

Table 3c: Regulatory regimes and main risk determinants for stakeholders

Regulatory model / **Main risk determinants**	***Free competition***	***Competition for the market***	***Monopoly with incentive regulation***	***Direct management***
Main risk determinants for consumers	Service failure Price discrimination Insufficient competition	Contract does not specify critical issues Failure to comply with contract Collusion between regulator and operator High concession fees	Cost pass-through Sudden price increase due to new investment	Low quality due to weak regulation
Main risk determinants for next generations	Underinvestment Environmental degradation	Underinvestment Low attention to resilience in order to keep prices down	Underinvestment Low attention to resilience in order to keep prices down	Underinvestment Environmental degradation
Main risk determinants for taxpayers	Bankruptcy of big operators requiring public intervention	Risk shifted on the public	Underinvestment leads to collapse of network and need for public subsidies	Under pricing
Main risk determinants for responsible entities	-	Incomplete contract Renegotiation clauses Information asymmetry	-	Weak control over managers
Main risk determinants for private operators	Market	Contract obligations Sunk costs Renegotiation clauses	Regulatory obligations Price regulation Unpredicted system's collapse	Unavailability of funds
Main risk determinants for suppliers	Market concentration Market power of operators	Market concentration and market power of operators Vertical integration	Pressure for operational cost saving shifted onto suppliers	Outsourcing leads to de-qualification of workers

Chapter VII

Analysis of the Environmental Implications of the Scenarios

Authors[*] *Nadine Herbke, Britta Pielen, Jessica Ward and R. Andreas Kraemer*

INTRODUCTION

This chapter analyses the environmental implications of each of the six scenarios developed within the EUROMARKET project (see chapter V) on the basis of a set of sustainable water management criteria. Each country and culture may perceive the importance of environmental aspects differently, resulting from varied expectations as regards water. Whatever its cultural role, water provides a multitude of services in the areas of public health, agriculture, environment and biodiversity, transport, and general recreation. Most importantly, water resources provide the basis for our existence and are finite. The Water Framework Directive formulates harmonised goals for all European Union Member States and states, "*Water is not a commercial product like any other but, rather, a heritage which must be protected, defended and treated as such*". The Water Framework Directive sets quality objectives and requires limitations of emissions (see chapter III). Guaranteeing a sustainable use of water resources is therefore a key priority, and this concern lies at the centre of this chapter. It focuses particularly on water protection and pollution prevention measures, i.e. long-term measures requiring little technology but a change in the behaviour impacting water resources (for instance farming practices).

The following paragraphs will present an overview of the methodology used for the analysis. The chapter then moves on to the results of the analysis of each scenario's implications on the protection of water resources and water quality as well as on drinking water related issues. The analysis also focuses, when necessary, on the repercussions on other uses and sectors not directly related to the drinking water and sanitation sector.[1] The synthesis then identifies the most determinant factors across scenarios with potentially great environmental impacts. The conclusion finally addresses the question of whether one scenario can be recommended from an environmental perspective. It should be noted that such an analysis clearly cannot provide for definite predictions of environmental impacts. The assessment of implications consists rather in identifying potential trends.

---

[*] This chapter is based on Deliverable 7 of the EUROMARKET project. This chapter would not have been made possible without the contribution of UNESCO-IHE – Institute for Water Education (Marco Schouten and Martha Garcia). Additional thanks go to two Ecologic Senior Fellow who provided valuable input and assistance: Ingmar von Homeyer on EU policy and Markus Knigge on trade policy.

[1] The industrial sector (chemical, metal and power industry and the food and beverage industry with cooling water and process water uses) and the agricultural sector (irrigation, farming practices).

Methodology

The reference date for the analysis of the scenarios is the year 2020, as referred to in the scenarios described in chapter V. The analysis presented in this chapter is a short version of the analysis conducted for Deliverable 7 of the Euromarket project. The complete study draws on results of the deliverables D1-D5 completed within Euromarket. The analysis is based on principles of sustainable water management and more specifically on indicators derived from these principles. In addition, examples of comparable, actual situations illustrate the environmental implications of the scenarios. These local examples serve an **illustrative purpose** only, since the main focus of this chapter is to analyse the environmental implications of the scenarios at EU level. Finally, we considered available literature on the analysis of environmental implications.

From principles to indicators

The indicators we chose and used to analyse the environmental implications of the six scenarios were developed on the basis of the principles of sustainable water management. These principles are elaborated on in various documents at international and European level.[2] Chapter 18 of the Agenda 21 – the outcome of the 1992 Rio Summit on Sustainable Development – is of particular importance as it lays down objectives for the protection of fresh water resources in terms of quantity and quality, including water pollution and prevention control, as well as for drinking water supplies and sanitation[3]. The study settled on the following set of key principles of sustainable water management:[4]

1) The **precautionary principle** implies that in order to protect the environment, a precautionary approach should be widely applied, meaning that threats of serious or irreversible damage to the environment or human health, and unknown risk must be avoided. Lack of full scientific certainty should not be used as a reason for postponing cost-effective measures to prevent environmental degradation.
2) The **prevention principle** requires action to be taken to protect the aquatic environment at an early stage. It is not a question of repairing damages after they have occurred, but rather of preventing those damages from occurring at all. This principle is however not as far-reaching as the precautionary principle.
3) The **source (reduction) principle** states that environmental damage / emissions of pollutants should preferably be prevented or reduced at the point of source, rather than by using 'end-of-pipe' technologies.
4) The **integration principle** stipulates, according to Article 6 of the EC Treaty, that environmental protection requirements must be integrated into the definition and implementation of EU policies and activities referred to in Article 3 (including, among others, agricultural and fishery policies), in particular with a view to promoting sustainable development.
5) The **resource minimisation principle** states that the direct and indirect consumption of resources and energy is to be continually reduced or should be minimal. Water uses should shift away from non-renewable to renewable resources.
6) The **regionality principle** and the principle of local provision imply that regional resources and habitats are to be protected and spatial environmental externalities avoided. Each region[5] should as far as possible resolve its water related problems while making use of its own resources.
7) The **reversibility principle** states that water management measures must be modifiable, and their results must be reversible.
8) The **polluter-pays / user pays principle** states that expenses resulting from pollution and the use of resources are to be charged to the polluter / user.
9) The **co-operation and participation principle** captures the issues of democratic control and local co-determination and states that all actors concerned or affected by a decision have to be consulted and involved into the decision-making process.

---

[2] Among others the Agenda 21, the EC Treaty, the Water Framework Directive, the Drinking Water Directive, the 6th Environmental Action programme.

[3] The chapter describes seven different programme areas: (1) Integrated planning and management of water resources, (2) Assessing water supply, (3) Protecting water resources, water quality and aquatic ecosystems, (4) Drinking water supplies and sanitation, (5) Water and sustainable urban development, (6) Water for sustainable food production and rural development, and (7) Impacts of climate change on water resource.

[4] Adapted from Kraemer and Kahlenborn (1998), and Kraemer (1998); precautionary principle and prevention principle from EEA Glossary.

[5] The term "region" does not refer to administrative boundaries but rather to a hydrological region, which can encompass, for instance, a sub-basin.

10) The **intergeneration principle** stipulates that the timeframe of observation with regard to water management plans and decisions must correspond to the timeframe of effects. Therefore, the interests of future generations have to be taken into consideration in the decision-making process.

These principles provide a basis for the study's understanding of sustainable water management but are in themselves not sufficiently concrete to be operationalised. We therefore developed a set of indicators on their basis. The indicators focus on elements of a sustainable water cycle specific to the protection of water resources and quality as well as to sustainable drinking water supplies. They can be understood as questions posed to assess the scenario's impacts:

Protection of water resources and quality: Is the water supply realised at local level, i.e. are the resources and customers in the same hydrogeological basin? Does the scenario leave room or provide incentive for the realisation of voluntary water protection measures, such as contractual agreements on land-use? Which incentives does the scenario set regarding the abstraction of water resources, in quantitative terms?

Sustainable drinking water supply: Which incentives does the scenario set as regards the quantity of water consumed? Does the scenario encourage a wide application of the minimisation principle, i.e. a minimal use of chemicals within treatment processes? Can the quality and the security of the drinking water supply be ensured? What incentives does the scenario set as regards investments in, and upkeep of, infrastructure, especially to reduce leakage and water losses?

The following matrix provides an overview of the relation between principles and indicators.

Criteria	(i) Protection of water resources and quality				(ii) Sustainable drinking water supply				
Indicators / Principles	Local water supply	Water source protection	Voluntary environmental protection	Quantity of water resources used	Quantity of water consumed	Application of minimisation principle	Quality and security of drinking water	Quality of infrastructure	Reduction of leakage and water losses
Precautionary		x				x	x	x	x
Prevention		x	x			x	x		
Source reduction		x	x						
Integration		x	x					x	
Resource minimisation				x	x	x		x	x
Local provision	x								
Reversibility		x							
Co-operation & participation			x	x					
Inter-generation		x		x	x			x	x

Now that the methodological framework of the analysis has been set, we can proceed to the analysis of the six scenarios, namely the two versions of Delegation Contracts, Outsourcing, Regulated Monopolies, Direct Public Management, and Community Management.

DELEGATION CONTRACTS

There are actually two Delegation Contracts scenarios. The main difference between them is the rhythm at which bidding procedures must take place. Scenario 1 sees these procedures as taking place every ten to fifteen years, whereas scenario 2 presents an extreme situation wherein open tenders must be introduced every five years. This analysis focuses primarily on scenario 1. It is expected that the impacts identified under this scenario will potentially occur in a more extreme form under scenario 2.

As already mentioned, the central aspect of the scenario 1 is the relatively short planning horizon (10 to 15 years) given to concessionaires to recover their investments in the infrastructure expansion and maintenance, and to affermage contractors to receive a revenue on the funds provided for routine repair and replacements of short-lived components. This short contract duration reduces the water providers' incentive to include in their budgets aspects whose benefits unfold within a much larger timeframe. The potential neglect of these long-term aspects can have many intertwined implications with reinforcing effects.

A first aspect that presents advantages only in the long term is the implementation of precautionary measures to protect the resource. These often take the form of contractual agreements with farmers or land-users regarding the implementation of best practices on soils either surrounding a source or covering a source-relevant area. Best practices do not automatically lead to an improvement of the water quality but can do in a significant number of cases, if the type of soil permits it. If they are not put into practice, the chance is high that the quality of the water resource will deteriorate. In a self-reinforcing effect, poor water quality means that more time is needed for land-use change to have an effect and reverse the trend, thus reducing even further the incentive for water providers to consider such measures.

In this context, end of pipe solutions and technological or chemical treatments take on a growing importance, as shown by the Lyonnaise des Eaux example (Box 1). The short term planning horizon incites water providers to diversify the ways in which to increase their profits, and selling certain technologies to the municipality or including their use in contracts offer such an opportunity. This allows for a much more direct return on investment than precautionary measures do. However, and this is particularly true when chemicals are used, such treatments can have a negative impact on the water quality. The incentive to move away from precautionary measures and implement post-abstraction treatment thus can have a double impact on the quality of the resource.

Box 1: Lyonnaise des Eaux, Brittany: Nitrate polluted drinking water – Who is responsible?

Brittany is a region characterised with intensive agriculture. With some eight million pigs and over two million cows, it ranks among the first French bovine and porcine producing regions. (Cour des Comptes, 2002)

When Lyonnaise des Eaux started providing the town of Guinguamp (in the northern part of Brittany) with water in 1987, the nitrate values were well below the present legal threshold (AMF, 2001). Starting in 1990 however, these values gradually increased until eventually, in 1995, Collectif Eau Pure, an association gathering 176 water consumers, launched a boycott and refused to pay the water bills. Over the previous years, their drinking water had presented unacceptably high nitrate values nearly two days out of three (Bauby and Lupton, 2004).

These water consumers resorted to legal means and successfully prosecuted Lyonnaise des Eaux. The water provider then lodged a complaint against the French State for not having implemented the 1991 European Nitrates Directive. The State was convicted in 2001 and made to pay an important fine to the company for its image prejudice (Drobenko, 2001).

An interesting aspect of the situation is the incentive it created for the contracted water company. Indeed, to solve the drinking water quality problems, Lyonnaise des Eaux offered, unsuccessfully, to sell a new technology to the municipality of Guingamp (Bourblanc, 2005).

Not only the quality, but also the quantity of the resource may be affected by the treatment. Indeed, treating poor quality water can lead to rising water abstraction to ensure that sufficient amounts of drinkable water are produced. Hence, both the quality and the quantity of the water resource are put at risk by the incentives given by a short-term planning horizon. In extreme cases, the source can even be closed and abandoned for another, possibly extra-regional one.

The investment in the infrastructure maintenance and extension (the latter only in case of concession contracts), a second aspect requiring long-term interests and planning, may also suffer under the arrangements described in this scenario. The lack of investment in infrastructure becomes visibly problematic only after several years, perhaps even decades, of neglect. Even before that point however, water losses through leakage will increase. To continue providing sufficient amounts of water to the consumers, water abstraction rates will therefore have to increase, thus adding pressures on the source. At the other end of the service, wastewater pipes may also leak, leading first off to pollution problems, but also to rain water coming into wastewater pipes. This dilutes the wastewater, which as a result cannot be treated as effectively, creating as it is released further environmental problems.

In short, the strains on the water resource both in terms of quantity and quality can be numerous and in part self-reinforcing. This naturally has an impact on other sectors, such as the farming and the beverage industry.

However, the ex-post regulation provided by an independent authority could alleviate the expected implications of the delegation management on the environment. The monitoring of the service quality through performance indicators can potentially influence the companies' objectives towards more environmental efficiency of the drinking water and wastewater processes. This depends largely on the selection and definition of the indicators used and the competence of the related authorities.

OUTSOURCING

The Outsourcing scenario envisages that in the year 2020 the European water and sanitation sector will be featuring a large amount of involvement of external parties to perform specialised tasks on contract basis. As the contract defines the terms on which the designated tasks are executed, it plays a central role in the analysis of the scenario's implications on sustainable water management. We therefore structure our analysis of the environmental implications and their interdependencies according to three sequential contracting stages, being the implications that are related to the pre-contracting stage, to the contract design stage and to the contract implementation stage.

Pre-contracting stage: The first stage of contracting relates to the pre-contracting stage when a competitive tendering phase is carried out to select, ideally, the party that proposes the highest value for money. Assuming that the importance of sustainable water management is recognised by the contracting party, and that it has been properly addressed in the terms of reference for the bidder, this competitive process might indirectly lead to positive implications on the quality and security of drinking water supplied as the selected bidder will theoretically be the one that has addressed the environmental requirements (along with other issues) best.

Contract design phase: Once the bidder is selected, the contract is drafted. Designing the contract requires that future risks be identified and addressed in the contract. It will not be possible in the case of long-term contracts to predict all future risks. The risk in this case is that, if something unforeseen happens, the contracted party chooses to solve the matter in an unsustainable manner. Due to this emergence of incomplete information at the moment of contract design, a large number of indicators of sustainable water management might be negatively implicated, such as the realisation of voluntary environmental protection measures, local water supply, quantity of water used, quality of infrastructure, and the quality and security of drinking water supplied. In case the contracting party tries to mitigate these problems of incomplete information by insisting on the establishment of a proper monitoring system, the above mentioned negative implications might be largely counterbalanced. The contract will in such cases force the contracting parties to clearly state the contract objectives and how these are monitored via environmental indicators. The creation of such a monitoring system of environmental indicators can have a positive impact on the protection of resources.

Contract implementation phase: Once signed, the contract is implemented. From then on, the environmental implications can be either positive or negative, depending on the local situation. The first element that determines if the contract implementation is environmentally sustainable is the success of the previous phases. The selection of the right bidder and a proper contract design will facilitate the emergence of positive environmental implications. But aside from the implications from these previous phases, the degree of environmental sustainability of the outsourcing contract will also depend on the extent to which the partners are able to build a sound relationship. A relationship with a relatively high level of trust will enable both contract partners to attune their actions to unwritten demands. Such level of understanding and trust will likely have a positive impact on the realisation of voluntary environmental protection measures. In this respect long-term contracts make it more likely than short-term contracts that such a level of trust between the partners establishes itself. It needs to be noted however that if the level of trust between the two partners is too extreme, it might in turn have a negative impact on the environment. In such a situation, the dependence of the contracting party on the contracted party might become too large, since the contracting party distances itself from the day-to-day operations. What is more, the contracted party might be tempted to abuse his freedom of operations, possibly resulting in paying less attention to long-term, sustainable, yet (in the short-term) more expensive solutions. Moreover, if the contract is long, the contracting party might also find out too late that the contracted party has managed the resources in his charge in an environmentally unsustainable manner. This would provide an increasing possibility that irreversible negative implications arise on the protection of water sources and quality.

The Outsourcing scenario solely envisaged the outsourcing of exclusively non-core activities. In this respect, the scenario as such is of limited relevance for the environment. The environmental impacts rather depend from the general context in which outsourcing takes place, as both public and private entities in different regulation systems can choose to outsource part of their activities.

REGULATED MONOPOLIES

The analysis of the environmental implications focuses on the first pole of the scenario in which private water monopolies are subjected to high-powered benchmarking with centralised regulation. The implications of the regulated monopolies scenario depend to a great extent on the regulative system and the indicators used for setting prices and regulative investments. As a result, the analysis could not identify obvious environmental impacts but only point to potential risks depending on the options chosen for regulation. At the forefront of these determinant options lies the definition of the economic level of leakage. Indeed, the economic level of leakage can be set in a number of ways and cover a range of aspects related either exclusively to the water provider and its budget, or based on a more holistic approach and including costs carried by society as a whole, such as environmental damages. In the first case, companies have a minimal incentive to realise extensive maintenance and related investment actions in the pipe system in order to reduce the water losses beyond the defined economic level of leakage. In the second, the calculation of environmental costs has to strike a fine balance between the needs of society as a whole and the ones of the company to remain functional and competitive. This approach presents the advantage of taking into account extensive leakage compromising the availability of potentially rare drinking water resources. Thames Water presents an example of the far-reaching impacts of leakage (see box 2).

Box 2: Thames Water – Leakage problems

Thames Water provides drinking water and sewage services to London and across the Thames Valley. With water losses representing a quarter of all leakage in England and Wales, the private company has the highest leakage rate of all the English and Welsh water companies and has been a cause of increasing concern for the water company regulator Ofwat. Indeed, the company loses 28% of the water it puts into supplies (Minting, 2000; Webb, 2003). The specific context of London presents many challenges to address the problem of leakage. The dense urban structure renders repairing mains difficult and costly. Replacing pipes in London, for instance, leads to disruptions. To avoid closing streets to traffic too often, co-ordination with road repair works would be necessary but is not always possible. In addition, urban expansion, a growing number of tall buildings (requiring high water pressure) and increasing water consumption exert further pressure on the infrastructure (Levett, 2002). In parallel to the slow renewal of its infrastructure, Thames Water proposed the building of a reservoir in Abingdon which would cover an area of about 10 km2. Such a reservoir would increase Thames' supplies by 5 to 10% and would not be running before ten or fifteen years after construction starts (Minting, 2000). The construction of the reservoir would have great environmental impacts. Among them: traffic increase during the construction phase, changes of groundwater levels and flow during the two-year period needed to fill the reservoir, loss of agricultural land and habitats, and the creation of new habitats to name but a few (Levett, 2002). Interestingly, faced with increasing water demands, Thames finds the reservoir solution more economical than the replacement and maintenance of its infrastructure, or the use of demand management measures such as metering (Levett, 2002; Minting, 2000).

A second important question is whether the regulative framework extensively considers precautionary measures for the protection of water resources and encourages water providers to seek out (co-operative) agreements with land-users or other agri-environmental measures to reduce water pollution. If this is not the case, the companies could fail to include these measures in their annual investment programme and it is unlikely that land-users will apply stricter standards than foreseen by national and European law. Instead, water providers could favour the implementation of (post) treatment technologies in order to meet quality standards. Blending, a process by which high quality water is mixed with water of insufficient quality to reduce the concentration of unacceptable pollutants or chemicals before it is distributed for general consumption is one such option. It induces high abstraction rates. In addition, diffuse water pollution would continue to increase under such conditions. This would imply higher costs among others for the industries dependent on water quality, such as the beverage industry, and for farmers, who rely on abundant, clean water resources. Finally, whether the regulation authority foresees a wider use of water meters will also induce varying degrees of water abstraction.

DIRECT PUBLIC MANAGEMENT

The Direct Public Management scenario presents various driving forces that pull, as regards the environment, in opposite directions. Some could lead to a sustained commitment to water and environmental protection while others could result in environmental degradations. In direct public management, the town-governing body, and therefore, indirectly, the town's citizens, plan and make their own decisions regarding, among other, water management. In addition, municipal representatives will tend to live in the area where they have political responsibility or were elected. They therefore have a vested interest in finding long-term solutions to secure the town's water provision. The long planning horizon provided by municipal ownership of the water provision infrastructure offers a favourable framework for such a far-sighted approach. Precautionary measures, including co-operation with land-users, can be more easily envisaged than in a context where returns on investment must be realised rapidly, as is the case for private firms (see box 3).

Box 3: Munich Stadtwerke – Voluntary environmental measures

The Munich Stadtwerke (SWM) are entirely owned by the municipality of Munich and supply the city and its surrounding region with water as well as energy and public transportation.
The Munich drinking water is abstracted from three areas that are predominantly used for agricultural purposes. To secure the water quality, the SWM has been consistently buying plots of lands in the catchment areas. These plots are then either reforested or leased under very strict conditions. Despite these measures, water pollution continued to increase due to extensive agriculture. Although the pollutants stayed well below the legal limits, the SWM decided to act preventively and launch a co-operation programme with farmers.
Now, some hundred farmers take part in this programme and apply strict organic farming methods on 2.500 km2 around the main water abstraction area of Mangfalltal, thus improving the quality of water, meat and produce. SWM supports the farmers with a little over 230 Euro per hectare, which for the consumer translates in a price increase of half a Euro cent per cubic meter of water.

Accountability is a further aspect encouraging a long-sighted approach to water management. Water quality, together with its price, belong to the issues that come into play when citizens elect their representatives, and this can give an incentive to provide high quality water at an acceptable price. However, the impacts of water management decisions and water protection measures become evident many years later, so that the original decision-makers most probably cannot be held accountable once the damage is apparent. In short, vested interest and commitment of elected and/or responsible officials, ownership structure allowing for a long planning horizon and, to a certain extent, democratic accountability all seem to provide incentives for a sustainable, long-sighted water management. The Vienna Waterworks case provides an example of how collective interests for high quality water can translate into an official long-term strategy, as the declaration of the Vienna Water Charter (see Box 4).

Box 4: Vienna Waterworks – Source protection

The Vienna Waterworks is part of the Vienna municipal government, and provides the population of Vienna (just over 1.7 million people) with water originating to 90% from the surrounding mountains. Some 600 km2 around the sources are designated as source protection zones, 50% of which are owned by the city of Vienna.
In 2001, the city of Vienna gave water constitutional protection and further strengthened this focus on safe water with the Vienna Water Charter. This Charter emphasises the aspects of quality and sustainability of water protection, provision and use over financial profits.
In addition to the waterworks, another division of the city administration plays a central role in implementing key points of the charter: the office for forestry and agriculture. Indeed, this office is responsible for the sustainable care of the forests and arable land covering the source protection areas, and works in close co-operation with the waterworks. All soil uses, such as forestry, tourism, hunting and fishing can thus be co-ordinated with the needs of water protection. The water is filtered through ecologically balanced forests and reaches the consumer without any treatment other than a disinfection with a minimal dose of chlorine dioxide.

Further driving forces, however, clash with this disposition. Geographic factors can severely limit a municipality's willingness to implement long-term water source protection measures, if the land it can protect within its boundaries filters water used by another community, for instance. In this case, unequal repartition of costs and benefits between communities could mitigate the readiness for adopting a long-sighted approach. In addition, and as in any democratic society, a municipality consists of citizens and groups with conflicting interests. Any decision regarding choices of investments, land development and land use has to strike a balance between these interests. What is more, strong power groups such as agricultural lobbies can wield political clout to impede or exert influence on the implementation of measures they are unfavourable to, for instance the setting up of best-practices programmes related to land use around water sources.

COMMUNITY MANAGEMENT

In the year 2020 of the Community Management scenario, users are frequently involved as owners or decision makers in their local water sector. This increased public participation can have various positive effects. It can for instance contribute to a better communication between different water-user groups, such as households, industrial users and farmers, and potentially help towards a co-operative balancing of various needs and demands with the actual resources. Community involvement, direct democratic control and local co-determination can lead to a higher awareness of pollution problems, and a greater acceptance of voluntary environmental protection measures, including changes in land-use practices, especially as participants and decision-makers have a vested interest, as users, consumers and local inhabitants, in long-term clean water resources. It is also conceivable that community management results in a high focus on the local or regional scale to organise water supplies, which is an important aspect of sustainable water provision. The German Water Associations provide an example of the advantages of user-participation (Box 5).

Box 5: German Water Associations – Benefits of involving the community

An estimated 1,200 Water Associations in Germany are active in water supply, including the supply for agricultural use. These Water Associations are self-governing consortia of property-owners, companies, public-law corporations and other interested parties for land and water management purposes, sometimes covering large land areas. Members control the association according to democratic principles. As relevant for our analysis of possible environmental implication, the functional self-government in Water Associations has proved to be effective in several aspects relevant directly or indirectly to the analysis of environmental implications. One is that the involvement of community members results in a high degree of political legitimacy. Secondly, the structure of water associations allows users to contribute their knowledge of local conditions and their potential co-operation in the spirit of the principle of shared responsibility. Finally, small-scale structures give meaning to the otherwise abstract term 'public benefit' in the eyes of those responsible: the 'public' is the population of a village, town or city (Kraemer and Jäger, 1995).

A more negative aspect of the involvement of users in the management of water services can result from the potentially insufficient qualified expertise necessary to manage the service provision adequately on a sustainable environmental basis. In particular the longer-term implications of current action might be overlooked in view of the higher level of complexity and in-depth understanding that it requires. Since the users are themselves involved in the decision-making, the community might for instance put too much emphasis on satisfying immediate demands of users for sufficient quantity and good quality drinking water at an affordable price, and possibly overlook the long-term consequences implied by the decision. This could be particularly true as regards investments in infrastructure or the decision on the type of treatment necessary for drinking water. Short-term focused decisions could also result from the various distribution of political clout of water-user groups within the decision-making bodies. Farmers and industrial users would benefit from their higher professionalism and political organisation, as well as their high local economic importance, and thus potentially more or less successfully impose decisions favourable to them. This could be tampered by the fact that negative long-term environmental implications would also affect the agricultural and industrial sectors. If farmers do not have sufficient access to surface water to irrigate their lands due to water pollution or scarcity, their crop yields will diminish. This problem might even be made worse if groundwater tables are lowered as this increases the dependence of the farmers on surface water to irrigate their lands.

SYNTHESIS

This section provides a synthesis of the analysis of the six scenarios. The analysis identified a certain number of determinant factors with potentially great environmental impacts. These are *inter alia* the length of the contract, the regulative framework in which companies operate and the level of monitoring.

Length of the contract: At least two scenarios give companies a limited time-frame to plan and invest: In the Delegation Contract scenario, bidding procedures take place every 10 to 15 years, in a second version of this scenario even every five years[6], and though the Outsourcing scenario does not specify any time frame, contract duration could vary greatly from short to long. This factor has a considerable influence on companies' decisions. Indeed, the water sector is characterised by the fact that it requires important investments in infrastructure and its maintenance and by a cost structure with 80% fixed. The investments do not translate into benefits before a long period of time (sometimes as long as 50 to 100 years) and the rates of return will tend to remain low. This means that companies with a short contract will have little interest in long-term investments in infrastructure maintenance and extension (the latter only in the case of concession contracts where the operator is responsible for this task), especially if they intensively look for cost reduction opportunities. A short contract linked with investment responsibilities therefore leads to a reduction in infrastructure quality, especially after a long period without appropriate maintenance. It can cause economic problems when the contract expires if the infrastructure has become so bad that the municipality needs to step in. Furthermore, water companies working within a short time-frame will most probably neglect precautionary measures, such as agreements with farmers to promote less polluting land-use and keep nitrate rates low in the long term, to favour end-of-pipe solutions, such as technological treatment of nitrate-polluted water.

The direct public management and the Community Management scenarios on the other hand offer a quasi-unlimited time frame in which to plan. Implementing precautionary measures then becomes economically sensible, despite the fact that the benefits of such measures appear only after several years, provided that the municipality has appropriate funds for the investment. However, political issues take on an important role in these scenarios. Indeed, the municipality or the decision-making body will need to achieve a balance between various interests and priorities in a transparent

[6] See Euromarket Deliverable 5, Scenario 2. The analysis of this scenario was left out of the present chapter for the sake of clarity.

manner in accordance with the principles of good governance. Urban, industrial or agricultural development needs to be implemented in parallel with an appropriate level of water protection.

Regulative framework: The indicators used for setting the prices and regulating investments in the Delegation Contract and the Regulated Monopoly scenarios will determine the environmental implications that are to be expected. Indeed, the regulative framework defines which investments can be recovered, for instance through an increase in consumer prices. If this framework does not leave sufficient flexibility for infrastructure maintenance or does not consider precautionary measures, the water providers will have little incentive to include these measures in their annual investment programmes. This is especially the case if the regulative system is based on the so-called economic level of leakage. The economic level of leakage is usually defined as the level at which further leakage reduction costs the water supplier more than to produce water from another source. In this context, water operators have no incentive to invest in infrastructure to reduce the leakage level below economic level of leakage. This increases the tendency to exploit alternative sources, potentially leading to a higher use of chemicals and energy, both being required for the production (and transport) of drinking water. The economic level of leakage can however be defined in an holistic way and also include environmental and resource costs as well as costs born by society as a whole.

Level of Monitoring: Finally, the level of monitoring water providers is subjected to can have considerable consequences. The analysis of the Delegated Management and Outsourcing scenarios found that water providers could potentially keep certain information undisclosed until the end of their contract, possibly leading to irreversible environmental damages. A monitoring that includes criteria of security of supply, drinking water quality and water resource protection could ensure that contractors do not take advantage of their freedom of operation to the detriment of water resources.

As regards the repercussions on other sectors, one should particularly focus on the agricultural sector. Indeed, agriculture plays a double role in the water sector: On the one hand, it benefits from and needs abundant clean water resources at low costs. On the other, it belongs to the users with the greatest impact on water resources through high abstraction and the use of such potentially polluting substances as fertilisers and pesticides. A balance needs to be found between the sector's long and short-term interests, keeping in mind the difficulty of depolluting water resources once the damage is done.

CONCLUSION

We can now turn to the question of whether one particular scenario presents clear advantages from an environmental perspective. One of the key aspects concerns the incentives that the scenarios' institutional and management framework sets in terms of sustainable water management. The assessment of the implications provides directions and trends for each scenario, and on this basis we can say that the Delegation Contract scenario does not provide, in itself, adequate incentives for the implementation of long-term oriented water protection and infrastructure maintenance measures. The impacts naturally depend on the contracts and on whether environmental standards are part of it. The procurement process and especially the key points of the contract defined between both parties play an important role in order to secure an adequate investment in water protection measures and water infrastructure modernisation over the whole contract period. However, the complexity of environmental aspects renders the inclusion of such element in a contract difficult. Flexibility is therefore necessary, and could be achieved through the periodic adaptation of the service contract.

The Direct Public Management and Community Management scenarios offer a higher incentive for the realisation of precautionary environmental measures in the long term provided that the municipality plans to carry out the water supply and sanitation services over a long planning horizon. Continuous information of the municipalities is a key aspect to prepare them for the challenges of water management issues such as new environmental and quality standards and setting political priority under tight public funds. The latter is a problem that municipalities are increasingly confronted with.

A long contract duration in itself however does not suffice to provide incentives to implement sustainable water management measures. Monitoring, or some form of supervision, is an essential element for both the delegated monopolies and the outsourcing scenarios to ensure that information asymmetries between the water or services provider and the contracting municipality remain minimal. Only thus can potential water-related environmental problems theoretically be acted upon before they become irreversible. Specific targets laid out in the contract may help to influence the contractor's behaviour. The authority responsible for the supervision of the water company must therefore still be able to hark back to professional and well-informed staff in order to act on the same level. Potentially strong political influence of organised, professional groups such as farmers or industrial users should also be addressed. Transparency at all stages of the decision-making process is necessary to increase the fairness of decisions and a balancing of interests.

One should also address the fact that the implementation of best practices regarding land-use involves payments to or financial compensations for the land-users, and this may seem to go contrary to the polluter-pays principle laid out in the Water Framework Directive. However, these financial compensations would generally not be provided by the State but by the water users. In addition, voluntary co-operation agreements have so far proved to be the most effective solution to diffuse sources of pollution, especially coming from the agricultural sector. These agreements can take a wide range of forms and are particularly easy to implement in small basins, where the relation between land-use and pollution is easier to demonstrate.[7] Their effectiveness may however vary according to, among other, hydro-geological factors.

As the few illustrative examples presented in this chapter hinted, the provision of water services in the EU takes place in very different contexts. Vienna for instance benefits from an extraordinary geographic situation and receives its water mainly from mountains, thus ensuring a high quality at a comparatively low cost. Thames Water on the other hand provides water in Europe's most populous city, a factor that weighs heavily on the infrastructure. No single system will possibly take account of such varied situations. Similarly, the legal and institutional systems for water supply and sanitation services vary widely between the EU Member States. To take account of all these differences, the responsible authority at local level must keep its ability to choose between different management systems including mixed management forms such as publicly owned undertakings under private law and public-law corporations.

In short, it is not possible to derive overall environmental implications from the scenario analysis and thus favour one single scenario from an environmental perspective, especially since some scenarios can exist in parallel (e.g. outsourcing can take place regardless of the management form). Despite the overall institutional and legislative framework, the length of the planning horizon defined by the responsible authority is one key factor. The authority has to identify first a clear view of how to guarantee a safe and sustainable water supply and sanitation before deciding on the management framework of the services.

REFERENCES

AMF. Association des Maires de France (2001) Pollution de l'eau potable dans la région de Guingamp (Côtes-d'Armor): l'Etat condamné pour n'avoir pas empêché les agriculteurs de polluer l'eau potable aux nitrates. 3 May. [www.maire-info.com/articles/archives.asp?param=525]

Bauby, Pierre and Sylvie Lupton (2004) Country Report France. in Euromarket. Work Package 4 (Phase 2): Analysis of the legislation and emerging regulation at the EU country level. Deliverable 4. 21-64.

Bourblanc, Magalie 2005: Personal Communication. 17 February 2005.

Brouwer, Floor, Ingo Heinz, and Thomas Zabel (eds.) (2005) Governance of Water-related Conflicts in Agriculture – New Directions in Agri-environmental and Water Policies in the EU, Dordrecht/Boston/London: Kluwer Academic Publishers.

Drobenko, Bernard (2001) Note sous CJCE, 8 mars 2001 et TA Rennes, 2 mai 2001, société Suez Lyonnaise des Eaux. Revue Juridique de l'Environnement. **3**, 448-457.

European Commission 2000: Directive 2000/60/EC of the European Parliament and of the Council of 23 October 2000 establishing a framework for Community action in the field of water policy, OJ EC L 327, 22.12.2000

Cour des Comptes France (2002) La préservation de la ressource en eau face aux pollutions d'origine agricole. Le cas de la Bretagne. Paris. [http://lesrapports.ladocumentationfrancaise.fr/BRP/024000109/0000.pdf]

Kraemer, R. Andreas (1998) Public and Private Water Management in Europe. In Selected Issues in Water Resources Management in Europe (ed. Francisco Nunes Correia), 319-352. Eurowater 2. Rotterdam: Balkema.

Kraemer, R. Andreas, and Walter Kahlenborn (1998) Regional Sustainability through Land and Water Management in Germany? In Agriculture and Sustainable Land Use in Europe (eds. Enid M. Barron and Ilga Nielsen), 25-55. The Hague: Kluwer International.

Levett, Roger (2002) In the National Interest? Government Proposals for Planning Major Infrastructure Projects. The National Trust. [www.nationaltrust.org.uk/main/news/national_interest.pdf]

Minting, Peter (2000) Thames still banking on Abingdon Reservoir. [www.edie.net/news/news_story.asp?id=1779]

Webb, Chris (2003) Leakage, a capital offense. [http://www.edie.net/library/view_article.asp?id=355]

[7] For more information, please refer to Brouwer et al., 2003.

Chapter VIII

Analysis of the Social Implications of the Scenarios

Authors[*] Sylvie Lupton and Pierre Bauby

INTRODUCTION

This chapter aims at analysing the social implications of the different European scenarios for the evolution of the water sector. Little research on social implications has been achieved up to now (apart from the PSIRU), and in particular the link between water management models and their implications regarding employment and consumer welfare. In Europe, many studies have been analysing the differences in water pricing between delegated management and public management for example (Carpentier and al, 2004; Ménard and Saussier, 2003 ; IFEN-SCEES, 2001).

In this chapter, we shall first develop the theoretical background regarding social implications and the definition of the main indicators. When analysing the implications of each scenario, the adjective "social" includes two different and complementary dimensions: First, it concerns the social corpus of the sector, meaning the workforce that contribute to its existence and its internal dynamics (employees, managers, etc…); Second, it also concerns the relationships of the water and wastewater sector with society, between users and providers. In this regard, we have decided to restrict our research to main criteria, which are access and employment. For each criteria, we will define and describe the main social indicators selected.

The second part of this chapter will analyse and develop the social implications of each scenario. In order to illustrate the different scenarios, case studies have been developed such as those of *Berlin, Paris, Nantes, Rouen, the OFWAT case study, Grenoble, Yorkshire and Dŵr Cymru from Wales,* which will be mentioned in this chapter. The methodology is essential to be able to analyse and compare different management models and their social implications. The question of comparison between countries is excluded according to us, as different social policies exist. Moreover, preferences and social values are different. That is why we shall focus on different local case studies to have a clearer and truer picture of possible impacts. We have selected three means to collect data and analyse social implications: general statistics (on employment…); existing literature on social impacts of different water management models; and case studies (based on interviews). For each management mode, we will try to consider several case studies to take into account a certain institutional diversity.

[*] This chapter is based on Deliverable 8 of the EUROMARKET project. This chapter would not have been made possible without the valuable help and contribution of the IHE-UNESCO team (Marco Schouten, Harshad Dhande and Meine Pieter Van Dijk).

SOCIAL IMPLICATIONS AND THE DEFINITION OF SOCIAL INDICATORS

The analysis of the social implications of each scenario includes two different and complementary dimensions: workforce and employment issues regarding water companies, and the relationships of the water and wastewater sector with society, between users and providers. These two criteria (access and employment) will be taken into account in this research. In this part, we will therefore explain why we have limited social implications to these two criteria and then develop the main indicators used to measure these criteria.

What social implications mean for the water supply and sanitation sectors?

Social implications may have a very large meaning but we have decided to restrict our analysis to two main criteria, which are access and employment. We indeed consider that these issues are the most important one to consider at the European level and there are major issues of debate surrounding these criteria in Europe, as we will see in the following section. Beforehand, it is important to clarify important concepts, which are related to our general understanding of social implications (i.e.: equity and equality) and how they relate to water and our criteria.

Equity and Equality as two central concepts of social implications

Our understanding of social implications was circumscribed around two main concepts: equity and equality. Equality refers to the *same treatment* in dealings, quantities or values (Gray, 2001). For water consumers, what interests us is to analyse: equal access to water (do all consumers have access to drinking water distribution networks? Is this compulsory or not?), equality of prices. This can also concern workers (equal rights for workers, equal opportunities).

Equity refers to *fairness: equal treatment for equal cases* (Boulding, 2004). Equity integrates the principle of proportionality. This may require different treatment or special measures for some people or groups. One can distinguish two forms of equity (Fauconnier, 1999):

1) Vertical equity (or distributional equity): vertical equity considerations prevail in redistribution programs where the affluent may pay more for a given good than the poor, and where some form of cross-subsidization occurs. This concerns affordability compared to an income level, and the existence of social funds or special rights (or no rights) for poor households (existence or not of water cuts for unpaid bills).

2) Horizontal equity: this corresponds to the "benefit" principle, whereby different individuals who receive the same amount of benefit from any good or service should pay the same price for it. This can concern price differences in water bills for the same service (comparing for example delegation contracts and public management through *régies*). It also concerns *geographical equity*: the even distribution of services across different geographic location (urban versus rural and centre versus periphery).

Applying these concepts to water supply and sanitation services, we have decided to restrict our understanding of social implications to two main criteria, i.e.: access and employment.

Justification of the main criteria

Access

The question of access is clearly a major issue in the area of water management and has obviously important social implications. At the international level, an increasing number of actors and movements have brought about during these last years the question of the establishment and guarantee of the right of access to water as one of the essential human rights. For the first time, the Committee on Economic, Social and Cultural Rights of the United Nations Economic and Social Council (ECOSOC) defined a "right to water" in 2002. Furthermore, access to water and sanitation is one of the key priorities of the United Nations Millennium Development Goal.

At the European level, the question of access to drinking water and sanitation is important, notably for sanitation with the EU wastewater treatment directive. For drinking water, the EU objective was to have a rate of connection close to 100%. In fact, this objective has been made compulsory by the European Union since the early 1980s (European Communities 1980 and 1998). Nevertheless, the rate of connection differs among countries, as figure 1 illustrates.

The major debate at the European level is not access per se but rather the notion of universal service. The notion of universal service defines a series of requirements of general interest that WSS should comply with in all the European Community. The definition and guarantee of a universal service enables to maintain, for all users and consumers, the accessibility and the quality of services during the transition from a situation of promotion of services under monopoly to that of markets open to competition.

The question of universal service is linked to the definition of services of general interest. The services of general interest refers to three complementary objectives: 1) the guarantee of access to each inhabitant of essential rights and services; 2) guarantee, at each level of organisation of society, economic, social and territorial cohesion and develop solidarity; 3) take into account the long term and sustainable development in its economic, social and environmental dimensions. These three objectives all have a link with the users, consumers and citizens, and more generally with society. The first obligation (guarantee of access) refers to the *rate of connection of water services*, and the *access of water services to poor households* (*affordability* of the water bill and existence of *social funds).*

Furthermore, the White Paper, published by the European Commission on the 12th of May 2004 on Services of General Interest, defines a series of principles (9 in total) that could have an impact on the water and sanitation sector if they were implemented. For social implications, three are important:

- *Ensure cohesion and universal access:* The access of all citizens and enterprises to affordable high-quality services of general interest throughout the territory of the Member States is essential for the promotion of social and territorial cohesion in the European Union. In this context, universal service is a key concept the Community has developed in order to ensure effective accessibility of essential services.

The application of this principle could lead to an EU definition of a "universal service of water and wastewater", ensuring access to all the EU territory. This comes back to the notion of access of wastewater and water services (rate of connection, affordability and social funds).

- *Maintain a high level of quality, security and safety*: the security of service provision, and in particular the security of supply, constitutes an essential requirement, which needs to be reflected when defining service missions. The conditions under which services are supplied also have to provide operators with sufficient incentives to maintain adequate levels of long-term investment.

This principle is important for this chapter as it highlights the importance of security of supply for water supply and sanitation services.

- *Ensure consumer and user rights*: these principles concern access in particular the access to services, including to cross-border services, throughout the territory of the Union and for all groups of the population, affordability of services, including special schemes for persons with low income, physical safety, security and reliability, continuity, high quality, choice, transparency and access to information from providers and regulators. The implementation of these principles generally requires the existence of independent regulators with clearly defined powers and duties. These include powers of sanction (means to monitor the transposition and enforcement of universal service provisions) and should include provisions for the representation and active participation of consumers and users in the definition and the evaluation of services, the availability of appropriate redress and compensation mechanisms and the existence of an evolutionary clause allowing requirements to be adapted in accordance with changing user and consumer needs and concerns, and with changes in the economic and technological environment.

This principle could lead to a series of new measures on consumer and user rights, regarding access to all citizens whatever their social status (and the prohibition of disconnection and self-disconnection throughout the EU).

The section above has defined the main principles at the European level concerning questions related to access. It should be emphasized that there is no common social policy across Europe and each European member country has its specific social policy. For example, each country developed its own mechanisms to maintain the affordability of drinking water in a context of price increases. To supply drinking water to all is explicitly mentioned as an objective in Germany, England & Wales and Belgium, as well as the prohibition of supply cuts (limited to the Brussels-Capital Region in Belgium). Affordability is also achieved through mechanisms of solidarity, i.e. the constitution of *solidarity funds* financed by a withholding on the consumer's bill and intended to make up for the water bills that people cannot afford. This mechanism is developed in France, Belgium, and in a lower extent in England & Wales. Disparities are also important concerning *social tariffs*. The Flanders Region of Belgium is unique in offering the first cubic meters for free (15 m^3/capita/year in Flanders). Otherwise, some social tariffs exist in Spain (licence fee much smaller for poor households in Barcelona), Sweden, Germany (social services provide income support on an individual basis) and England & Wales (affordability measures for water bills). However, these measures are dissimilar and often developed at the local level either by the municipalities or the operators themselves.

The following table summarizes the existing social policies in different EU countries. As the table shows, some countries like Finland and the Netherlands have no social policy regarding poor households, whereas in other countries like France and Austria, social tariffs or social funds exist.

Table 1: Different social policies in some European countries

Countries	Social measures (funds, tariffs)	Disconnection policy
AUSTRIA	Social tariffs exist for poor and old people.	Prohibition to cut off water supply to a domestic costumer for non-payment.
BELGIUM	In Flanders, a minimal provision of water for free is set. In Wallonia, a social fund for water is used to help people in financial difficulties to pay their water bill.	Disconnection is possible. Due to the federal system, regions are responsible for the water sector; there are 3 different regulations. In the Flemish region, complete disconnection is prohibited. In Wallonia, no general legal prohibition, but companies need local authority approval. In the Brussels region, no prohibition to disconnect, and companies need adjustment from the court.
ENGLAND AND WALES	By late 1998, ten private water companies out of the 27 then existing in England and Wales had made arrangements through which customers could receive charitable help if they faced disconnection of their water supply for non-payment.	Disconnection has been banned for domestic customers since the 1st of July 1999.
FINLAND	No social tariffs	All customers can be cut off from water supply due to non-payment
FRANCE	Existence of social funds	Companies can cut off the water supply for non-payment.
GERMANY	Community welfare centres deal with households in financial difficulties	Water suppliers may disconnect every non paying customer after a fixed time of two weeks if it can be deemed that the costumer will pay in the future.
HUNGARY	Direct cash subsidies are given to help poor families with water bills.	
NETHERLANDS	No special social tariff.	The company can cut off the water supply for non-payment, whoever the customer may be.
POLAND	The municipal or district *Social Support Centre*, which is responsible for giving financial aid to poor households, may help pay the water charges as part of housing assistance.	
PORTUGAL	There exist specific tariffs for social/charity institutions.	The company can cut off the water supply for non-payment, whoever the costumer may be.
SPAIN	Social tariffs exist and depend on each municipality.	Most customers can be cut off from water supply due to non-payment.
SWEDEN	No social tariffs.	The water supplier can cut off the water supply to customers under the following condition: there must be no risk for the costumer's health
SWITZERLAND		Disconnection for non-payment is rare. In any case, water supply necessary for vital requirements must be delivered to bad payers.

Source: S. Lupton, Euromarket, WP8 (2005)

Let us now turn to the justification of employment as our second main criteria for social implications.

Employment

The question of employment is another core issue at the European level. Since the 2000 "Lisbon strategy", the EU fixed itself quantitative and qualitative objectives regarding employment. In effect, the Lisbon strategy has assigned the European Union a new strategic goal for 2010: to become "the most competitive and most dynamic knowledge-based economy in the world, capable of sustainable economic growth, with more and better jobs and greater social cohesion, with the respect of the environment" (Lisbon Strategy, Lisbon European Council, 2000). The strategy has an objective regarding growth (growth rate of 3% per year), and for the first time a goal for full employment in 2010, with a medium employment rate of 70% (60% for women and 55% for wage-earners between 55 and 64 years old). At present, the EU has still to face important rates of unemployment in some countries. Some new member states suffer from important unemployment levels: Poland's unemployment rate is 18.2%, against 16.5% in Slovakia (Infobase Europe, 2005). Unemployment is also a priority in elder member states like France (the unemployment rate is around 10%), Greece (10.5%) and Germany (9.6%) (Infobase Europe, 2005).

More specifically linked to the water supply and sanitation sectors is the issue of the nature of employment contracts. The evolution of the sectors towards a more competitive market has important consequences on the nature of employment contracts. Under direct public management, the employers were considered as civil servants and their employment was viewed as very secure. With the changing managerial mode towards more flexibility, the traditional rights in this sector are, to a certain extent, currently been called into question.

It is interesting to see that in terms of employment, several water companies are starting to adopt new social initiatives. We have focused on the biggest one such as Suez, Veolia and RWE as access to this information is much easier compared to smaller companies. Of course, this does not mean that smaller companies do not have special social measures concerning employment. These measures are in fact strongly encouraged at the European level (e.g.: the Green Paper on the social responsibility of firms (COM (2001) 366), or at the international level (e.g.: the *Global Compact* of the United Nations in 2000). At the international level, Suez for example has developed an International Social Charter, which lists fundamental rights and principles such as equal opportunities, non-discrimination, respect of human rights, and personal development and has created in 1998 an International Social Observatory. At the European level, Bouygues has developed a European Social Charter. In fact, one can distinguish at least three main issues in these voluntary initiatives.

Equal opportunity between men and women in employment is one illustration of these initiatives. In fact, these measures are also regulated by a new EU directive on the equality of treatment between men and women (23rd of December 2002), that implies an absence of discrimination based on sex, regarding access to employment, training, professional promotion, and working conditions. For example, *Veolia Environnement* supports the principle of equal opportunities between men and women. The proportion of women in the total workforce is stable at 19%. In 2003, women employees who attended a training programme increased in proportion to men (+15% against +12%). Women are more present in RWE Thames Water compared to Véolia, as they represent, in 2003, 25.8% of employees. Access to work for handicapped people is certainly another major issue. In France, a partnership convention was signed between *Générale des Eaux* and AGEFIPH (*Fonds pour l'Insertion Professionnelle pour les Personnes Handicapées*- National Fund for the professional integration of handicapped people) for 2002 and 2003. This has lead to the recruitment of 34 handicapped people. Finally, the nature of employment is the third major issue. In Bouygues, 97% of workers work on permanent contracts of employment. Another example is the Group Suez. Since 1993, the Group has insisted on the employment of young workers, and created 21 000 jobs in this regard.

Overall, these new social performance indicators have enabled to increase the transparency of these companies' social policies in terms of: the number of core employees, the training regarding occupational health and safety, the number of women working, and also the percentage of handicapped workers in the water companies. However, there is a lot of important data missing in order to have a systematic analysis of the social measures of these companies. For example, it is difficult to find information regarding non-core employees (therefore the extent of outsourcing remains unknown). Moreover, the number of employees does not give an indication on comparative data like the number of employees per thousand connections. Therefore, although increased transparency is much appreciated, it has not necessarily led to enable analysis and comparisons between companies (apart perhaps for the number of women working and percentage of handicapped workers employed).

Current research on access and employment

After having identified access and employment as the two major issues in terms of social implications at the European level, we will now look at the various scholarly research that has been done on these two issues. One must add that this section for both access and employment is based exclusively on England and France and cannot be extrapolated to all EU countries.

Access

In terms of access, one can distinguish two main types of research: one type of literature focuses on the possible link between access and private sector participation while another type of literature looks at the possible correlation between management modes and pricing, which has of course an incidence on affordability.

Regarding the *rate of connection to drinking water supply network*, there is no immediate link between coverage of drinking water and sewerage and different management modes: a World Bank study (Clarke, Kosec and Wannsten,

2004) on developing countries shows no evidence of any public or private management impact on coverage of drinking water and sewerage.

As for affordability, several studies suggest that prices are higher in delegated management models compared to public management ones. The following table on the situation in France clearly illustrates this point.

Table 2: Water price average (in euros/m3) in France in communes with collective sanitation

	Drinking water supply		Collective Sanitation	
	Management mode		Management mode	
	Régie	*Delegation and mixed companies*	*Régie*	*Delegation and mixed companies*
Communal organisation	1.03	1.26	1.01	1.44
Intercommunal organisation	1.19	1.58	1.45	1.56

Source: IFEN-SCEES, (2001)

This table shows higher prices for delegation but one must also add that:

1) Delegation contracts concern larger communes (IFEN, 2004)
2) Delegation could concern more complex cases requiring heavier investments (Ménard and Saussier, 2003)
3) Price differences are also due to local specificities like water pollution, and technology used.

Another study based on the production and the distribution of drinking water confirms these results (Carpentier et al., 2004a). Carpentier et al. (2004b) focus on the price of drinking water in communes of less than 10 000 inhabitants. However, one should note that the communes that have chosen delegation were also confronted with difficult operating conditions from a technical and/or financial point of view, which can also explain higher prices.

Table 3: Average price of the drinking water service in France

	Average Price[1] In French Francs/m^3	**Number of communes**
Total of the sample	7.88	3 135
Delegated management	8.85	1 793
Régies	6.58	1 342
Difference delegation/régie	+ 34%	

Source: Carpentier and al. (2004b: 2)

The findings of these researches are related to France, and cannot be extrapolated to other countries. Let us now look at the scholarly research on employment.

Employment

According to a study made by the International Labour Organisation (1999), private sector participation in the WSS sectors does not lead to significant employment reductions like in the electricity and gas sectors, One should note that the WSS sectors is only partially liberalised compared to these other sectors. Another important finding of this study is that the nature of employment contracts considerably changes after private sector participation with an increase in non-core employees and a decrease in the number of core employees (IL0, 1999). If one looks at the case in England and Wales, one can see that privatization did not lead to a decrease in the *overall* number of employees as the following table illustrates:

Table 4. Water company employees in England and Wales from 1990 to 1996

	1990	1991	1992	1993	1994	1995	1996
Non-core employees	1237	4290	10 906	14 340	18 933	19 983	24 239
Core employees	41 126	41 504	41 679	40 402	39 105	37 385	34 174
Total	42 363	45 794	52 585	54 742	58 083	57 368	58 413

Source: ILO, (1999: 2)

[1] The average price corresponds to the price per cubic metre paid by consumers corresponding to a consumption of 120 m^3.

However, the nature of employment has considerably changed. When looking more precisely into data derived from UK companies' annual reports, one can clearly observe job cuts of core employees. Hall and Lobina (1999) noted an overall fall of employment of 21.5% for 10 regional companies[2] from 1990 to 1999 (representing 8 599 core employees). Some of these job cuts are due to mergers and takeovers, and sub-contracting. Moreover, as well as loosing their jobs, "*water workers have also suffered by companies deliberately eroding employee rights. Northumbrian, for example, systematically put more and more of their employees on short term contracts of 1 year 11 months, in order to be able to dismiss them at will without the employees concerned being able to make us of the protection of unfair dismissal legislation. This is likely to have affected employee morale and standards of work*" (Hall and Lobina, 1999: 6-7). The evolution of employment is also dependent on other factors, notably investments in drinking water and wastewater infrastructure. According to the Directorate General Environment of the European Commission[3], the urban wastewater treatment directive has generated considerable investments that have lead to the creation of new jobs. In France for example, we can observe a considerable increase in the number of employees between 1993 and 2002: the total employment in the water sector increased from 31 207 to 39 446, which means a 26% employment increase in 10 years. The main source of increase is in the wastewater treatment sector (77% increase compared to a 17% rise in the drinking water sector). Of course, in absolute numbers, catchment, treatment and distribution of drinking water creates more employees than the wastewater treatment sector.

Regarding equality of access to men and women, the employees in the water sector are mainly men and this regardless of the management mode. According to ILO (2003), in the utilities sector, most women work in clerical jobs and few reach middle management levels while even fewer hold top-management positions. The gender imbalance is at unacceptable levels, and much needs to be done to improve the situation.

Social indicators for access and employment

In order to look at the possible social implications regarding access and employment, this last part will define a number of indicators that will be used for evaluating these implications on the scenario and the illustrative examples.

Access

- *Access to drinking water services (rate of connection)*

As we have already mentioned, rate of connection is usually high in EU member states. Nonetheless, it is important to recall that this indicator is a major issue in very isolated areas. In some new member states, this problem is also accurate in urban areas (see Latvia and Lithuania, European Commission, 2003)

- *The price of water*

The price of water affects the affordability of water especially for poor households, and it was therefore chosen as a social indicator. Price increases can pose problems for poor households.

- *Existence of social funds*

As we have seen in the previous part, social funds have been implemented in different European countries. These funds are used to help poor households to access water services.

- *Equal or unequal distribution of water quality according to different sectors of population* This indicator looks at the equal or unequal distribution of water quality in different sectors of the population. This could be especially interesting for the community management scenario: rich population could have an independent community management with better quality than the rest of the population (this belongs to the problem of geographical equity developed above).

- *Tariff principles*

Are there special measures regarding tariffs on water and sanitation services, destined to facilitate access to poor households? The OECD report stresses on the fact that a progressive system of tariffs (by layers of consumption) can have important potential advantages as the first layer is free or the tariff is reduced according to the size of the family (OECD, 2003). This type of tariff system is progressing throughout the world.

- *Same price for same quality*

This corresponds to the "benefit" principle, whereby different individuals who receive the same amount of benefit from any good or service should pay the same price for it (same price for same quality). This can concern price differences in water bills for the same service (comparing for example delegation contracts and public management

---

[2] These ten companies are Anglian Water, Hyder, Northumbrian Water, Severn Trent, Southern Water, South West Water, Thames Water, United Utilities, Wessex Water, and Yorkshire Water. A Decrease in number of employees was observed in all companies, except Northumbrian Water and South West Water.

[3] Working report reference : EC 4739/M/1145-0, September 2000

through *régies*). This indicator has only rarely been considered, due to the difficulty in collecting data on this subject. However, the case study of Nantes (France) illustrates the efforts made by the *agglomération de Nantes* to impose the same price and same quality requirements for all consumers in the communes of the *agglomeration.*

Employment

- *Number of staff employed*

The number of staff employed is an important criteria. In this chapter, we will look to what extent the different scenarios may have (or not) an impact on the number of staff employed.

- *Nature of contract (temporary/long term)*

The major question behind this indicator is again to see if some scenarios have an incidence on the nature of employment contract.

- *Equal rights of workers between men and women, between levels of qualification and responsibilities*

We will see to what extent equality of treatment between men and women may (or may not) change according to the different scenarios.

- *Social policies of companies*

This indicator will be used to see what kind of social policies have been developed by these companies in the different case studies for the different scenario.

- *Working conditions*

Working conditions constitute also an interesting indicator. We will look in the next section to see to what extent working conditions may be altered according to one or another scenario.

EVALUATION OF THE INDICATORS ON THE DIFFERENT SCENARIOS

Regarding implications of different scenarios, one must insist on the fact that some indicators were not taken into account, as they do not vary according to scenarios. The *access of water and sanitation services to all (rate of connection)* has been developed in the theoretical background section of this report: there is no immediate link between coverage of drinking water and sewerage and different management modes. However, this does not mean that the coverage rate could not vary and improve (or not) in 2020. This indicator had therefore not been developed regarding social implications of different scenarios.

As for the *existence of social funds,* and existence of *tariff principles*, this also depends on national social policies that exist in some EU member states, and do not in others. Therefore, the existence, or not, of social funds and tariff principles has nothing to do with the different case studies (except for scenario1), and will therefore not be extensively developed. Regarding scenarios, scenario 1 is the only scenario representing a significant change regarding social funds as there is an EU social policy with compulsory social funds.

Regarding *social policies of companies*, what we can observe is an increasing transparency regarding statistics on employment, on the *quality of contracts* (permanent and short term contracts), equal opportunities, and training on health and occupational safety. What we learn is that there is a high percentage of permanent employment contracts in multinational companies. However, no data is given regarding outsourcing and non-core employees.

To summarize, the following indicators were selected to evaluate the social implications of scenarios. Regarding access, the *price of water* (used to evaluate the affordability of the service to consumers[4]), *geographical equity* and *same price for same quality* will be selected. Regarding employment, the *number of employees*, the *nature of employment contracts*, the *equality between men and women* and finally the *working conditions* were selected as indicators that vary from one scenario to another and must be analysed. Some indicators will not be integrated on the analysis of implication of *all* scenarios: for example *same price for same quality* has been analysed in the case of local regulation of price and quality in a given agglomeration of communes (example of Nantes) that has enabled the same price for the same quality both for *régies* and private companies (delegated management).

[4] Affordability can also be evaluated through the existence of social funds and social tariffs. However, the latter do not depend on scenarios but on the existence of national policies as we mentioned in the introduction. Affordability will therefore be evaluated through the water price, as the latter changes across scenarios.

SCENARIO 1: DELEGATION CONTRACTS AND STRONG REGULATION

Access

A range of social indicators is positively implicated due to the high level of regulation that this scenario incorporates. The existence of a strong regulator, in particular price cap regulation as illustrated in the case of Ofwat, allows not only limited increases in water prices, but also high quality standards, and the encouragement of the participation of consumers. A particular indicator related to access that is relevant for this scenario is the *price of water*. The cases of Grenoble and Berlin demonstrate that prices tend to increase with delegation contracts compared to public management. Also the data on average prices in France, represented in Figure 3, illustrate a higher average water price for delegated and mixed companies compared to the public management companies. However, it needs to be noted that the reasons for this price difference cannot be simply ascribed to the different management mode. Straightforwardly comparing water prices is in general already difficult due to the dependence on local variables as the availability of good quality water resources and the difference in infrastructure requirements, but in this case a comparison is even more difficult to make. For example, delegated companies serve on average larger service areas compared to public modes. Also, it can be assumed that delegated management contracts are more often established in areas where the management of the service provision is complex and with a lot of investment requirements, while the areas with relatively low requirement in terms of management remain in the public realm.

Also due to the high level of regulatory interference, a social indicator as *the rate of connection* is positively implicated, as operators will need to oblige that inhabitants have the right to access and that each individual consumer can demand this right even before a court. Disconnection of customers by operators is prohibited, it is even forbidden for consumers to disconnect themselves from the network. Also *the existence of social funds* is compulsory and enables poor households to also have access to water services.

In general, one could state that the negotiating power and the level of control that the local public authority is able to exercise over the delegated party, determines to a large extent all social implications related to access. The benefits of such are clearly illustrated in the case studies of Nantes, Rouen en Paris. In these cases, it was found that the stronger the local authority is able to operate and the better it is able to create a very competitive environment, the more it is able to induce the delegated companies to mitigate potential negative social implications.

Employment

Related to the indicator of the number of staff employed, the experiences with a change in staff due to a shift from public management to delegation contracts (and vice versa) vary according to member states' policies. In France, the law prescribes that in case of such change from public management to delegation (and *vice versa*), the employees are also transferred to the delegated party. However, in other EU member states changes in the number of employees can be observed, such as in Berlin or in Gdansk. In these cities a reduction in employment can be noticed as productivity rates of personnel increases after the delegated management contract has come into effect.

Regarding the indicator of the *nature of contracts*, the change from public management to delegation contracts may not change significantly the nature of contracts. Large companies that work as delegated management party, as Suez and Veolia, employ the vast majority of their workers under permanent contract, respectively 95.9% of Veolia workers and 95.6% of SUEZ workers. The same goes for the indicator of *equal opportunities between men and women,* where a shift from public management to delegation contracts does not suggest a large change. In the whole water sector women are not well represented regardless of management modes. Also in the large companies that act as delegated contract partners, as SUEZ, Veolia and RWE, women are underrepresented, respectively 13.6% of women in Suez, 21.9% of women in Veolia and 25.8% of women in RWE.

Summary social implications for scenario 1

Regarding access, the high level of regulation that marks this scenario could positively influence numerous indicators, as the price of water, the rate of connection, and the existence of social funds. Nevertheless, especially for the implications on the water prices of a shift towards delegated management, it needs to be noted that there are various examples illustrating the tendency of price increases after delegated management contracts come into effect. It cannot be generalised that these price increases can be ascribed to the change in management mode; numerous other factors complicate a sound comparison. As for employment in scenario 1, unless regulation includes the prohibition of job cuts, employment will have a tendency to decrease in countries where no legislation protecting workers exists (like in France).

SCENARIO 2 DELEGATION CONTRACTS AND EXTREME COMPETITION

Access

This scenario might implicate negatively numerous indicators related to access. The only regulation mechanism that this scenario incorporates is through competition law applied to each contract. This regulation is limited as there might be an oligopoly of some big companies that control most of the market. Even if there is competition and oligopoly, competition can be limited (some companies keeping the same niches like in France), and can be more of a renewal of contracts with the same existing company.

The *rate of connection* will be negatively implicated as for all households unpaid bills will lead to disconnection of water in countries where this is allowed. Also the *existence of social funds*, the *geographical equity* and the tariff principles might be negatively implicated due to the total reliance to market forces and the absence of regulatory control and oversight. Services will concentrate on good customers (cherry picking). The participation of consumers is not encouraged.

Prices may vary significantly after each management contract, according to the influence of competitive pressure from frequent tenders. Here again, all depends on the negotiating power and control of the local authority, that can in certain cases define tariffs enabling a minimum service, and social policies according to the policies existing in each member state. In fact, the changes in prices are highly dependent on the competitive character of the contract awarding. Prices may vary significantly after each delegated management contract, depending on the level of control that local authorities are able to exert upon companies. If extreme competition for the market exists, the price may not necessarily rise and can even decrease.

Employment

Especially the indicator of *number of staff employed* is of importance for this scenario. The reduction in employment figures is due to the pressure towards increasing job productivity. Such pressure for increased productivity will be even higher in an extreme competitive market.

Summary social implications of scenario 2

Regarding access and employment there is a more extreme situation as compared to scenario 1, as the mitigating effect of a regulatory regime is absent. There will be a higher tendency to an increase in prices and a tendency towards job cuts, as this also concerns delegations contracts. The role and responsibility to local authorities becomes more eminent as they are the ones that should exercise a mechanism of price control and employment guarantees in the contracts.

SCENARIO 3 OUTSOURCING

Access

Our analysis of access is directed towards the efforts that water providers make to realize that their services are increasingly accessible to their prospective users. The accessibility of water and sanitation services depends largely on the affordability of the services to the prospective consumers. The issue of affordability relates to *the price of water* and the *tariff principles* that the operators formulate, and also to the degree that national policies are willing to subsidize lower income groups through *social funds*. For the scenario of outsourcing several elements play a role:

Organising the provision of water and sanitation services through outsourcing contracts might facilitate a more *even (or uneven) distribution of water or water qualities across different geographical areas or groups of the population*. As in outsourcing only a small part of the business process of the water provider is contracted to an external party, it can be imagined that the act of outsourcing can have implications on the equal distribution. In the case where only a specific geographical region or group of population is benefiting (or not) from the work of the outsourcing company, the equity distribution becomes violated. The service levels are not generally equal for all users, possibly triggering feelings of unfairness. Depending on the success of the outsourcing party compared to the public utility the group of people directly affected will feel privileged or victimized.

Water providers are mostly motivated for reasons of cost savings to engage into outsourcing. These savings are usually in labour costs, which is the outcome of flexible working hours and short-term contracts for employees. Outsourcing allows firms to offer services more efficiently and effectively without risking large investments in labour. Another reason for the cost savings is due to competition between private contractors to provide services cheaply. Also cost savings can be achieved by outsourcing arrangements through accessing modern and more cost efficient technologies and expert knowledge. The replacement of manpower by labour saving machinery will reduce working hours. Another implication is that with outsourcing, the utilities can concentrate on their core activities and provide services in a more efficient manner. All of these cost savings might indirectly implicate *lower water prices*, increasing the affordability for consumers of water services.

Employment

Two factors play an important role that determine if and to what extent, there are implications on *the number of employees*. These two factors are the duration of the outsourcing contract and the transfer of employees. When the duration of the contract becomes extended and the employees from a public utility are transferred to the outsourcing party, the chances of reduction in the number of employees over a long period of time are higher, rather than when the duration of the contract is short. The reason most often cited for downsizing is that the public utilities are overstaffed with low productive staff, five to seven times compared to other utilities and also have a more hierarchical structure (Haarmeyer, 1998). The outsourcing party would always like to be more cost effective by having a less hierarchical structure and productive staff, because the decisions are taken faster and productive staff is enabled to take more initiatives and responsibilities.

In many cases of outsourcing, when there is a transfer of employees internally from one department to another to perform different functions, usually after a period of training, there are risks of a "brain drain" in key positions where specific know how is required. Most companies believe that the primary risk of long-term outsourcing resides in a loss of technical know how (ILO, 2001). The core objective of outsourcing is to realize cost savings – crucially labour costs (Thomas, 2003). Another reason for outsourcing is that the public utilities cannot generally lay off workers from the job due to many reasons. Through outsourcing arrangements, utilities found an alternative way to get rid of inefficient work force, by transferring them to the outsourcing party.

With increased reliance on outsourcing, the replacement of manpower by labour saving machinery is another significant reason for job cuts. The point mentioned above is very well illustrated by the case of Dutch water industry. Considering that the Dutch water industry is quite stable in terms of size, significant labour productivity gains are made by the water companies at the expense of employment. In the 15 years between 1985 and 2000, employment in the companies has dropped by nearly 20% from 8,500 to 6,803 (Van Dijk, 2003). There are many examples all over the world, where there is a decline in the number of employees especially in water sector through outsourcing. In the United Kingdom, during the 1980's there has been an increase in outsourcing and the use of contractors, which resulted in a decline in number of core employees by almost 17 percent, from 41,126 employees in 1990 to 34,174 employees in 1996. However, over the same period, there has been an enormous increase in the number of non-core employees from 1,237 in 1990 to 24,239 in 1996 (ILO, 1999). There is further evidence of significant work reduction due to the outsourcing of activities in the water industry in some other countries like the Czech Republic, in which there is a reduction of almost 34 percent of employees. The increase in outsourcing can be a result of the shift from the socialist regime to a more capitalist regime in the early 90's.

With the introduction of the private sector comes an associated issue of the 'employer and employees' relationship. The relationship or *nature of contract* depends on the human resources strategies adopted by these companies (Hall, 1997). If there is any transfer of employees, it can lead to more flexible, fixed term employment contracts. These fixed term contracts will then lead to further possibilities of having strained relations between employer and employees, and between different groups of employees. This can lead to a high degree of employment uncertainty among workers. If there is no transfer of employees, then the nature of contract between employer and employee is of permanent nature. The employer in most cases will be a public entity and workers will enjoy the benefits other than just the salary, like governmental retirement plans, government health insurance, holiday arrangements, and governmental policies for bereavement and maternity leave. In the illustrative case of Dŵr Cymru Welsh Water only 137 staff are directly employed by the water provider (a ratio of 8,760 connections per employee). These people are mainly concerned with administrative functions, as debt and treasury management, and reporting to Ofwat. All the remaining input from human resources comes through outsourcing arrangements. In the annual report of 2005, the chairman of Dŵr Cymru Welsh Water acknowledges his concern about the uncertainty and concern for staff rights due to this extensive form of outsourcing (Dŵr Cymru, 2005).

The outsourcing of activities often results in changes, although in varying proportions and degrees, in terms of salary and other *working conditions*. When outsourced, the companies generally adopt their own wage policies to be more cost effective and achieve greater productivity along with being more efficient. Salaries differ depending on the type of job carried out in the enterprise and there are major differences between salaries paid to different categories of people depending on the experience, the status and the level of qualification. Although there is a lack of data for an analysis to be made of salary differences between manual workers, engineers and managerial staff, the example of the United Kingdom does highlight some aspects of this phenomenon. With increased outsourcing there was an increase in the salaries of the directors of the water and electricity enterprises at a time when there were massive redundancies for lower staff. This is just an illustrative example where a discrepancy in the policies adopted by the companies towards different workers is shown, but this might not be the case all the time.

The competitive pressures for better performance and productivity in limited working hours may lead to increased psychological pressure on employees. This may have negative implications in terms of *working conditions*. One of the aspects of the working conditions is also health and occupational safety of workers, though this is often an underrated issue when considering outsourcing. As stated in the ILO report, in addition to the classical hazards associated with most large industrial plants, workers may also face dangers of occupational illness caused by working in proximity to unidentified pollutants, drowning, underground work, work in confined spaces, risks of explosion and working with under high pressure (ILO, 1999). When the activity is outsourced, one of the obvious targets of the new employer company is to reduce expenses and to be more profitable which is at the expense of allocating fewer funds towards the health and safety of workers. Although this will depend on the length of contract as for short-term contracts the employer will have less incentives to do so. But on the contrary when the length of the contract is long the employer will invest to protect the trained employees.

While considering *job security*, the *working conditions* in public enterprises are generally quite favourable and these enterprises have highly formalized internal structures with a high level of employment security. They also benefit from the availability of the internal labour market. Regarding the impact of outsourcing, there are changes in the employment relationships: full time secured jobs will be changed into part time jobs. Much of the shift from these formalized and secured benefit structures towards flexible and unsecured employment has taken place in relation to the outsourcing activities (ILO, 2001). Moreover, when an activity is outsourced, the workers of that department are generally transferred to outsourcing, which leads to a reduction in the strength of a unit for negotiations with management. The smaller unit formed due to outsourcing is also not in a position to negotiate with management and this may lead to compromises in the decisions taken regarding job security, decent wages and better working conditions.

Summary of social implications outsourcing

On the basis of above analysis one can conclude that the impacts of outsourcing depend on the scope of the outsourcing activities and the degree to which any cost savings will implicate water prices. In general terms, one could conclude that if the water provider succeeds in establishing cost savings due to outsourcing part of its operations, this will positively implicate in the long-run the affordability of the service provision to the consumers. The existence of such positive implications is based on the assumption that the tariff is to a certain extent a reflection of the realised costs of the water provider. Nevertheless, if the result of the outsourcing is that the collection and billing operations will be separately managed via an outsourcing contract and regarded as a means to maximise the revenue stream, this will negatively implicate the affordability of the service provision. This negative implication is not due to an increase in tariff, but to a reduced leniency of the water provider towards its customers in collecting the bills.

One could state that outsourcing arrangements can lead to a reduction in the number of employees, the risk of brain drain and the loss of job security, to name a few. This can be due to the private sector wanting to realize cost savings by having fewer employees, but more productive ones. A possible reduction of the number of employees might also be due to the motive of the public utilities to use the outsourcing arrangement as an alternative means to layoff excess workers. With this reduction in employment, there is also a possible risk of 'brain drain' of experts due to outsourcing to the new company for better remuneration. In outsourcing, with private sector involvement, the job security is less important, as the employment contracts are generally temporary and short-term ones. Moreover, outsourcing has implications in terms of the level of unionisation. With increased outsourcing, there is a reduction in the negotiation power of unions as generally employees are transferred to the private sector leaving fewer employees in unions. An interesting example to analyse outsourcing is Dŵr Cymru from Wales, as this water and sewerage company outsourced 80% of all of its activities. Another case that to a certain extent illustrates outsourcing is Yorkshire Water, that planned to outsource a large part of its operations although this was prevented by the regulator Ofwat.

SCENARIO 4: REGULATED MONOPOLY

Access

For this scenario, issues of access depend on the orientations and choices of the regulatory authorities, even if the existence of compulsory benchmarking creates the conditions of an efficient regulation. This leads to contrasted access configurations according to the nature of regulation, regarding social funds and disconnection policies, tariffs, prices and the participation of consumers that can be encouraged when the regulatory authority wants to exert important control. If benchmarking is compulsory, and high prices are sanctioned, then prices will be maintained or may even be reduced.

A very interesting and illustrative example can be found in Scandinavia. Benchmarking developed since 1995 by the 6-cities Group. The 6-cities Group consists of Copenhagen, Gothenburg, Helsinki, Malmö, Oslo and Stockholm. All these cities have between 250 000 and 800 000 inhabitants. In the 6-cities Group, the water and wastewater utilities are all under public control. Benchmarking was developed in order to compare the efficiency of publicly operated water utility systems with private alternatives (Stahre and Adamson, 2002). Regarding affordability, an important part of the metric benchmarking concerns costs of the water and wastewater services to explain the water bill to consumers. Metric benchmarking can be described as a quantitative comparative analysis that enables a utility to follow the performance of their system over time and to compare this performance with the performance of other similar utilities (Stahre and Adamson, 2002). The comparisons of the costs were structured as follows: 1) the whole business; 2) administration; 3) production of drinking water; 4) distribution of drinking water (including reservoirs, water towers, pressure pumps etc.); 4) collection of wastewater; 5) treatment of wastewater; 6) system improvements and renewal. When comparing costs, there turned out to be big gaps between the lowest and highest specific cost (expressed in €/m^3 billed). However, differences in costs cannot necessarily correctly inform customers about water prices as capital costs (depreciation and interest) are handled quite differently in the 6 cities. The capital costs differ by a factor of 2.7 among cities. Therefore, even when information is given to consumers, this will not allow proper comparisons of prices and costs. The fact that information is given to consumers may not always lead to correct assessments of the services, as costs can be calculated differently between companies, and higher prices of water can be explained for example by a better quality of water, better maintenance and low leakage rate. Therefore, even if benchmarking is established, it has to integrate very different dimensions of the service that may not be sufficiently taken into account by customers. Price differences may not be sufficient for consumers to evaluate different water services and act accordingly. Therefore, benchmarking may not lead to changes regarding the price of water, as consumers may not be able to properly evaluate services and put pressure for change.

Employment

Regarding employment, if benchmarking is compulsory and there are sufficient credible threats (for example consumer discontent) for companies that under-employ, employment will be maintained. On the contrary, if benchmarking is not compulsory and operators have no obligations and credible threats, employment may decrease. However, when looking into detailed examples like the one of the 6-cities Group, the results of benchmarking may depend on the choice of indicators and consumers may also prefer not to have an important number of employees per 1000 connected inhabitants, as they can think this increases costs and induces the price increases. It can be also true that a very low number of employees per 1000 connected inhabitants may also not be desired by consumers that prefer to have a minimum number of employees in order to guarantee the quality of the service (maintenance, leakage rates…). But two questions can be asked: can benchmarking facilitate evaluation? Can consumers engage themselves enough to put pressure on different companies in order to impose change? The example of the 6-cities Group in Scandinavia is illustrative. Benchmarking also concerned the number of employees per 1000 connected inhabitants, which enables an easier comparison between companies. This indicator has not been taken by Suez, Veolia and RWE, and does not allow comparing companies easily as only the number of employees is given. For the six cities, there are important differences: Gothenburg has the lowest number of employees per 1000 connections (0.5 employees per thousand connections), and Copenhagen has the highest level (1 employee par thousand connections) Stahre and Adamsson (2002). However, this data can be misleading, as all cities do not have their own personnel to perform all duties. One city for example is connected to a regional water supply system, and outsourcing (which is not counted in these statistics) can be important. It is therefore necessary to have a more detailed analysis of the organizational structure. If one evaluates the indicator of in-house work (which can be expressed as the ratio between cost for own personnel and the total cost for operation, maintenance, administration and improvements), this can give a more precise picture of employment in these companies. According to Stahre and Adamsson (2002), the in-housework varies between 30 and 35%: this means that 65-80% of the work (material and external services) is carried out by other parties.

Here again, if benchmarking exists regarding employment, it must also integrate data on the extent of outsourcing in order to correctly evaluate the number of employees per 1000 connected inhabitants. Otherwise, comparisons between companies may be largely insufficient. Consumers can be mislead regarding the extent of employment in different companies, and the establishment of proper indicators like the number of employees per 1000 connected inhabitants including both in-house employees and out-sourced employees, would enable a solid comparison between companies.

The effect of benchmarking could therefore be limited as consumers have limited rationality: not only do they have to analyse existing information given by benchmarking, but they could have difficulties in understanding and treating information according to their knowledge, and may not properly evaluate data given by benchmarking indicators (Simon and al., 1992).

Summary implications for scenario 4 Regulated Monopoly

Few examples of case studies on benchmarking in the EU exist. One is that of the Dutch water sector in which benchmarking is used. In this benchmarking, the performance of water companies is compared, using performance-indicators on water quality, costumer service, environment and finance (Van Geel, 2004). The aim of this benchmarking is threefold: 1) increase transparency; 2) give an account to the public, the shareholders and the Board of Directors; 3) generate information to improve performance.

However, the impact of this benchmarking has not yet been sufficiently studied. Implications depend on the extent of benchmarking (if there are sanctions or not), and the importance of diffusion of information. It does not mean that because there is more information on price, quality, employment... that this information is centralized and easy to compare (the example is that of social performance indicators of private companies).

SCENARIO 5 DIRECT PUBLIC MANAGEMENT

Access

The scenario provides that tariffs are based on cost recovery and that there are no direct subsidies. This may have an important impact on *water pricing.* Important modifications can arise in price setting after each election. In some cases, the operator of the service is very poorly controlled, whereas in other cases, the public authority can exert strict control, by for example promoting the participation of consumers like in the case of Grenoble. Regarding public management, the case of Grenoble is an illustration of the importance given to accessibility to the services, while prices are usually given less importance. In Grenoble, the two successive changes in management modes led to notable evolutions of prices: a rise of about 50% between 1991 and 1995 with the change to delegation contracts, and then a decrease of 8,5% between 1996 and 1997, during the change to a *régie,* which has lead to a stabilisation of prices ever since. However, one cannot generalise from this tendency for all changes in management modes. Direct public management is not necessarily always less expensive than other management modes like delegated management. Moreover, the existence of some control by the organising authority can help harmonise prices be they from delegated management or public management, like the cases of Rouen and Nantes illustrate. In Rouen, in 2003, for 120m^3 (before tax), the price was between 109.04 euros to 190.70 euros in the agglomeration. The *agglomération*'s objective was to better control private companies and increase competition. One objective was to increase negotiating power with companies in order to have a unique and harmonized price in 2012 that should be fixed at the lowest existing price. This will lead *to geographical equity*: same price for the same service. In this scenario there are no social obligations at European level, but this also depends on local particularities (existence of local social funds or not).

Employment

Regarding *the number of staff employed,* it is usually assumed that direct public management employs more workers compared to other modes of management. This really depends on the implication of the local authority to properly evaluate the service and increase its performance. Regarding employment, the case of Berlin illustrates that public management employs more personnel than delegated management. One can remark the important job cuts with the delegation of water management to private companies. All in all, 2637 jobs were suppressed in a ten-year period between 1996 and 2006, representing 39.1% of the jobs in 1996 before the delegation process. However, in countries like France with a legislation that obliges no change in employment, employment remains unchanged. The French legislation (*Code du travail article 122-12*) obliges that a juridical modification of the employer (for example a passage from a *régie* to a delegation contract) does not lead to any change regarding employment contracts. This however is not necessarily always positive: although the French law protects employees, it does not necessarily enhance efficiency according to the French Competition Council, as it does not permit to chose more efficient management modes and reduce costs (Conseil de la concurrence, 2000). Moreover, over-employment is also possible, without the possibility in France to enhance efficiency in the water management sector.

As for *the nature of contracts*, the change from delegation contracts to public management may not change significantly the nature of contracts, if one refers to the social performance indicators shown for Veolia and Suez: in Veolia, 95.9% of workers are under permanent contracts in 2003; for Suez, 95.6% of workers are under permanent contracts. However, one should note that in direct public management, the nature of contracts are very secure (i.e. civil servants). For *equal opportunities between men and women*, the implications will be very limited as the whole water sector, regardless of management modes, has little representation of women in the sector.

Summary social implications scenario 5

The above analysis shows that the main social implications of scenario 5 are related to the number of staff employed. Although highly dependent on local circumstances, a tendency can be noticed that private management modes put more emphasis on reaching high labour productivity rate compared to direct public management. As for indicators concerning access, a lot of effort is undertaken to guarantee a high level of accessibility to the service.

SCENARIO 6: COMMUNITY MANAGEMENT

Access

Access to water and sanitation services depends largely on the *water price* that is asked of the consumers to pay for the service provision. *The tariff principles*, including the possible subsidies for target groups, and the collection policies that the operators formulate determine to a large extent the affordability of the services. In the community management scenario, the community itself is responsible for setting the tariffs and organising the way it wants the bills collected. It can be hypothesised that this type of tight relationship between the provider and the users will soften the tariffication and collection policies. Unless it is found by the community as a whole that a high tariff and strict collection policy are extremely necessary, the community will be motivated to keep low tariffs and find a proper mechanism to protect the parts of the community that have difficulty in paying for it. For example in the illustrative case of Dŵr Cymru Welsh Water, its non-profit structure allowed the company to give customer bill rebates worth about £ 23 million per year. No water and sewerage company in England is giving a price rebate in this way by charging less than the regulatory price cap. To anticipate properly to the specific demands of identified target groups within the community, generally poor people are provided with a facility of having *social tariffs* or cross subsidization. Moreover, there can be a positive implication on social tariffs with an increased level of accountability, as the members of committee or boards will be held responsible for the provision of basic services to all the people in the community. In doing so they may institutionalise a social tariff although there are examples in which income support to poorer households is sometimes channelled through social welfare systems or linked to housing benefit programmes, and there are numerous examples of local authority assistance arrangements. In the example of Dŵr Cymru Welsh Water, a Customer Assistance Fund, is set up designed to help customers to meet their water bills, while for the elderly and disabled additional services are offered as large print or Braille bills and bottled water in emergencies for less mobile customers. Also Dŵr Cymru Welsh Water is active in community engagement by establishing education centres, and the installation of water coolers in schools.

Organising the provision of water and sanitation services through community management arrangements might facilitate a more *even distribution of water or water qualities across different geographical areas or groups of the population*. For the scenario of community management two different elements can be identified linked to the equity in water provision: (1) within the community, (2) outside of the community.

Within the community, it can be hypothesised that engaging into a community management set up will positively implicate the *equity in water provision*. As members of the water provider are in direct touch with the needs and emotions of the larger community, they will be better able to provide a fair distribution of services over the different geographical areas or community sub-groups. Also in the communication of the water provider it will be more likely that the distribution of services is perceived as fair. People within the community will not feel easily victimized or privileged compared to other groups within the community.

The equity issue provides a different picture once comparing the distribution of services of the community to its' surrounding neighbours. Often it can be seen that the better organised communities that are able to establish a community management model are communities that have strong social ties and have the financial and human resources to make it work. Such communities organise themselves possibly at the cost of surrounding poorer and less able areas that are excluded from this partnership. This exclusion can be intentional because the rest of the community does not regard their inputs as valuable to the success of this management mode, but it can also be unintentional, just because the poorer areas are not able to participate in this partnership. These outside areas still need to rely on more traditional ways of water provision, as for example through public management. So, if the community management model is restricted to the more wealthy areas of the overall population, this might mean that it will have negative social equity implications if we regard the differences between the area in which community management takes place and the surrounding areas. The community that is included in this management will make sure, by every means, that their water services are of good quality and offer low prices, even when that would go at the expense of their surrounding (poorer) neighbours. For example, water abstraction will be used to its' fullest for the community without taking primarily into account the implications this might have for neighbouring communities. A situation of sub-optimalisation will arise since there is no holistic overview of all of the area.

Employment

When community management is seen as one of the management options, one can analyse implications from two standpoints. The first situation is when the workforce in community management is basically the local people who manage everything for the community. This is the typical community management model. The people are generally volunteers or they are people within the community; and the community appoints them. These people are always compensated to do the job. The advantage of having such a workforce is that being from the community, the employment opportunities are generated within the community. Since the communities own the system they are more committed towards the management of the system and are easier to control. There are also certain negative aspects, like more chances of cheating and poorly managed systems, and a lack of specified knowledge. Involvement of volunteers does influence the number of employees. There will be no need to hire the people to do the job from outside the community. In this type of community management, external agencies in the form of experts' advisory body play an important role in terms of capacity building. It is for those involved in the management of the system who will benefit. These experts can be government, community workers, NGO's or even private sector depending on the intervention needed for capacity building. There are examples where NGO's played an active part in the provision of the services to communities.

In the case of community management, job security is generally high. It is one of the basic principles of community management that people in the communities have to be involved in the management of most of the work, which gives them job security. With this sense of responsibility and commitment, the status of employees does not change generally in the case of community management, as the employees are mostly chosen by the communities to do the job. The second situation is when community management outsource the operation and maintenance of the system. Outsourcing and the number of employees will depend on the size of the community and the complexity of the WSS system. For example in the case of Dutch water boards, the size of the water board has many implications in terms of participation and also the process of selecting representatives of different stakeholders. With an increase in the size of water boards, people are not aware anymore of the candidates for the elections in the water board assembly. This has also a negative implication in terms of participation, as only 20% of the people participate in the voting process for selection of stakeholders' representatives.

When the private sector is involved in community management, the risk (in terms of breach of contract) involved is higher as most community management forms do not have a legal status (Schouten, 2004). Under such circumstances, it is difficult to manage the contract with private parties without a legal status. The implications on workforce is analysed based on illustrative case from Wittenbach in Switzerland, where the operation and maintenance of the system is given to a private party and that private party is one of the members of the community. This is the case where employment is generated within the community but since the size of the community is small and the system is not very complex the workforce required is also very limited. The indicator like the degree of unionisation has less important implications, as it does not fit in the overall concept of community management. But corporate social responsibility as an indicator is very important as it protects the social values of the community management and has implications in terms of democratic elections, involvement of stakeholders in decision-making and increased level of accountability towards community.

Summary social implications scenario 6

From the above analysis, it can be assumed that the scenario of community management has positive implications on numerous indicators as related to access and employment. The involvement of the users in the service provision seems to relate positively to strive for affordable and accessible service provision. Some interesting observations are made related to the equity of the service provision. A negative implication might arise for the surrounding areas as the better able communities will manage to co-operate effectively together potentially at the cost of lesser able surrounding communities.

In the World Water Vision that was announced at The Hague in March 2000, a tri-partite alliance between government, civil society and the private sector was propagated, to seek "more community participation [in order to] provide a sense of ownership and empowerment to the stakeholders."(WWC, 2000). In spite of the felt urgency to establish community management in the water sector, as exemplified by the World Water Vision call, still there are currently only few cases truly representative for the community management scenario. An interesting example to analyse community management is the rejected proposal of the Kelda Group to mutualise Yorkshire Water. Mutualising means that the shares of the company are sold to the consumers of the company. Another illustration is the case described for Dŵr Cymru in Wales, that is owned by a not profit entity. In this entity members that are selected based on being representative of the service provision area execute the control and supervision of the water company.

CONCLUSION

This chapter aimed at analysing the social implications of the different scenarios. Two criteria have been taken into account in this research: access and employment. These two criteria constitute a major concern at European level: on the one hand, the definition of universal service clearly raises the question of the conditions of the right to water for all; on the other hand, the question of employment, and more particularly the nature and type of job contracts in a context of reforming of utilities, is subject to subsequent changes and it consequently raises many important questions.

Social implications are different from one scenario to another for indicators such as geographical equity (equal or unequal distribution across geographical areas), the nature of employment contracts, and the number of employees. As we saw in case studies like Rouen and Nantes, the local authority can induce public or private operators to harmonise prices across different areas, increasing geographical equity.

Regarding *access*, some indicators do not vary when looking at the case studies, like the access of water and sanitation services to all (rate of connection), the existence of social funds and tariff principles and equal opportunities. They depend on national policies and local decisions, except for scenario 1 in which there is an EU social policy, and member states are obliged to establish social funds, and disconnection for unpaid bills regarding poor households is prohibited. As for *employment*, social performance indicators have been published by multinational companies since 2000 and this provides interesting information on equal opportunities, and the number of employees. However, no information is given on non-core employees. Regarding the nature of employment contracts, these are mostly permanent employment contracts in private companies and public companies. However, the extent of outsourcing and the nature of employment contracts regarding non-core employees are not known to us, as statistics regarding this topic are tricky to find. Regarding the number of employees, there are many cases when the change from direct public management to delegated management led to important job cuts. This of course can increase productivity, but it would be interesting in the future to determine what the optimal level of employment is: when can one assess that a company over-employs or under-employs, and what are the consequences of employment on the quality of service (for example, the maintenance of networks can depend on the number of employees per connections)? Regarding equal opportunities, women are poorly represented in the water sector whatever the management mode be it public or private management, therefore, this indicator will not vary across scenarios.

Regarding **scenario 1**, a general EU social policy makes a significant change in the affordability of water services to poor households, taking the example of some member states like England and France. As for **scenario 2**, prices may vary significantly after each delegated management contract. Regarding poor households, unpaid bills will lead to water cuts in countries were this is allowed. Regarding **scenario 3**, cost savings through outsourcing may implicate lower tariff levels or a more gradual increase of tariff levels, meaning a higher affordability of water. As for employment, there may be a decline in the number of employees. Employee rights may deteriorate with short-term employment contracts. **Scenario 4** is characterized by contrasted access configurations according to the nature of regulation regarding social funds and disconnection policies, tariffs, and prices. If benchmarking is compulsory and there are sufficient credible threats for companies that under-employ, employment will be maintained. However, the actual number of employees is not always given as the extent of outsourcing is not published. The effect of benchmarking could therefore be limited as consumers have limited rationality: they could have difficulties in understanding and treating information according to their knowledge, and may not properly evaluate data given by benchmarking indicators. As for **scenario 5**, regarding affordability, there will certainly be an effort to keep the prices low. Regarding employment, the nature of contracts is very secure (i.e. civil servants). Finally for **scenario 6**, tariffs will be kept low, leading to a better affordability of water.

More needs to be achieved regarding research on the social implications of different scenarios to understand the changes regarding access and affordability to consumers. As Megginson and Netter (2001) note, economic literature often ignores the impact of private or public management of services on the welfare of consumers. Moreover, more research is needed on social implications of water management regarding employment, and more precise data is needed to understand the extent and impacts of outsourcing and non-core employment.

REFERENCES

Bakker K. (2001), "Paying for Water: Water Prices and Social Justice in England and Wales" Transactions of the Institute of British Geographers 26(2), pp. 143 - 164

BIPE/SPDE (2005), Les services collectives d'eau et d'assainissement en France. Données économiques, sociales et techniques, BIPE/SPDE.

Boulding K. (2004), The meaning of equity, The Progress Report. Available at http://www.progress.org

Bouygues (2003), Informations environnementales et sociales. Available on the Bouygues internet site.

Bröstrom S. (2004), Disconnection of domestic customers for non payment of water charges in the EU and the EFTA countries, Update of EUREAU report from June 1998, EUREAU, April.

Carpentier A., Nauges C., Reynaud A. and Thomas A. (2004a), Effets de la délégation sur le prix de l'eau potable. Une analyse à partir des résultats de la littérature sur les « effets de traitements », *Rapport interne*, INRA.

Carpentier A., C. Nauges, A. Reynaud and A. Thomas (2004b), « Une mesure de l'effet de la délégation sur le prix de l'eau potable en France », INRA Sciences Sociales. Recherches en Economie et Sociologie Rurales, n° 2/04, July, pp. 1-4.

Clarke G., Kosec K. Wallsten S. (2004), "Has Private Participation in Water and Sewerage Improved Coverage? Empirical Evidence from Latin America", World Bank Policy Research Working Paper, n° 3445, November.

Conseil de la Concurrence (2000), Avis n° 00-A-30 du 4 décembre 2000 relatif à une demande d'avis de la Fédération nationale des collectivités concédantes et régies sur la convention collective des entreprises des services d'eau et d'assainissement.

Corton M.L. (2003), "Benchmarking in the Latin American water sector: the case of Peru", Utilities Policy, vol. 11, pp. 133-142.

De La Motte (2005), WaterTime case study- Gdańsk, Poland, January. Available at www.watertime.org

Dore M., Kushner J., and Zumer K. (2004), "Privatization of water in the UK and France-What can we learn?", Utilities Policies, vol. 12, pp. 41-50.

Ernst J., (1994), Whose Utility? The Social Impact of Public Utility Privatization and Regulation in Britain, Milton Keynes, Open University Press.

European Commission (2003), European consumers and services of general interest, Qualitative Study, Eurobarometer.

Gray D. (2001), Equal opportunity, available at http://www.jcu.edu.au/office/eandsd/principles/equality.html

Green J. (2001), "England and Wales Water Industry Privatisation- A Case study", Water Aid, available at http://www.wateraid.org.uk/pdf/psp_england.pdf

Fauconnier I. (1999), The Privatization of Residential Water Supply and Sanitation Services: Social Equity Issues in the California and International Contexts, Berkeley Planning Journal, 13, pp. 37-73

Haarmeyer D. and Mody A. (1998), Worldwide Water Privatisation: Managing risks in water and sanitation, London, Financial Times Energy.

Hall D. (1997), Restructuring and Privatisation in the Public Utilities – Europe, Public Service International Research Unit, University of Greenwich, London, UK

Hall D. (2000), Impact of electricity privatisation on industrial relations – lessons from UK and Hungary, PSIRU, London, University of Greenwich.

ILO (1999), Managing the Privatization and Restructuring of Public Utilities (Water, Gas and Electricity), Geneva, 12-16 April. Available at the following address: www.ilo.org/public/English/dialogue/sector/techmeet/tmpu99/tmpure1.htm)

ILO (2001), The Impact of Decentralization and Privatization on Municipal Services, Geneva, ILO.

ILO (2003), Challenges and Opportunities facing Public Utilities, Geneva, ILO.

IFEN (2002), Les Comptes Economiques de l'Environnement en 2002, Paris, IFEN, Ministère de l'Ecologie et du Développement Durable.

INFOBASE EUROPE (2005), Unemployment in the EU Member States, Factsheet n° 078, March, available at www.ibeurope/Factfile/78unemp.htm

Holdsworth M. and Simpson R. (2004), Future water and Sewerage charges 2005-10, Response to Ofwat's periodic review 2004 draft determinations, London, National Consumer Council.

Lanz K. and Eitner K. (2005), WaterTime case study – Berlin, Germany, available at www.watertime.org

Lee T. (1996), "Alternatives for private participation in the provision of water services" Natural Resources Forum 20(4), pp. 333-341

Lisbon European Council (2000), Presidency conclusions, Lisbon Strategy, 23 and 24 March.

Lobina E. and Hall D. (2001), "UK Water Privatization- A Briefing", PSIRU.

Mairie de Paris (2004), L'Eau de Paris. Des Clés pour comprendre.

Maksimović, Č. and Tejada-Guibert, J.A. (2001), Frontiers in Urban Water Management. Deadlock or Hope, London, IWA Publishing.

McCommon C, Warner D and Yohalem D (1990), Community Management of Rural Water Supply and Sanitation Services, Washington D.C, The World Bank.

Megginson W. et Netter J. (2001), « From State to Market: A Survey of Empirical Studies on Privatization", Journal of Economic Literature, Juin, vol. 39, n° 2, pp. 321-389.

National consumer council (2004), Water for all. Campaigning for affordable and sustainable water supplies for all consumers, NCC.

Nantes Metropole (2005), « Regroupements communaux et transferts de compétences, quel mode de gestion choisir ? », 6ième Carrefour des Gestions Locales de l'Eau, Réunion du 3 février 2005, Direction Générale des Services Urbains, Mission Intégration des Services Publics.

Newbery D. (2000) Privatisation, restructuring and regulation of network utilities, Cambridge, MIT Press.

O'Connell-Davidson J. [1990], "The Commercialisation of Employment Relations: The Case of the Water Industry", Work, Employment and Society, 4 (4), pp. 531 – 549

OECD (2003), Social Issues in the Provision and Pricing of Water Services, Paris, OECD.

OFWAT (2003), Customer Charter. Our standards of service, OFWAT, Watervoice, Birmingham, July.

Outsourcing Journal [2001], Accessed through outsourcing-journal.com.

RWE (2001), Personnel Report, RWE. Available at:

RWE (2003), Personnel Report, RWE.

Sansom K., Franceys R., Njiru C. and Morales-Reyes J. (2003), Contracting Out Water and Sanitation Services Volume 1, Guidance Notes for Service and Management contracts in Developing Countries, WEDC, Loughborough University.

Sansom K., Franceys R., Njiru C. and Morales-Reyes J. (2003), Contracting Out Water and Sanitation Services Volume II, Case Studies and Analysis of Service and Management contracts in Developing Countries, WEDC, Loughborough University.

Saunders P. and Harris C. (1990), "Privatization and the Consumer", Sociology, 24, pp. 57 – 75

Schouten T. and Moriarty P. (2003), Community Water, Community Management: From System to service in rural areas, ITDG publishing, London, UK

Schouten T. and Moriarty P. (2004), Scaling up the Community Management of Rural Water Supply, Waterlines, Vol. 23, No. 2, pp. 2 - 4

Seidenstat P., Haarmeyer D., Hakim S. (2002), Reinventing water and wastewater systems. Global lessons for improving water management, New York, John Wiley and Sons.

Seppala O. T., Hukka J. J. and Katko T. S. (2001), Public Private Partnerships in Water and Sewerage Services: Privatization for Profit or Improvement of Service and Performance?, Public works Management and Policy, Vol. 6, No. 1, pp 42-58

Simon H., Egidi M., Marris R., et Viale R. (1992), Economics, Bounded Rationality and the Cognitive Revolution, Aldershot, Edward Elgar.

Smets H. (2001), Mise en oeuvre du droit à l'eau potable dans les pays de l'OCDE, February, OIEau, Cartel.

Spulber, N. and Sabbaghi A. (1994), Economics of Water Resources: From Regulation to Privatization, Natural Resource Management and Policy Series, Boston, Kluwer Academic Publishers.

Stahre P. and Adamsson J. (2002), Performance Benchmarking. A Powerful Management Tool for Water and Wastewater Utilities.

Suez [2003], Rapport d'Activité et Développement Durable.

Thomas D. (2001), "Welsh Water: role model or special case?", Utilities Policy, vol. 10, pp. 99 - 114

Thomas S. and Hall D. (2003), Restructuring and Outsourcing of Electricity Distribution in EU, PSIRU, London, University of Greenwich.

Thomas S., Hall D. and Popov V. (2002), Gas Industry Liberalisation, Restructuring and Employment in the European Union, Public Service International Research Unit, University of Greenwich, London, UK

Van Dijk M. P. (2003), Liberalization of Drinking Water in Europe and Developing Countries, UNESCO-IHE Institute for Water Education, Delft, The Netherlands

Vanclay F. (2002), "Conceptualizing social impacts", Environmental Impact Assessment Review, vol. 22, pp. 183–211.

Van Geel P. (2004), Innovative practices in the Drinking Water Supply in the Netherlands. Private business, Public owners, UNEP/GMEF, Jeju, Republic of South Korea, March 29-31.

Vickers J. and Yarrow G. (1988), Privatization: An Economic Analysis, London, MIT Press.

Chapter IX

Liberalisation and private sector involvement in WSS: the European experience

Authors
Antonio Massarutto[*], *Elisabetta Linares*[**] *and Vania Paccagnan*[***]

INTRODUCTION

Regulatory reforms in the last 30 years have provided a massive shift of the boundaries between state-controlled and market activities throughout the world; in the field of public utilities (PU), this has been achieved through new forms of regulation aiming at creating suitable conditions for competition to occur (Armstrong *et al.*, 1994). This trend has affected most network industries (energy, transport, telecommunications), as well as environmental utilities, including water supply and sanitation (WSS).

WSS services are now provided under a high variety of organizational systems throughout the world, with many different and alternative arrangements for involving the private sector (World Bank, 2005; 1999). While being commonly recognized as an activity that requires state intervention due to a multiplicity of market failures, they are also increasingly being considered as an elective terrain for private sector initiative (Rees, 1998). This occurs either because of the evidence of public sector flaws in the management of WSS (and, more broadly, of PU); or more simply, because the public sector faces constraints (financial, technological etc) that limit their capacity to invest and manage, particularly, but not only, in developing countries (Massarutto, 1993; OECD and World Bank, 2002).

International institutions such as the World Bank, the OECD, the WTO and the IMF have strongly pushed in favour of increased private sector participation (PSP), sometimes providing moral suasion, sometimes imposing it as a precondition for supporting financial aid to developing countries. The strategy for the achievement of Millennium Development Goals, with its compelling targets concerning water, is for the most part relying on private sector initiative (Winpenny *et al.*, 2003).

In more recent times, however, this attitude has been strongly criticized. Empirical evidence is definitely not supporting the claim of superior performances of the private sector, while it also demonstrates many failures in private provision of WSS (Lobina and Hall, 2003). In fact all experiences show that introducing competition in the WSS is quite difficult, and even when this occurs in some market segments, state regulation and direct management of other phases continues to be fundamental; it also shows that WSS is a very risky business and it is difficult – and above all increasingly costly – to ask the private sector to carry on this risk (Bayliss and Hall, 2002). The international literature

[*] Università di Udine and IEFE, Università Bocconi.
[**] IEFE, Università Bocconi.
[***] IEFE, Università Bocconi and DP, Università IUAV.

 Water and Liberalisation: European Water Scenarios. Edited by M. Finger, J. Allouche & P. Luís-Manso. ISBN: 1843391139. Published by IWA Publishing, London UK

assessing WSS performance is not conclusive with respect to the preferred strategy for PSP, showing that public sector–based solutions may still have an important role to play (Massarutto, 2005).

The present chapter provides a discussion of the main regulatory issues characterizing PSP in the WSS sector and develops an assessment model aiming at evaluating the outcomes of alternative scenarios, considering environmental, economic and social aspects.

Section 2 sketches out the main driving forces that bring along PSP and the emerging likely scenarios, with special focus on the European WSS market, namely technological complexity, financial needs and the division of competences between operators and planning authorities.

Section 3 develops an analysis of the economic literature and of existing international experience. Section 4 recalls the six alternative management models described in chapter 5. They can be represented through three archetypes derived by considering various institutional settings aiming at providing regulation in terms of ownership of water resources, division of responsibility in the WSS provision, infrastructure ownership, legal status of the WSS operator and so on. They are the Delegated Management (DM), the Regulated Monopoly (RM) and the Direct Public Management (DPM):

Section 5 contains the proposed liberalisation assessment model. It is based on a definition of social welfare that encompasses economic and social aspects. In other words, we will not only refer to economic efficiency, but we will consider the value creation for all the stakeholders involved and the distribution of this "net dividend" among them. We conclude that, even if different management models have different impact on total costs (i.e. the extent of the "net dividend"), nevertheless distributional impact is highly correlated with the allocation of risk among stakeholders, which is in turn dependent on the regulatory framework. We will finally provide the regulatory policy implications.

DEVELOPMENT OF THE WSS INDUSTRY AND DRIVERS OF PSP

For a long time, WSS have been basically local issues, managed quite simply by local communities with relatively small efforts and correspondently low input of technology and investment. Human settlements used to be conditioned and constrained by the availability of water resources as well as by the proximity of rivers or other "natural" ways to dispose of wastewater. Availability of water represented one of the most important "limits to growth" for human settlements (Massarutto, 1993).

Of course this statement is true only in general terms. Throughout history, powerful examples of water infrastructure still capture our admiration for their skilful technological solutions and the massive effort provided for realizing them. Urban development always called for the creation of waterworks aimed at making available enough water for an increasing and increasingly demanding population. The diffusion of sanitary standards during the 19th century in parallel with massive urbanization has been identified as the main driving force that led to the institution and overall diffusion of WSS services as we know them today, and the starting point of a specialized water industry (Correia, 1998).

Until recent times, however, the matching between water demand and water availability used to be searched through an "extensive" management model based on physical infrastructure, long-distance transfers, dams, reservoirs, aqueducts and other similar devices whose main philosophy was that of bringing in "from elsewhere" water that was not available locally, because of quantitative limits or inadequate quality. On the other hand, the problem with wastewater management was mostly that of safely getting rid of it in order to avoid consequences on public health; once wastewater could be brought far from urban settlements, it could be given back to the environment without too many concerns for the final quality of watercourses.

Technology for this kind of "extensive" water management systems is quite simple and developed on the basis of scientific knowledge that was already in place at the time of ancient Romans or during the Middle Age; the main problem lies in the availability of economic resources for realizing and later operating and replacing water infrastructure. As Barraqué (1995) puts it, what is "scarce" is not water, but money. In many countries, this model has lasted until very recently; many developing countries are now engaged in a similar strategy.

This feature helps us understand why the typical organizational pattern for this kind of water management system relies highly on centralized institutions and on the role of the state as the principal provider of economic means.

The most important sources of inputs to water management were therefore construction works and engineering of long-distance supply systems, dams and reservoirs. Very often, the necessary investment for these water resources management facilities was carried on by the State and financed from the public budget. The water policy network is dominated by "iron triangles" between water users, construction industry and public administration spending money in order to achieve consensus (Bressers et al., 1995). The role of the private sector in this model is basically that of supplying public works and construction.

At a certain point, the "extensive" model enters a crisis: available water resources are too distant and too costly to gather; construction of new water systems meets increasing social opposition; environmental quality of local water resources becomes a problem in itself; and more generally, urban development and social demands become so widespread and conflicting that the extensive model is not practicable anymore without severe external effects that turn out into increasing social costs.

This turning point starts a new phase of water management that we can label as "intensive": the focus is concentrated on integrated management of locally-available resources; technology becomes more complex and specialized in qualitative rather than quantitative issues; stronger and stronger emphasis is put on information, quality control, pollution prevention. This transformation of the object of water policy has obvious effects on the structure of the water industry, determining strong pressures towards PSP and thus generating as a feedback effect the need for new governance mechanisms (Barraqué, 1992).

New water policies impact on the equilibria of the water industry basically through four channels, each on its way leading to PSP, unless the public system is able to provide the required functions directly and effectively.

First, the increase of technical complexity of water management forces local water management systems to delegate decisions to professionals having the necessary expertise. This leads to the creation of specialized water management organizations, whose belonging to private or public sector follows a good deal national paths and traditions. On the other hand, specialization and increased complexity also means that less and less critical inputs can be produced "on site" and more and more are incorporated into "technological objects" and managerial skills that local operators will select and purchase on the market (Clausen and Rothgang, 2002; Prost, 1999). WSS operators are required to be in the best position to dialogue with external suppliers of technology, and this in the end might represent a strong incentive towards corporatisation and an increasing recourse to the market for the provision of inputs. The value chain of water services includes more and more value added that is produced by specialized firms on the market rather then directly by WSS operators (Kraemer, 1998).

In some countries, like France, the need for professional skills has led quite early to delegation of operation to private companies, instead than developing the same skills in-house[1] (Barraqué, 1995; Menard and Saussier, 2002). This is not an obligate route, however: Italian and German municipally-owned water companies, as well as the British Water Authorities before privatisation are well-known examples of WSS undertakings developed within the public sector, though with some sort of entrepreneurial autonomy. On the other hand, it means for sure that the public management system has to incur in fixed costs whose relevance in the service economy is growing higher. The more sunk costs are required in order to acquire and maintain appropriate managerial skills, the more these undertakings are forced to search for strategies aiming at keeping the average cost low (Massarutto, 2005). This is achieved for example by an increase of territorial size, in order to acquire economies of scale, or by diversification, through the creation of multiutility operators (Cozzi and Massarutto, 2002). Of course, this also means that in order to achieve greater efficiency, WSS undertakings should also become increasingly autonomous from political power and oriented towards a business-like behaviour.

At a certain point, the trade-off might result unsustainable: maintaining acceptable levels of efficiency is possible only by increasing degrees of freedom and entrepreneurial autonomy, until public control on activities becomes only nominal. The sale of shares to the market, often forced by budget constraints, completes the process, until companies finally behave as private companies (Drusiani *et al.*, 2004).

[1] Peculiar institutional mechanisms might influence this development; for example, in the French case, delegation to private companies was often made necessary by the difficulties that the French law opposes to the creation of local publicly-owned companies and by the difficulty for municipalities to have access to credit (Barraqué, 1995).

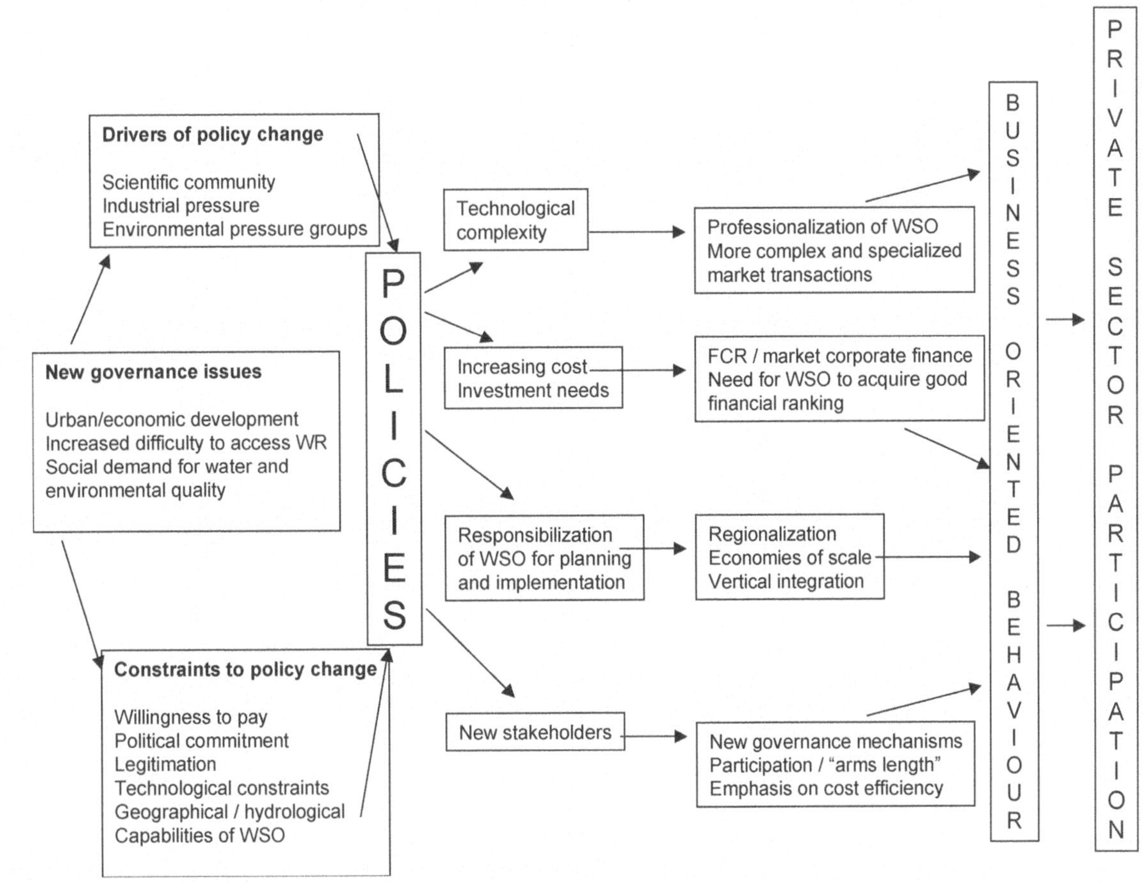

Note: WR = Water Resources; WSO= Water Service Operator; FCR= Full Cost Recovery

Figure 1: Drivers of private sector involvement in the WSS industry

The second channel is the financial one. The increasing cost of water services in the present phase is paralleled by the crisis of public finance and the imposition of upper limits to taxation and government budgets (Winpenny et al., 2003; Oecd, 2000). Self-finance through full-cost recovery becomes necessary: tariffs and prices paid by consumers will serve as a cash flow for sustaining market-based finance, for which the WSS operator becomes the intermediator.

Once again, this does not necessarily lead to private operation of services, since the public sector might have the capacity to access financial markets at even more favourable conditions than private companies; on the other hand, what matters for financial markets is the reputation of water companies and the guarantee of their faithfulness to a business-oriented philosophy (Buckland and Zabel, 1998).

The decision to involve the private sector can be therefore "political", in the sense that governments and local authorities might wish to sell property in order to give a signal that eventual losses in the future will not be transferred again on the public budget; privately-owned companies, from this point of view, will need to achieve a good balance between revenues and costs and therefore will not engage in new expenditure unless they will be able to finance it through prices paid by consumers[2] (The World Bank, 1997).

The third channel regards the division of competences between WSS operators and water resources planning. In the traditional model, water management is mostly concerned with operation, while infrastructure is planned, financed and realized in the frame of central government policies (Massarutto, 1993).

[2] The history of British Water Authorities is enlightening. When they were created in 1973, they were required to operate on a cost recovery base; however, charges used to be kept low for political reasons, while significant new investment had to be carried on; so the WA experienced a substantial rise of their debt, what was possible only because financial markets treated this debt as if was warranted by the central government. At the moment of privatising the Water Authorities in 1989, the British government had to write off this debt and absorbe it into the public debt, otherwise the whole operation would be a failure (Barraqué, 1995; Rees and Zabel, 1998)

In the new paradigm the water industry is required to play a more important and responsible role, with substantial degrees of freedom; one important consequence is that WSS operators should become responsible of taking investment decisions and therefore acquire both suitable geographical scale and vertical integration.

In the new division of competences, public functions are more concerned on regulation of water uses, setting quality objectives and enforcement, while WSS companies become responsible for the integrated management of infrastructure together with service operation. This is, once again, a mighty force that changes the adequate territorial size (from municipal to intermunicipal or regional dimension) and, above all, multiplies the tasks that operators have to perform and the level of economic risk that it is required to bear.

Finally, the fourth channel regards the increased conflictuality of the water policy arena and the emerging need to establish an "arm's length" regulatory system. WSS operation, while taking care of a service of general interest, cannot claim to represent the whole of the public interests concerned with water. In order to guarantee a credible enforcement of environmental and quality regulation, autonomy of WSS undertakings from water policy is an important prerequisite (Massarutto, 1993). Once again, this can be achieved within the public sector; yet many examples tend to demonstrate that some sort of PSP of the WSS significantly increases the degree of antagonism between regulators and water companies and thus of regulatory effectiveness[3].

ANALYSIS OF LIBERALISATION AND PSP OPPORTUNITIES IN THE WSS SECTOR

Following the analytical approach developed in the Euromarket project (see chapter 1 and van Dijk *et al.*, 2003), we can identify in the WSS sector 4 different sets of transactions among relevant actors, each of whom raises specific governance issues. Each of these sets of transaction identifies a particular submarket, namely the "market for drinking water and sewerage services" (given by the transactions between customers and the service provider); the "supplier markets" (i.e. the transactions between the WSS operator and its inputs suppliers); the "market for water abstraction and discharge" (i.e. the transactions between the WSS operator and the "owner" of the resource) and the "market for delegated contracts" (i.e. the transactions between the responsible entity and the WSS operator). All these submarkets are extensively described in chapter 1, so we will not go into more details here. What is worthy to note is that each of these transaction layers can, in principle, be opened to (some) competition such as:

a. outsourcing, public-private partnerships in DBFO (Develop Build Finance Operate), market for procurement, market for corporate ownership of water companies for what concerns the supplier market;
b. water markets and tradability of water use rights in the water market;
c. eligibility of final customers and local communities for autonomous service provision in the operator market.

The degree of competitiveness of the WSS industry cannot be reduced to the degree of competition occurring only in the customer market; quite often, increasing competition in this segment leads to fewer opportunities to compete in other segments since industry tends to react with increased concentration and vertical integration. As a consequence, liberalisation opportunities should be analysed considering all the submarkets described above. This because interrelations between transaction layers are, nonetheless, important and should be adequately considered. In many cases, introducing competition in one segment leads to fewer opportunities to compete in other segments, or might hinder the capacity to achieve water policy objectives that pertain to those segments.

The starting point of the analysis is the market failures that characterise the WSS provision. They are summarised in Table 1 below (Cowan, 1997; 1994; 1993). The pervasiveness of these market failures in all the different submarkets make it crucial to set up a regulatory system that is able to tackle them.

In order to understand which PSP models are suitable for the WSS sector, we can refer to the "policy roadmap to regulation" presented in figure 2 (adapted from Braeutigam, 1989), identifying the main variables to be checked in order to understand if PSP can be made compatible with the presence of market failures (Massarutto, 2005).

[3] The British case is again enlightening. Before privatization, the Water Authorities were responsible both for WSS operation and for administering water policy. The level of compliance with the standards that they themselves had established was estimated below 15%, and no prosecution did ever occur. After privatization, compliance levels have rapidly increased to close to 100% (Summerton, 1998).

Table 1: Transactions in the WSS value chain and related market failures

Market	Description	Regulatory issues / market failures
I.	Transactions between the WSS operator and public entities holding the responsibility for service provision	Incomplete contracts Transactions costs Sunk costs Information asymmetries
II.	Transactions between the WSS operator and suppliers of inputs along the value chain	Vertical integration Cost of capital for long-run undertakings Principal-agent relations in procurement
III.	Transactions between WSS operator and entities holding the property rights on natural resources	Externalities Long-run sustainability of water management systems Transactions costs in the trade of water rights
IV.	Transactions between WSS operators and final consumers	Natural monopoly Public good dimensions (e.g. health issues) Accessibility and affordability issues Resilience and flexibility

Generally speaking, the starting point is to evaluate the viability of introducing competition in the market. If it is possible, there is no need for regulation and the sector can be liberalised. At the contrary, when competition cannot be immediately introduced, it is of interest to detect whether there are some major economic losses due to the monopolistic features of the industry. If not, we do not need regulation and the market can be liberalised, once that some qualitative parameters have been defined. If the answer is yes, then some alternative forms of introducing competition can be considered, i.e. competition for the market. The next step is thus to verify if some form of competition can be introduced in some segments of the market. If competition for the market is viable (i.e. there are no massive transaction costs and information asymmetries), then competitive tenders can be introduced. Otherwise, in all the cases where the contracts cannot be specified, some form of tariff regulation is needed, i.e. price caps, marginal or average cost pricing.

Having this map in mind, in order to understand and check liberalisation opportunities, the main WSS features to be considered are (see also chapter 6 of the book) the *dimensions of general interest*, the risk dimensions and the WSS specificities that could make difficult to introduce competition in this sector:

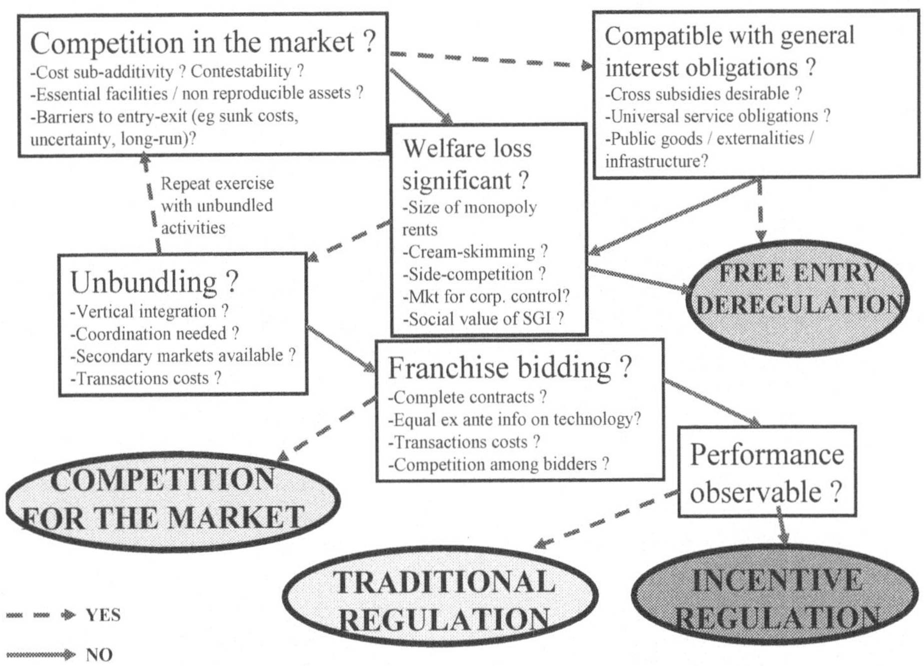

Figure 2: The roadmap for regulating services of general interest: the main questions to verify in order to assess suitability of alternative liberalisation models

For what concerns the dimensions of general interest, they are widespread and fundamental, including public good dimensions, externalities and equity issues. With this respect, the evaluation of WSS performance includes many different aspects:

- as a network industry, WSS operator should be able to guarantee its customers value for money and avoid exploiting its monopoly position;
- as a service implying consequences in terms of public health and quality of urban life, it should ensure adequate quality standards;
- as a utility that deals with a vulnerable natural resource, it should respect all the requirements arising from environmental policy at different levels (local, river basin, regional or national level);
- as an essential service necessary for human life, WSS should be accessible and affordable;
- as an infrastructure having a very long economic life and entailing substantial sunk costs, it should guarantee long-term resilience and adequate investment levels in order to maintain the value of assets over time.

Moreover, introduction of competition in the case of WSS sector should take into account that (Rees, 1998; Beecher, 2001):

- *Unbundling* between phases (raw water extraction, drinking water production, distribution, wastewater collection and sewage treatment) is possible, though both services remain natural monopolies even if isolated;
- *Competition for the market* is well possible, but the sector fails to satisfy requirements for perfect contestability. In particular, information asymmetries and transactions cost protect incumbent against new entrants and are the cause of incomplete contracts along the I market. For this reason, tenders cannot be based on price only, unless they are effectuated only for very simple and short-termed tasks, what would ultimately require that the public be responsible for long-term asset management and planning. Conditions are better satisfied, in turn, once operation and asset management are treated as separate businesses and/or if private commitment in asset management and development and its assumption of risk is limited and corresponded by adequate compensation;
- *Regulated monopoly* is an adequate solution provided regulators are strong enough so as to reduce information asymmetries and quality regulation is detailed enough and credible. Efficient capital markets (for takeover), systematic benchmarking and yardstick competition are fundamental. Concerns on long-term sustainability are nonetheless raised, since the regulatory system is concerned on short-term and depreciation of existing assets not guaranteed;
- Limited potential for *competition in the IV market* due to physical indivisibilities in the WSS network, economies of scale in service provision and negligible importance of variable costs; also, limited interest for common carriage, since essential facilities represent the most part of the total cost.

Risk dimensions are correlated in particular to the very long life of assets. Being investment sunk for a long time, guarantee that costs are actually recovered depends on the credibility of commitment of public entities to ensure water company viability as well as from the actual willingness and ability to pay of water users (table 2). Market risks are present in the segments open to competition.

Different patterns of risk allocation are possible. In principle, risk can be borne by a combination of subjects, namely company shareholders (lower profitability); actors along the supply chain (e.g. reduced level of employment or guarantees for workers); consumers (higher tariffs, lower quality of service); other water policy stakeholders (negative externalities); public sector and taxpayers (need for subsidies to be financed out of general taxation); future generations (disruption and uncompensated depreciation of natural capital and/or artificial assets; lack of resilience). Risks can be allocated in different ways through regulatory devices. In each institutional model, pricing rules, tenders and contracts, quality regulation and similar instruments can be adopted to obtain the desired outcome (OECD, 1997).

For instance, minimizing WSS costs today could also mean that costs are simply transferred to the next generations, in terms of bad maintenance, insufficient replacement, inadequate protection of vulnerable natural resources, poor development of the system, inadequate investment in information systems and human capital, etc. Financing capital expenditure under public mechanisms or with guarantees provided by the public sector might reduce the cost of capital in the short term, but also encourage an excess of leverage, favoured by the expectation of investors that the state will impede eventual defaults (Barraqué, 1999). This implies a higher risk for taxpayers in the long run. Increasing the risk borne by company shareholders (e.g. via fixed-price contracts or strict enforcement of performance standards) reduces the willingness of private companies to invest and engage in sunk costs, long-term involvement and to accept responsibility for the system. The remuneration requested by private capital (expected profit, interests) is a growing function of the scope of responsibility (which can range from ordinary maintenance and operations only to asset management and development) and of the subsequent risks borne by investors.

As we can see, opportunities for PSP are well present, yet severely constrained (Oecd, 2004). Nonetheless, once the policymaker is interested in boosting PSP, he has to face a crucial trade off. On the one side, given the monopolistic and other "market failure" feature of the WSS and the limited scope for unbundling, the state is required to fulfil complex regulatory tasks and acquire an adequate contractual power. On the other side, however, a higher regulatory capacity reduces interest for private sector to participate. In general, even if a higher regulatory capacity increases transparency, a strong regulator will try to extract monopoly rents from the WSS operator. Moreover, if the regulatory framework is not predictable and stable, this will increase regulatory risk. Solutions to this trade-off are not simple. An important reason of this difficulty is the interdependence of the four markets: solving for market failures in one of them is well possible, but it tends to create the case for more fundamental market failures in one or the other markets[4]. Moreover, regulatory reforms aimed at improving the overall performance are hardly possible without a substantial reallocation of costs, benefits and risks. This process has winners and losers; the decision on how costs, benefits and risks are to be allocated is a political choice, since an objective and scientifically founded determination of an "optimal" solution is not possible.

Table 2: Principal typologies of risk in the water industry and their main determinants (*Source*: adapted from Pricewaterhouse&Coopers *et al.*, 2004)

Typology of risk	Determinants
Performance risk	Need to meet demand or quality standards or universal service obligations
Operational	Status of knowledge on infrastructure and investment needs; Performance standards to be guaranteed and allowed flexibility margins; Patterns of water demand in relation with tariff structure (e.g. water consumption if tariffs are based on volumetric charges)
Tariff	Perspective tariff dynamics vs. operators' commitments; Cost-based (RoR) vs. price caps; Cost pass-through and guarantees of minimum total revenue
Capital risk	Expectations on capital remuneration; Measurability of risk exposure, sizing of facilities, appropriateness of technological choices; Guarantee of the recovery of all the sunk costs at the end of the contract
Competition	Contract length; Barriers to entry and exit; Post-contract arrangements (e.g. sunk costs compensation after contract termination)
Regulatory	Degree of clarity and coherence of regulation. Financial viability of operators and recovery of sunk costs Stability and predictability of legislative framework Discretional margin for regulators Provision for new environmental / quality regulation in the contract. Transfer of the corresponding cost into tariffs?
Municipal	Financial standing of municipality and capacity to respect financial obligations
Political commitment	Commitment of regulators to maintain prices at a level that covers quasi-rents and not only marginal costs. Political attitude towards private companies; Political commitment to ensure financial viability of operating company
Instability	General economic situation, currency issues, financial rating of the country etc
Civil society	Social attitude towards private companies and pricing policies
Country interest	Expectations on GDP growth, water market development, long-term perspectives

[4] For example, the economic theory suggests that in order for competitive tendering to be efficient, tenders should be based on single performance indicators (e.g. price), contracts should be complete (they should foresee all potentially relevant events) and all relevant information should be shared by participants to the tender (also meaning that this information should be handled by the public authority and communicated to all bidders). It also suggests that short-term contracts are more suitable for this purpose. This model can be applied more easily for single activities (e.g. managing a wastewater treatment plant) than for the integrated service as a whole: but this means that in order to maximise the scope for competitive tendering, all the more risky and uncertain activities (such as AMD) should be provided by the public sector.

The consequence of the trade-off described above can be described more clearly with the aid of figure 3, representing the value chain of WSS services (Kraemer, 1998). In the figure we distinguish a *service core sphere* (responsibility, asset management and operation), a *market sphere* (activities to be supplied in order to provide WSS services) and a *regulatory sphere* (activities that are performed by the public sector). Different layouts are used for representing transactions in the different markets. For what concerns the service core sphere, delegating investment responsibilities to private operators requires long-term contracts, flexibility of obligations and other provisions to lower the market risk; this is not compatible with writing complete contracts, unless frequent contract renegotiations are envisaged. The applicability of tenders based solely on cost is also extremely difficult. Conditions for maximizing tender efficiency (short-term, complete contracts etc) require the responsible entity to be able to perform coordination tasks and ensure long-term planning of the system. Combining far-reaching tenders with short-term and complete contracts is extremely difficult, and in any case entails extremely high transactions costs for the tendering process, contract enforcement, litigation, management of handover between incumbents and new entrants etc.

Regarding the "market core", separating different phases and functions can help to single out market segments that are more suitable for competition than integrated systems; however, this entails transactions costs and places on the responsible entity enormous planning responsibilities. Separation of operation and capital asset management might reduce the risk for the private sector and thus the need to remunerate it; in turn, it might favour dualism (capital investment not coordinated with operational exigencies) or foster unnecessary investment by the operator (since it does not carry the corresponding risk, and can profit in various ways from over-investment). Guaranteeing current and future performance of WSS is more easily achieved if investors are released from the operational risks and allowed to transfer costs onto prices and/or if extra costs are borne by the responsible entity. However, this also reduces incentives to efficiency and possibly encourages over-investment, gold-plating, over-staffing.

Finally, considering the regulatory sphere, increasing the effectiveness of regulation requires an increase in the discretional power and resources used by regulators; this can result in higher regulatory costs and in the risk of arbitrariness, against which appropriate institutional counterbalances should be made available.

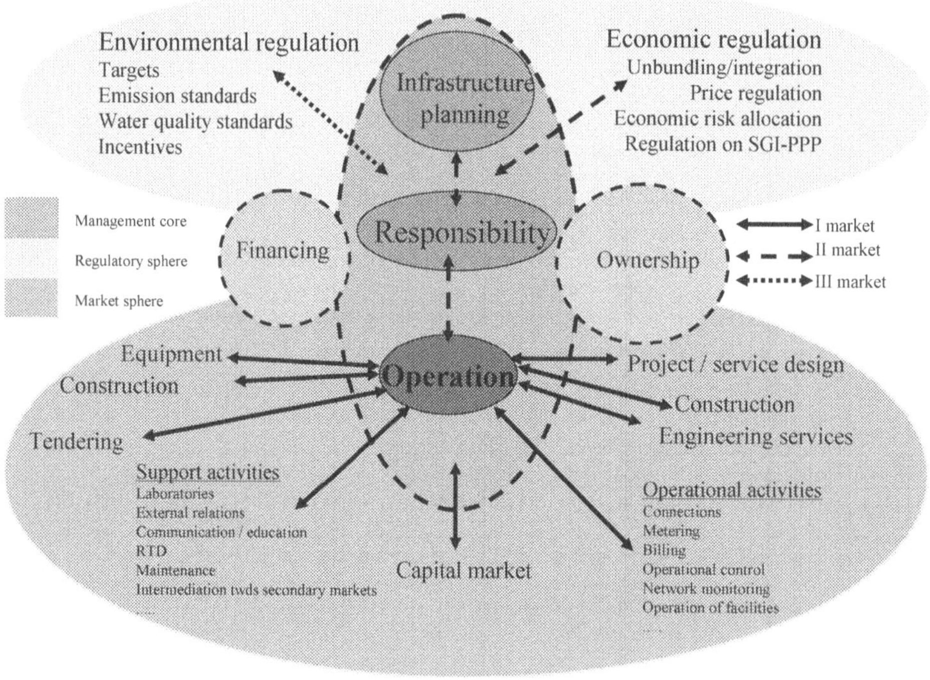

Source: Adapted from Kraemer (1998)

Figure 3: Organisation of the WSS value chain

To sum up, theoretical analysis and empirical evidence underline two main trade-off occurring. The first one concerns primary and secondary markets. The more the primary market is based on market transactions (tenders, contracts etc), the more the industry seems to tend towards vertical integration in the secondary market; vice-versa, when the primary market is based on the incumbent operator's monopoly, this shows a greater tendency to purchase inputs from the market and outsource (Kraemer, 1998). This can be explained by considering that local monopolies are not necessarily reaching a scale that is optimal for all phases. On the one hand if historical, institutional and political reasons force local monopolies to small scales, their capacity to perform activities in house will be reduced, and the likelihood of a recourse to the market higher. This will be more probable, if price regulation does not allow operators to transfer costs on consumers. On the other hand, competitive tendering in the primary market increases risks; vertical integration helps private operators to reduce the risk by increasing the economic margin of the transaction and by raising information barriers against potential new entrants or even against regulators[5].

The second trade-off concerns risk assumption, with particular reference to asset management development (AMD). Here a reversed mechanism seems to apply: the more the private sector is involved in operation and faces some market and operational risk, the more it tends to refuse responsibility on AMD; vice-versa this responsibility is more easily accepted when these operational and market risks are reduced (e.g. because contract terms are much longer or even eternal, because contractual clauses allow to pass extra costs onto tariffs). This can be easily understood by recognizing that AMD contracts are exposed to significant transactions costs arising from information asymmetries and opportunistic behaviour. On one side, once the investment is sunk, responsible entities might be tempted to impede price increase over the marginal cost, thereby appropriating the "quasi-rents" and making it impossible for private companies to recover the fixed cost (Noll et al., 2000); on the other side, the company might easily capture regulatory decision and force technological and investment choices (Lobina and Hall, 2003).

In fact asset ownership, management and development are almost never definitively allocated in the service core or in the external spheres; rather, they are posed somewhere in-between, with a continuum of solutions based on sharing competences, responsibilities and risk. The more they are allocated on the private sector, the higher is normally the time horizon and the correlated risk premium (OECD, 2000); on the other hand, when these competences are separated from operation and remain in the public sector, contracts are more easily released for operation only. The public sector can later on adopt other arrangements for PSP in that particular phase (such as DBFO), while pursuing autonomous strategies for operation.

ALTERNATIVE MODELS FOR ORGANISING WSS SERVICES PROVISION

The existence of a multiple dimension of market failures helps understand why so many alternative solutions have been experienced throughout the world. Fundamentally, alternative solutions differ on the basis of the answers to the following questions (figure 3):

- Who is the owner of water resources? On what basis are abstraction and discharge rights assigned and what degrees of freedom are available for eventually trading these rights in favour of other subjects?
- Is there a legal responsibility for providing services (i.e. is there a subject who is responsible to guarantee that service is provided at certain conditions)? Is this subject a public entity or a water company?
- Who is the owner of assets and infrastructure used for providing the service?
- On what basis is the operator chosen? Is it a public entity or a private firm?
- What are the main sources of economic risk, and how is this risk shared between company shareholders, public entities (taxpayers) and consumers?

By considering these analytical dimensions, we can distinguish three main alternative models, whose main features are resumed in table 3 below[6].

We stylised the analysis by considering only three (instead of six) scenarios, because the two delegations scenarios can be considered as two extremes of a continuum. Moreover, outsourcing and community management models do not constitute two different archetypes, but they only contribute to change some limited features of the main models. For example, in the regulated monopoly or in the direct management model, outsourcing places some operation activities in the market sphere instead of in the management core sphere.

Figure 4a, 4b and 4c (see the annex) represent each model in terms of the value change structure analyzed in figure 3 above. Transactions between the different components of the value chain are represented with different layouts in order to distinguish market transactions (normal lines), regulatory transactions (bold lines) and hierarchical transactions (belonging to the same sphere without linking rows). In the delegated model the tasks referred to core sphere (in terms of infrastructure planning and ownership) are responsibilities of the public authorities, through environmental and

[5] Industrial economics normally considers transactions costs as the main explanatory factor for vertical integration. We suggest in turn that market risk reduction is a more powerful driving force in this case; transactions costs in the II market do not appear to be significant, even if it should be emphasized that efficiency in procurement is directly linked with knowledge and technical capabilities of operator.

[6] Kraemer, 1998, refer to them as French, British and German model, because these countries represent their most typical example; even though, an evolution is taking place in each of them, while other countries exhibit similar or alternative combinations. We prefer to consider the three models as archetypes

economic regulation. The WSS financing responsibilities are normally split between public authorities and operators. Finally, market sphere regards mainly operation of the WSS system.

In the regulated monopoly the management core sphere is bigger than in the previous case, encompassing infrastructure planning, ownership and operation responsibilities. Nonetheless, the market sphere retains certain financing responsibilities.

Table 3: WSS management models and their most important features

	Delegated Management	Regulated monopoly	Direct public management
Main example	France	England and Wales	Germany
Other examples (often involving variants)	ITA (few), ESP, East European countries		NL, SWE, ITA AUT, GRE
Ownership of water resources	Public Variant I (US): water rights can be privately appropriated and transferred Variant II (some limited cases in Europe): licensed users can trade use rights under public supervision		
Responsible entity	Municipality, ev. voluntarily or compulsorily associated	Water company	Municipality, ev. associated
Ownership of water companies	Private	Private and public	Public (responsible entity); Often multiutilities; Minority shares for private partners or other public bodies
Legal form of operator	Private company with concession contract; Variant: specific purpose company jointly owned by municipalities and private companies	Private-law company contract and being owned by a parent company	Public law arrangements with different degrees of autonomy and governance structure; Private law companies (with obligation for unbundling in case it is also active on other markets)
Ownership of assets and infrastructure	Responsible entity, either directly or through dedicated private law companies (poss. mixed during the contract lifetime)	Operator; Variant: assets owned by public entities and/or consumers' associations	Responsible entity; Variant: users' associations and individual consumers
Choice of operator	Tender (based on a mixture of economic offer and "beauty contests"); Variant (ITA): tender for partner in mixed venture companies, with majority shares in public hands but operational decisions fully delegated to private partner	Direct (usually resulting from partial or total privatization of previously publicly-owned firms); Variants: municipality may retain shares and even the majority of shares; governance rules protect the private	Direct
Structure of the industry along the value chain	Vertical integration aimed at maximising the value added produced in-house and minimize contract, operation and market risks	Vertical integration on make/buy considerations; "Contract companies" usually produce only activities that require sunk costs that are specific for that contract; other activities are purchased from the parent company and/or from the market	Vertical disintegration; Market for procurement; Joint ventures and PPP for specific initiatives (DBFO); Variant/trend: Increased outsourcing of service activities as well as procurement; Horizontal integration of distribution with gas/electricity; Sewage treatment with waste management; Dedicated companies for raw water production
Patterns of competition	Competition for the I market	No competition in the I market; Inset appointments (possibility for new customers to choose their preferred operator with ev. bulk supply from the main one); Competition for company ownership in the stock exchange market; Yardstick competition (benchmarking)	Competition in the II market; Procurement and outsourcing

Finally, in the direct public management model planning authorities hold responsibilities in terms of infrastructure planning. The market sphere is more limited compared with the regulated monopoly one.

Each model provides different solutions to the trade-off discussed in the previous section and entail different regulatory issues that are summarized in table 4.

Table 4: The WSS models of management with the related regulatory issues

	Delegated Management	**Regulated monopoly**	**Direct public management**
Regulatory system	Quality standards and environmental standards exogenously imposed; No price regulation (tender + direct agreements between responsible entity and private partner); Voluntary instruments (e.g. service charters)	Quality regulator; Environmental regulator; Economic regulator (price-caps on 5 year base); Service charters; Variant: service contracts	No explicit regulation; Soft-regulation and benchmarking for providing comparative info; Variant: public sector contracts; service charters
Main regulatory instruments	Tenders; Contracts; Performance standards	Price-caps and performance standards enforced by independent regulatory authorities	Hierarchical control; Benchmarking; Voice
Main regulatory issues	Contract completeness; Regulatory capture in the renegotiation phase; Dealing with sunk costs after termination of contracts; Conflict resolution; Barriers to competition in the next bids due to information advantages of incumbent	Site specificities in the cost function; Transfer prices and in-house transactions; Easy to parameter OPEX and cost of assets, very difficult to assess AMD as a whole	Cost inflation and limited incentive to enhance productivity; Gold-plating and excessive investment; Less credible enforcement of quality targets
Non-economic factors	Corruption; Collusion between private companies and elected people (may be enhanced in case of PPP)	Reputation; Conflict btw regulators	Extra costs required for meeting political agendas of elected people; Corruption

Fundamentally, the Delegated model relies on tenders and contracts, the Regulated Monopoly one on regulatory authorities specifying service obligations and price limits, the Direct Public Management one on hierarchical control by the responsible entity.

ASSESSMENT

One of the reasons why an assessment of PSP (and more generally of regulatory reforms) is so difficult is the fact that performance involves a great number of critical dimensions regarding different aspects of individual and social welfare (Dore et al., 2004). A mistake that should be avoided is that of referring only to some of these dimensions and neglecting the others. In particular, many studies on privatization are limited to considering the outcome on managing companies (in terms of profitability, cost recovery, viability). In fact it is quite common (and also easy to understand) that privatized companies show better economic performance; however, this can be due to improved efficiency, but also to higher prices, reduced investment or quality levels, higher transfers from the public budget etc (Megginson and Netter, 2001; Renzetti and Dupont, 2003).

In the case of WSS and other environmental services, it is crucial to consider other aspects than economic ones, namely environmental and social dimensions (de Carli et al., 2003; Gleick, 1998; Dinar, 2000). In fact, WSS services give the collective access to the natural resource, by transporting it until the point where water is demanded and then discharged the same resource, through the sewerage service. It is straightforward that WSS have a strong environmental impact, both in qualitative and quantitative terms, which should be taken into account in any assessment exercise. Equally, the social dimension of WSS management is pervading. The access to clean water and sanitation is considered one of the Millennium Development Goals of the 2002 UN Johannesburg summit.

For the assessment of regulatory models and options for reform we have developed an approach that is based on a definition of social welfare entailing all possible stakeholders and adopted a procedure based on two steps:

- Assessing the "net social dividend" of regulatory reform: this is the net overall difference between service costs, before and after the reform, and regardless who actually pays for them. The net dividend could be positive (meaning an increase in overall efficiency) or negative; it can arise both from a reduction of x-inefficiencies (production costs closer to theoretical optimum) or from the elimination of deadweight losses
- Assessing how this net dividend is shared among stakeholders: this depends on regulatory arrangements, but is not completely neutral (in the sense that not all stakeholders can benefit from the same net dividend in each scenario). Net benefit for a given stakeholder may be negative even if the net social dividend is positive[7].

In table 5 we have summarized the main welfare components that should be considered in this assessment exercise, the related stakeholders and some indicators. As the reader can see, the assessment of different management models considers all the agents that are affected by the WSS management choices, either directly and indirectly. These indicators encompass economic, environmental and social aspects.

Having these indicators as a point of reference, we can attempt to evaluate the transition between different management models.

Table 5: Stakeholders of WSS and concerned outcomes of policy reform

Stakeholders	Main concerned outcomes	Indicators
Companies and shareholders	Profitability; Corporate value	ROI / EVA; Net asset value
Responsible entities	Positive externalities (e.g. public health); Synergies w/ public goods provision (e.g. rainwater management); Coherence with urban development patterns; Financial transfers (royalties, revenues from privatisation etc)	Water-borne diseases; Concession fee and royalties
Customers	Value for money; Affordability; Accessibility; Sizing / Capacity to meet peak demand	Annual cost per hh; % of water bill on hh budget; Service interruptions
Environment / other water uses	Achievement of water quality objectives; Compliance with environmental and health protection standards	Compliance w/ regulation; Contribution of WSS to water environment degradation
Suppliers of inputs (e.g. workers)	Volume of economic transactions with the water sector; Levels and quality of employment; Salaries; Technological development	Value added and its composition; Mean labour cost
Next generations	Maintain the system financially viable in the long term; Guarantee service functionality / resilience; Adopt state-of-the-art appropriate technologies	Actual investment / real depreciation; Degradation of water environment
Taxpayers	Reduce impact of WSS on general budget	Net contribution from/to the public budget

As stated above, the first step is to define the "net social dividend". It can be assessed by considering the difference in total costs occurring before and after the reform. Table 6 summarizes the main costs components to be considered and states a likely cost trend. Regarding operational costs, the WSS management model with the strong incentive to cost reduction is the regulated monopoly, whilst the one with the weakest incentive is the direct public management. Incentives in the case of the delegated management model are highly dependent on the awarding criteria. With respect to infrastructure costs, we cannot draw a general conclusion. Investments in infrastructure maintenance and renewal are carried out, provided that WSS operator is sure it will have them recovered through tariffs and during a time horizon compatible with the management one. The cost of capital is lower in the DPM model than in other models. Similarly, transaction costs are lower in the DPM model and increase as the degree of delegation increases (Correljé *et al.*, 2005).

To sum up, the three models perform differently with respect to the various components of costs. On the one hand, some reduction of operational cost due to increased productivity is likely to be stronger if competitive incentives are stronger. This is not simply due to PSP, but rather depends on the guarantee that the operator has on revenues. We can believe in any case that these incentives are higher in the RM (especially if price caps are adopted) and lower in DPM. Delegated model has some potential for cost reduction, but awarding criteria and contract negotiation patterns may

[7] For example, tariffs might increase even if service is more efficient, because of higher profits or higher investment.

favour instead some cost inflation that is profited by suppliers belonging to the same group as the operator. On the other hand, capital costs and depreciation costs are likely to be lower in the case of DPM, since long-term schedules can be adopted and interest rates are lower[8]. A further cost component on which regulatory models impact differently is transactions costs (that are presumably higher in the delegated model (tender, conflict resolution, consultants etc) and lower in the DPM).

It is worthy to note that the environmental impact assessment is implicit in the definition of infrastructure costs. As noted above, water resource availability is highly dependent on the investments done in infrastructure maintenance and renewal. The minimisation of the costs, as a consequence, should not be attained at expenses of environmental objectives.

The resulting net effect on costs is not easy to predict. Empirical observation is inconclusive, for two reasons: the first one is that service costs are not the same everywhere because of local specificities in the cost function; a benchmarking should be made on individual cost components, what is actually impossible with the available data. The second and more important reason is that different models correspond also to different way to calculate the cost and define levels of recovery (Massarutto, 2005)[9].

Table 6: Expected overall net dividend: drivers of cost increase and reduction

	Delegated Management	Regulated monopoly	Direct public management
Operational cost	Depends on awarding criteria: *low incentive* if tenders are based on beauty contests and/or protect incumbents; *high incentive* if based on fixed price	Permanent incentive to reduce costs particularly with price-cap based regulation	Overall weak incentive to reduce costs in order to avoid conflicts with trade unions; Quasi-market mechanisms (e.g. subsidy caps), price caps and appropriate management rewarding schemes can provide higher incentives, but lower than in the RM due to the reduced risk of bankruptcy; Threaten to privatize can be effective if credible
Infrastructure cost	*Lowering factor*: willingness to reduce conflicts with local people and avoid unnecessary investment; concessions provide permanent disincentive to invest if not explicitly foreseen in the contract *Increasing factor*: operator is vertically integrated with construction industry, consultancies, equipment manufacturing	*Lowering factor*: price-caps provide permanent incentive to minimize capital expenditure *Increasing factor*: cost pass-through and cost-based regulation provide permanent incentives to expand investment; market often requires shorter repayment schedules	*Lowering factor*: possibility to depreciate over longer time schedules *Increasing factor*: gold-plating, possibility to invest (enhanced if the WSS system is more autonomous from local administration and tariffs are based on FCR); lower incentive to make agreements with neighbouring services in order to share infrastructure and sunk costs
Cost of capital	High, proportional to risk effectively borne	Medium-high; risk that regulator underestimates it for keeping price low	Low, but constrained by public finance conditions, country rating etc
Transactions costs	High (tender, contract, enforcement, monitoring, conflict resolution)	Medium-high (regulatory agencies, reporting)	Medium-low (higher if some of the above remedies are adopted)

[8] This statement depends of course on the financial rating of the country (and of municipal authorities). In developed countries, governments can normally obtain better conditions since they do not have to face market risks; on the other hand in developing countries the country risk might be higher than the market risk, resulting in a lower rate for private companies. However, in this case it is likely that private companies will require a supplementary risk premium for the country risk (e.g. for preventing monetary inflation) and thus a higher remuneration.

[9] In the UK the regulatory asset base is represented by what private investors have actually paid for the assets; in Germany, it is the full reconstruction value, while in France it corresponds to the financial costs paid by municipalities (loan repayment). These different ways of calculating the RAB lead to huge differences in the final cost (Massarutto, 2004).

The second step in the analysis is the assessment of the distribution of the "net dividend" among stakeholders. It is worthy to emphasise that the distribution of the "net dividend" is dependent on the initial risk allocation among the stakeholders.

Table 7 analyzes in more detail the structure of economic risks that is likely to occur in the three models, considering all the relevant stakeholders. We can conclude that each model entails some kinds of risks that are typical of that model. Therefore, the way these risks are actually shared among different actors depends on the concrete regulatory instruments adopted and the set of incentives provided to actors.

Once again, a confrontation of table 6 and 7 shows the importance of the trade-off discussed above. More effective regulations (e.g. detailed contracts and quality specification, frequent tenders, credible enforcement etc) are also putting higher pressures on the private and increasing its risks (and consequently the expected returns on investment, which reflect on total costs). In turn, a looser regulation and a weaker incentive system (cost-based regulation, poor specification of duties, renegotiation of contracts etc) lead to higher profitability and lower pressure on cost reduction.

Table 7: Management models, economic risk and main expected outcomes on performance

	Delegated Management	Regulated monopoly	Direct public management
Main risk dimensions for the private	Market risk (tender) and recovery of sunk costs; Operational risk (initial information missing or wrong; emerging new issues during contract lifetime); Commitment of public authority to ensure cost recovery and viability	Regulatory risk; Takeover; Unforeseen investment; Public reaction forces regulators to keep down unpopular price increases	Limited to DBFO and market for procurement
Main risk dimensions for the public	Information asymmetries; Technological lock-in	Regulatory capture	Lower efficiency; More vulnerable to pressures from workers and consumers
Main risk dimensions for consumers	Collusion leads to extraction of monopoly rent shared by municipality and private company; Quality reduction if contracts are not fully specified and/or badly enforced	Higher cost of capital; Cost pass-through; Quality reduction corresponding to what quality dimensions are actually specified by regulations and service charter and enforced.	Lower credibility of quality standard enforcement may lead to deterioration of service quality
Main risk dimensions for future generations	Lack of incentives to invest in assets having a long term repayment schedule, unless this is specified in the delegation contract.	Underinvestment induced by unwillingness to raise tariffs in the short term	Underinvestment; Slowdown of environmental and quality expenditure due to public budget pressures
Main risk dimensions for suppliers / workers	Market power of operator face to suppliers; Vertical integration	Pressure for lower salaries – outsourcing and for staffing reductions	Higher competition on procurement and reduced profit margins for suppliers
Public subsidies and likelihood of self-sustaining WSS finance	Obligation for FCR; Mutuality systems financed by ear-marked taxes; Variant: public budget contributes to investment with specific grants	In principle no subsidies and obligation for FCR; new obligations only when tariff increase allow investment to be viable; Variant: public sector can assume part of the risk for long-term infrastructure renewal in order to guarantee against risk of bankruptcy	Water tariffs and charges intended as local taxes and aimed at long-run FCR; Variants: public accounting does not consider depreciation and capital costs; public budget finances investment
Patterns of risk allocation	Investment risk separated from operational risk	All investment responsibilities on the water company (variant: creation of specific purpose companies for the ownership of assets, also responsible for fund raising and owned by public or consumers); Responsibility for regulators to ensure industry viability; Price caps and cost pass-through in order to share risk of unexpected events with consumers	Entirely on the public; Cost-based tariffs ⇔ economic risk shifted to consumers and/or taxpayers; Some limited assumption of risk by private firms in DBFO arrangements

CONCLUDING REMARKS

As we have seen in previous section, it is impossible to conclude with respect to the desirability of any of the described alternative regulatory models; quality of regulation in all cases is more important a determinant for overall efficiency gains (net dividends) than the model itself; on the other hand, regulatory instruments have a decisive impact on the set of incentives, risk sharing and ultimately for deciding what social actors are most to benefit from reform.

Liberalisation and more generally PSP is often blamed for causing rising water bills. For instance, the empirical evidence clearly shows that after PSP prices almost always rise (Hall, 2001; D'Souza and Megginson, 1999; Florio, 2004; Gleick et al., 2002). Required returns and profit earned by private companies are repeatedly indicated as the main reason for this. Obviously, delegating any activity to the market implies that market operators earn a profit, and this comes naturally with the very idea of competitive markets and a capitalist economy. In principle, this is not negative per se, providing cost savings and increased efficiency are partly transferred to consumers and the profit margin equals the market cost of capital and does not incorporate monopoly rents. The question should not be whether it is just or fair for a private company to earn a profit for providing an essential service, but rather to evaluate whether water users and society as a whole are better or worse off after PSP. The answer to the last question is not straightforward. As we argued above, the WSS is a very capital intensive industry: while opportunities for productivity gain certainly exist, they could be overturned by private investors' remuneration so the final outcome is not so beneficial to consumers; in turn, public-sector based management and financial models could allow longer-term repayment schedules and lower cost of capital, and hence lower total costs. Price increases after PSP do not necessarily depend on profit. Rather, they often are due to discovering costs that were previously unaccounted for and the need (or the political choice) to recover them from the water bill instead of taxation. Raising prices at cost-recovery levels is also a way to be fairer to the next generation, particularly concerning the correct depreciation of assets value with the aim of restoring their capital value. Very often, PSP is not the cause of price increase; for example, sometimes the need to increase prices is at the origin of PSP since local authorities might not want to be responsible for raising the water bills. PSP is also a way to raise higher barriers against the temptation to once more pour public money into the system. Finally, price increases might be due to the need to finance new investment. The WSS industry is undergoing a dynamic phase, given the contemporary necessity to replace ageing assets and to develop them further in order to catch up with the demanding targets imposed by quality regulation and environmental policy.

In order to decide whether consumers and society are better off after privatization, the water bills are not the only element to be evaluated; the net effect on taxpayers (present and long-term) and on future generations as well as improvements in service quality and environmental performance should also be considered. One of the fundamental criteria set by welfare economics is thus violated: we cannot simply concentrate on overall efficiency gains, since the way they are distributed also affects critically the likelihood of their occurrence. This consideration has some important consequences for policymakers

The first one is that, instead of choosing *among* regulatory models, it would be wiser to leave all the models available for choice and create the conditions that allow municipalities to change from one model to the other. Reversibility of choice is a key issue, since the best way to increase efficiency – in an industry in which competition is neither easy nor frequent – is the threat to move to a different regulatory model (e.g. from DPM to RM, or from DM to DPM).

The second is that each model is more suitable for certain segments of the service and less suitable for others. On the one hand, dividing the industry in different segments (both horizontally and vertically) could be very helpful for the sake of designing appropriate incentive schemes and to reduce market risks. On the other hand, some diseconomies (coordination costs) may arise, especially if the public sector is not sufficiently skilled for providing coordination on its own. In particular, it seems fundamental that operation and asset management responsibilities are separated in order to reduce risks; but on the other hand it is extremely dangerous to separate decisions in these two spheres since this might very easily conduce to inefficient planning of assets. Starting a virtuous circle of skill accumulation enabling the public sector to provide these tasks is therefore of utmost importance, as well as finding instruments that allow to share the risks for asset management while maintaining operator's role in choosing technologies and shaping investment strategies.

The third lesson is that models' performance can be surely improved by adopting innovative regulatory solutions. The most promising of these solutions appear to be those in which different competitive patterns are introduced "laterally" (e.g. through benchmarking or quasi-markets) and those in which consumers have a role in assessing service performance and controlling both elected people and operating companies (through public participation to decision-making, self-responsibility for specifying certain service components or even through the ownership of managing systems, for example through consumers' cooperatives). Innovative models of competition, although insufficient and imperfect if considered alone, can also provide partial but meaningful incentives to efficiency, especially if combined together. For example:

- market for corporate control of operating companies (threaten of takeover of inefficient firms);
- non-price-based tenders (e.g. tenders based on reputation and trust);
- "yardstick competition" and benchmarking;
- competition through "reputation and embarrassment";
- participatory evaluation and mechanisms based on consumers' "voice";
- allocation of subsidies and government funds based on comparative performance ("quasi-markets");
- eligibility of certain consumers (or groups of consumers) for autonomous service provision.

The main policy recommendations are summarised in table 8 below.

In the WSS sector, many factors are promoting a greater involvement of the private sector. The main issue to tackle is not whether to involve the private sector but rather finding the most effective and efficient solutions to involve it given that WSS remains a service implying significant market failures.

Table 8: Strategies and options

	Delegated Management	Regulated monopoly	Direct public management
Main available strategies for improving performance	Soft regulation providing benchmarking and info; Separate operation from AMDP; Simplify awarding criteria, tender objects (e.g. single activities vs. integrated service) and contract duration; trade-off with extent of PSP Increase contractual power of local authorities (associations etc)	Yardstick competition and econometric benchmarking; Improve accountability through information, benchmarking, reporting and public participation; Impose outsourcing through unbundling (trade-off with coordination costs); Reduce risks by providing guarantees and/or by keeping some part of the risk in public hands	Separate operation and management from enforcement; Improve accountability through information, benchmarking, reporting and public participation; Private-law arrangements and contracts; Outsourcing and delegation of tasks
Potential for introducing more competition	Reduce duration of contracts (requires solutions to avoid sunk costs); Reduce size of contracts (requires that public authorities provide strategic planning of the WSS system); One-dimension bids based on economic performance (requires very detailed contract specification and separation btw operation and asset management)	Inset appointments; New customers; Allowing bigger consumers to bypass the utility	Outsourcing; DBFO
Potential for outsourcing	Low due to permanent incentives to vertical integration	High, but requires unbundling and transfer price regulation	High; can be further expanded by legal provisions and/or tight price regulation
Potential for community management and individual solutions	Depends on the degree of development of the system, technical availability of low-cost solutions (e.g. self-treatment of effluents) and local circumstances (e.g. population density). Suitable for less developed service areas (e.g. drinking water production where the system still relies on local water)		

REFERENCES

Armstrong, M., Cowan, S. and Vickers, J. (1994). *Regulatory Reform*. Massachusetts Institute of Technology.

Barraque, B. (1992), "Water resources planning: beyond the privatization debate", *Economia delle fonti di energia e dell'ambiente*, n.3

Barraque, B. (1995), *Politiques de l'eau en Europe*, La Découverte, Paris

Barraque, B. (ed.) (1999), *Water 21. Towards sustainable water management in Europe*, Final report to the European Commission, Bruxelles

Bayliss, K., Hall D. (2002), *Can risks really be transferred to the private sector? A review of experiences with utility privatization*, Psiru, University of Greenwich, www.psiru.org

Beecher, J. (2001), *Privatization, monopoly and structured competition in the water industry: is there a role for regulation?* In Holwarth e Kraemer, eds, Environmental consequences of privatising water services in Germany, Ecologic, Berlin

Braeutigam, R.R. (1989), *Optimal policies for Natural Monopolies,* in SCHMALENSEE, R. and WILLIG, R.D. (eds.), *Handbook of industrial Organisation*, vol. II, North-Holland, Amsterdam, 1289-1346

Bressers, H., O'Toole L.J., Richardson J. (1995), *Networks of water policy*, Frank Cass, London.

Buckland, J.; Zabel, T. (1998), Economic and financial aspects of water management policies, in Correia, 1998, Selected issues in water resources management in Europe, Balkema, Amsterdam

Clausen, H.; Rothgang, M. (2002), Innovations and Sustainability in the Water Services Sectors – Institutional Framework, Actors, and Policy Instruments, Paper prepared for the Workshop "Advances in Applied Infrastructure Research" Berlin, 12 October 2002.

Correia, F.N. (ed.) (1998), *Water resources management in Europe*, Eurowater, Rotterdam: Balkema.

Cowan, S. (1993), "Regulating several market failures; the water industry in England and Wales", Oxford Review of Economic Policy, vol.9 n.4

Cowan, S. (1994), 'Privatisation and Regulation of the Water Industry in England and Wales', Chapter 5 in Matthew Bishop, John Kay and Colin Mayer (eds), Privatization and Economic Performance, Oxford: Oxford University Press.

Cowan, S. (1997), "Competition in the water industry", *Oxford review of economic policy*, vol.13 n.1, 83-92.

Correljé et al. (2005), Deliverable 6: Analysis of the economic implications of the scenarios. Euromarket Project. URL: http://www2.epfl.ch/mir/page18246.html

Cozzi, G.; Massarutto, A. (2002), "Dalla municipalizzata all'impresa pubblica locale", Economia delle fonti di energia e dell'ambiente, 1.

De Carli, A., Massarutto A, Paccagnan, V. (2003) "La valutazione economica delle politiche idriche: dall'efficienza alla sostenibilità", *Economia delle fonti di energia e dell'ambiente*, n. 1-2.

D'Souza, J. and Megginson, W. L. (1999), "The financial and operating performance of privatized firms during the 1990s", *The Journal of Finance*, **54,** no 4: 1397-1438

Dinar, A. (ed.) (2000), *The political economy of water pricing reform*, Oxford University Press, UK.

Dore, M., Kushner, J., Zumer K. (2004), "Privatization of water in UK and France: what can we learn?", *Utilities Policy*, 12, 41-50

Drusiani, R.; Fazioli, R., Massarutto, A., Matino, P. (2004), *Regolazione e ristrutturazione del servizio idrico italiano*, CLUEB, Bologna.

Florio, M. (2004), *The great divesture. Evaluating the welfare impact of British privatisations*, 1979-1997, Mit Press, Cambridge Ma.

Gleick, P. (1998), "The human right to water", *Water Policy*, 1, 487-503.

Gleick, P., Wolff, G., Chalecki, E., Reyes, R. (2002), *The new economy of water. Benefits and risks of globalization and privatization of freshwater*, Pacific Institute.

Hall, D. (2001), *Water in public hands*, PSIRU, London.

Hall, David (2003), *Water and DG Competition*, PSIRU, London.

Kraemer, R.A. (1998), *Public and private water management in Europe*, in Correia, 1998, Selected issues in water resources management in Europe, Balkema, Amsterdam.

Lobina, E. and HALL, D. (2003), *Problems with private water concessions: a review of experience*. On www.psiru.org.

Massarutto A. (1993), *Economia del ciclo dell'acqua*, FrancoAngeli, Milano.

Massarutto, A. (2005), "Full cost recovery of water services: the European experience", Zeitschrift für Angewandte Umweltforschung, special issue on water pricing, forthcoming.

Megginson W., Netter J. (2001), "From State to market: a review of empirical studies on privatization", *Journal of economic literature*, 39, 2.

Menard, C., Saussier, S. (2002), *Contractual choice and performance: the case of water supply in France*, in BROUSSEAU, E. and GLACHANT, J.-M. (eds.), *The economics of contracts. Theory and applications*, Cambridge University Press, Cambridge, UK.

Noll R., Shirley M., COWAN S. (2000), *Reforming Urban Water System in developing countries*, SIEPR Discussion Paper 99-32, Stanford Institute for economic policy.

OECD and The World Bank (2002), *Private sector participation in municipal water management in Eastern Europe and Asia*, Conference write up, Paris, April 10-11

OECD (1997), *Regulatory impact analysis: best practices from Oecd countries*, Oecd, Paris

OECD (2000), *Global trends in urban water supply and wastewater financing and management: changing roles for the public and the private sector*, CCNM/ENV(2000)36/Final, Oecd, Paris (www.oecd.org)

OECD (2004), *Competition and regulation in the water sector*, DAFFE/COMP/WP2(2004)1, Oecd, Paris

PRICEWATERHOUSE&COOPERS, Office Internationale de l'Eau, Bipe (2004), *MEIF – Evaluation methods for investing in the water sector: forward-looking financial strategies and water pricing*, European Commission, 5th Framework Programme, Final report, http://www.meif.org

Prost T. (ed.) (1999), *R&D and Water Management Systems in a perspective of sustainable development*, European Commission, DgResearch, 4th Framework programme, Final report, Bruxelles.

Rees, J.A. (1998), *Regulation and private participation in the water and sanitation sector*. Natural Resources Forum, **Vol. 22** (N. 2), 95-105

Rees, J.A. and ZABEL, T. (1998), "United Kingdom", in Nunes Correia (ed.), *Institution for Water Resources Management In Europe*. Balkema, Rotterdam.

Renzetti, S. and DUPONT, D. (2003), "Ownership and Performance of Water Utilities", *Greener Management International*, **42** (Summer 2003), 9-19

Summerton N. (1998), "The British way in water", *Water policy*, 1, 45-65

World Bank (1997), *Private Sector in Infrastructures*, The World Bank, Washington DC, Usa

World Bank (1999), *Water, Competition and Regulation*. The World Bank Group. Washington D.C.

World Bank (PPIAF) (2005), *Approaches to Private Participation in Water Services: A Toolkit*, 504

van Dijk et al (2003), Deliverable2: Analysis of the European Water Supply and Sanitation Markets and its Possible Evolution. Euromarket Project. URL: http://www2.epfl.ch/mir/page18246.html

Winpenny *et al.* (2003), *Financing water for all*. Report of the World Panel on Financing Water Infrastructure (Camdessus Commission), World Water Forum, Kyoto. URL: http://www.gwpforum.org

ANNEX

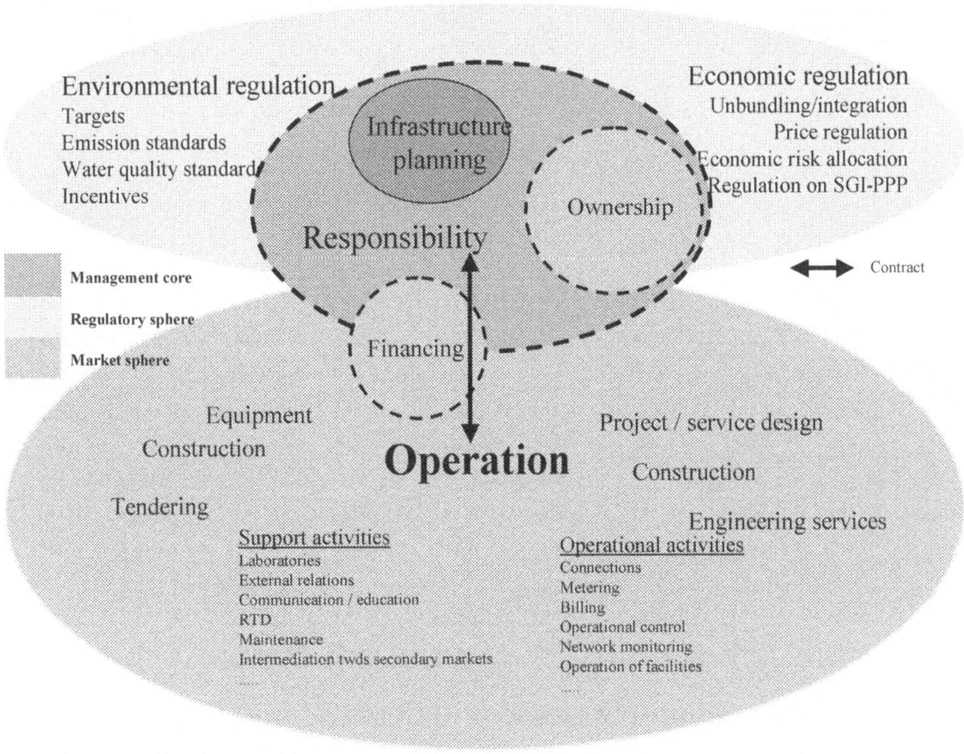

Figure 4a: Organization of the WSS value chain: Delegated model

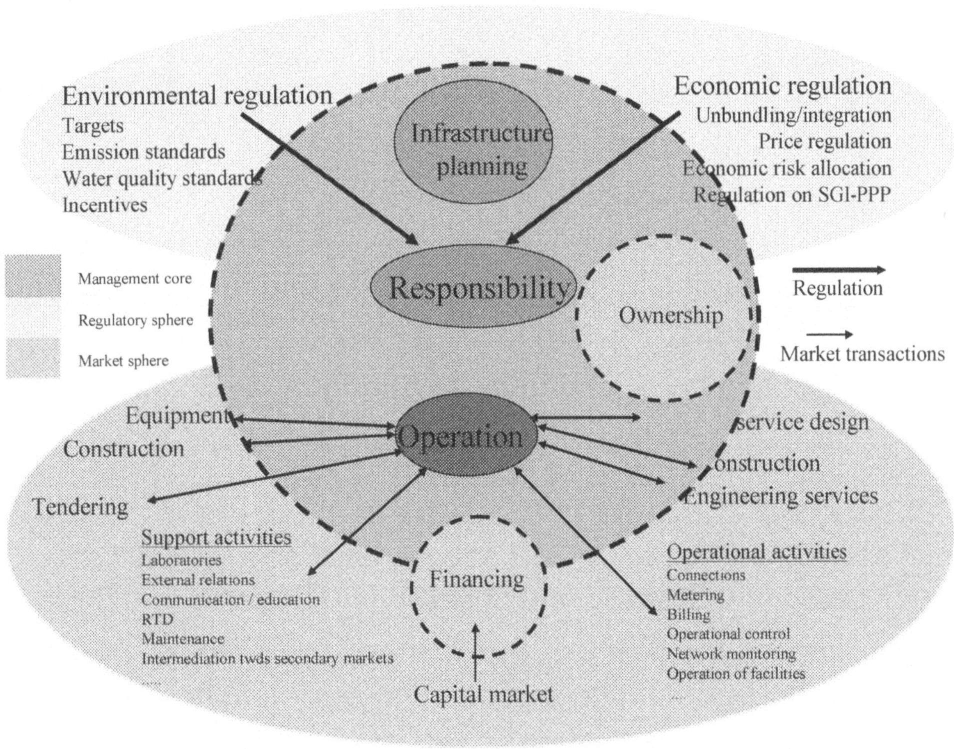

Figure 4b - Organization of the WSS value chain: Regulated Monopoly

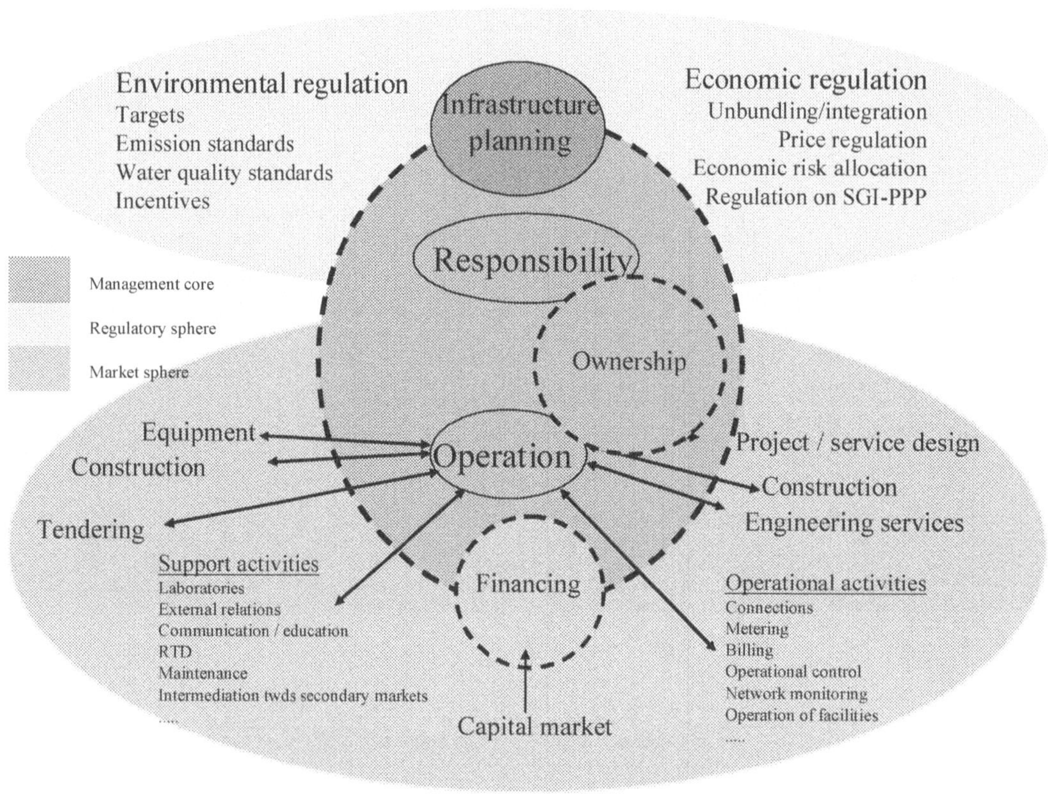

Figure 4c: Organization of the WSS value chain: Direct public management

Conclusion

Institutional and organisational dynamics and trade-offs: no single path for water

Jeremy Allouche, Patrícia Luís-Manso & Matthias Finger

The overall objective of this research project, and consequently of this book, was to study the likely evolution of the European water supply and sanitation (WSS) sectors in the foreseeable future (i.e., in the next 15 years). In order to do so, the book has (1) defined the actual structure of the European water supply and sanitation sectors, (2) analysed the existing and potential dynamics of the water sector, (3) constructed different scenarios, and (4) assessed their economic, ecological, social implications. This research project wanted to address two particular questions: firstly to see to what extent the water supply and sanitation sectors are distinct from other network industries; secondly to examine the possible forms of competition that could occur in these sectors in the near future. Both aspects have been extensively dealt with throughout this book and won't be analysed in this conclusion.

The aim of this conclusion is to evaluate the different modes of management in institutional and organisational terms. This assessment is difficult in that it requires a dynamic (rather than static) approach. Indeed, these modes of management constantly evolve in that they are influenced by three important drivers characterising the sector. The first one pertains to the evolution of the sector (i.e., the structure and forms of real or potential markets). The second one relates to operator strategy. Each operator reacts or anticipates business opportunities or threats. Finally, the third variable concerns regulatory functions and activities at various levels. The constant need to adapt to new sector developments and operator strategy, and the need to guarantee consumers a certain level of services makes regulatory intervention an important variable in apprehending the dynamic feature of the different institutional models. The current dynamic of the sector therefore needs to be understood according to these three drivers. Having in light this difficulty, we first set up an institutional framework, and then identify three criteria that address the long-term vision needed to evaluate different organisational models of WSS management. Finally, we assess each scenario according to these criteria and identify the main challenges of each of them.

INSTITUTIONAL FRAMEWORK AND MAIN CRITERIA FOR ITS ASSESSMENT

The institutional and organisational implications of each mode of management are assessed within the institutional framework specific for each end state. The main challenge of designing an *institutional framework* to water management is to account for the specificities of the sector, especially its merit and public good, as well as its strong natural monopoly characteristics. These specificities make the provision of water services the focus of special operational and regulatory attention. In particular, the fact that water services are of general interest does lead to the need for guaranteeing principles in terms of access, quality of service, and price.

The concept of institutions is used in a broad sense and includes all relevant actors, and in particular, the formal and informal rules governing their relationships (Finger, 2005). Institutions are viewed as codified power structures that shape the behaviour of the actors involved. Simultaneously, actors may be capable of reshaping the institutions within which they operate by forming strategic alliances (Crozier, 1963).

Organisations are social units (or groups of human beings) deliberately constructed and reconstructed in order to pursue specific goals (Etzioni, 1964). An organisation's structure is the framework whereby the desired goal can be achieved. It has three major components, namely complexity (set by the amount of aims), formalisation (pertaining to the rules an organisation has and it's reliance on these rules and procedures to direct behaviour), and centralisation (pertaining to the development of the decisions making powers within the organisation).

In the remaining of this chapter, we first present the general features of the institutional framework for the WSS sector, and then describe the three criteria identified for assessing the scenarios.

The institutional framework

The institutional framework for the WSS sector generally reflects a three-tier institutional structure based on three types of functions: policy making, regulation and operation. We identify the main institutional functions by deriving them from the operational and regulatory issues in network industries in general and the water and sanitation sector in particular. Policy making pertains to the definition of the medium/long term guidelines for the sector. Regulation refers to the controlling, as well as the enabling and facilitative roles exercised by the public authority (dependent or independent), on the basis of a legislative mandate, over activities that are generally regarded as desirable to society (see Selznick, 1985; Baldwin et al., 1999). Finally, operation concerns the activities and processes taken and implemented by the operator in order to provide water services.

The institutional framework developed for the sector is based on the transactions framework developed in chapter 1. To recall, four markets were identified:

- Market for drinking water and sewerage services (consumers receive water and sanitation services from the management entity in return for payment);
- Market for suppliers (private/public companies provide various inputs – bulk water, products, services and finance – to the management entity in return for payment);
- Market for water abstraction and discharge (the environment provides/receives raw water/wastewater discharges with payment possibly being made to environment agency to administer such controls); and
- Market for delegated contracts (the seller – i.e., the responsible entity for service provision - 'sells' the right to act as service provider for a geographical area to an operator, in return of a fee or profit from the service provision).

The institutional framework focuses on (1) the various types of actors and their relationships; (2) individual transactions and information flows; and (3) the broader policy making/legal context (as the institutional framework's foundation). It enables us to include the specificities of the water sector and understand the various implications of the different scenarios in institutional and organisational terms. It also highlights the constant dynamic of the sector.

The institutional implications of the scenarios are assessed on the basis of a set of criteria – stability, efficiency and legitimacy – derived from organisational, transaction cost, and governance theories. These criteria also enable us to integrate the long-term vision needed to assess the different models of management (here our scenarios). The following sections develop these criteria and define the corresponding indicators (based on Euromarket, 2005).

The stability criterion and its main indicators

The institutional framework of the water sector is considered a dynamic system of organisations that differ in terms of goals and which operate in changing environments. Each scenario hence corresponds to a specific set of organisations operating within a defined institutional framework, which is prone to be affected by change. In the specific case of the WSS sector, changes in the environment might stem from liberalisation pressures, water-related policies (e.g., stricter environmental standards), financial pressures, as well as ideological shifts. The environmental changes affect primarily the rules (i.e., institutions), yet via the rules the actors and their strategies are also affected.

If we are interested in analysing the institutional implications for the scenarios, stability emerges as an important criterion. The stability of the scenario pertains to the sustainability of its institutional framework in a context where organisations under permanent power struggles might redefine priorities and strategies and possibly, as a consequence, the rules of the game. For practical reasons, we have derived from organisational theories the following indicators to assess the stability criterion: coherence, balance of power and conflict resolution.

Coherence: Coherence pertains to the definition and allocation of the institutional functions, which means organisational separation of institutional functions. It is important that institutional functions do not overlap. For example, policy making, regulatory and operational functions should be distinguished and clearly defined. Moreover, operational and regulatory functions should not be dependent upon political interference. And finally, operational functions should be dependent upon the respect of the regulatory framework; and separated from political direction and control.

Balance of power: The WSS sector is a dynamic system of organisations that differ in terms of goals. It is therefore important to analyse for each scenario whether there are organisations that enjoy considerable discretion and protection from possible interventions which enables them to look after their own interests. In each of the three markets considered in the institutional framework (i.e., customer, supplier and resources markets), the study looks if there is one (or more) actor(s) more powerful than the others. A balanced system of power means that each organisation responsible for specific institutional functions in the system is checked by the others, so that one organisation is not the final arbiter of all powers.

Conflict resolution mechanisms: In the context of changing environments, organisations must adapt and change, which is likely to lead to a redefinition of priorities and (eventually) goals, as well as to a redistribution of resources. As a matter of fact, change affects the power relationships among the actors, and consequently it threatens the status of some actors. There is therefore a strong need to manage potential ruptures and crises (e.g., in the cases there is a new mission to pursue). We check for each actor/organisation which are (if any) the existent mechanisms to solve conflicts of interest with other actors/organisations.

The efficiency criterion and its main indicators[1]

Institutions seek to achieve efficiency – minimising comprehensive cost – inclusive of both production and transaction costs. The efficacy of economic institutions should be judged on a systemic basis considering both *ex ante* and *ex post* transaction arrangements.

To develop a complete set of efficiency indicators to evaluate the scenarios, we have focused on two levels: the nature of individual transactions (within our institutional framework), and the nature of the generic institutions of governance. The indicators are: level of transaction costs, adaptability and incentives to perform.

Level of transaction costs: In a general view, transaction costs relate to both motivation (to align actors' interests) and coordination (of agents' actions) costs. Governance is an exercise in assessing the efficacy of alternative modes of organisation – namely Markets, Hybrids and Hierarchies – to economise on transaction costs (Williamson, 2005). Hence hierarchies (e.g., public/private/social enterprises), Hybrids (e.g., delegation contracts) and Markets are alternative forms of organisation for managing the very same transaction.

In order to characterise the transactions according to their information problems/contracting hazards, one needs to focus on each individual transactions (e.g., between customer and operator; between customer and responsible body; between responsible body and regulatory body). The key information problems associated to transaction costs have been grouped into:

I1. *Information asymmetry:* creates adverse selection (leading to *signalling/screening* costs);

I2. *Non observability* of agents' action: creates a moral hazard (leading to costs of *monitoring and incentive/penalty schemes*);

I3. *Non verifiability* of information (leading to *audit* costs/costs due to *misrepresentation* when audit too costly);

I4. Information *uncertainty:* creates the need for incomplete contracts (leading to complex contract *construction, negotiation,* monitoring/ *renegotiation* and *dispute resolution* costs).

Adaptability: Uncertainty increases the frequency of adaptation. In the discrete governance forms, market and hybrid governance represent two polar types of adaptability (Williamson, 1991). Market governance supports high exit flexibility and low idiosyncratic investments adaptation, i.e. autonomy, whereas hybrid governance supports high modification flexibility and high idiosyncratic investments adaptability, i.e. co-operation.

Incentives to perform: According to Williamson (2000), a governance structure reshapes incentives (ex ante and especially ex post incentives). The academic literature argues that well-designed and well-implemented market-based instruments generate high-powered incentives to perform and require limited administrative controls. On the other hand, command-and-control instruments are frequently criticised because of their high administrative costs and low-powered incentives.

The legitimacy criterion and its main indicators[2]

In the context of this chapter, legitimacy of an institutional arrangement is not necessarily political regime legitimacy, but rather the "legitimacy of specific structures and procedures through which policy is being produced"

[1] The content of this section is based on a chapter written by David Musco in Euromarket (2005).

[2] The content of this section is based on a chapter written by David Aubin in Euromarket (2005).

(Scharpf, 1997). This means the "general capacity of particular policy-making institutions to produce policy choices that are likely to approximate the common good" (Scharpf, 1997). Legitimacy can be regarded along two ways: (1) input-oriented: legitimacy derives from the agreement of those who asked to comply; and (2) output-oriented: it refers to substantive criteria of *buon governo*, in the sense that effective policies can claim legitimacy if they serve the common good and conform to criteria of distributive justice.

Good governance can be regarded as a source of legitimacy. According to the Commission's white book on good governance (European Commission, 2001), "governance means rules, processes and behaviour that affect the way in which powers are exercised at European level, particularly as regards openness, participation, accountability, effectiveness and coherence" (considered as principles).

The two first principles concern the input-oriented legitimacy of public management, which depends on process openness and public participation (democracy understood as government by the people). In contrast, the three other principles focus more on output-oriented legitimacy (democracy understood as government for the people). In our assessment, two over the three indicators related to output-oriented legitimacy are already retained in the proposed assessment criteria of stability and efficiency, respectively coherence and effectiveness. As such, we chose to build the criteria of legitimacy on the three remaining indicators, one output-oriented (accountability) and two input-oriented (openness and participation).

Openness: In order to reach good governance, organisations among the institutional arrangement should work in an open and transparent way. The broad public should have access to reliable information and be able to scrutinize the decision-making process. Such openness requires active communication and information provided in accessible language. An active communication implies the availability of detailed information to the public and access to working documents and final decisions. It also means that citizens are aware of the procedures and can locate at any time which step the decision-making process has reached and the periods within which they can intervene (e.g. time schedules of the consultation processes). In addition, available information should be adapted to local preoccupations and needs.

Participation: The process should be formally institutionalised at all steps of decision-making and at two levels. First, participation must involve experts according to clear and transparent mechanisms within each organisation. Second, participation should be opened to the broad public. In addition, the range of participation goes from consultation to co-decision. We consider that a decision-making process is participatory if there is at least a consultation of the interested parties and the broad public.

Accountability: the respective roles of organizations in the decision-making and implementation processes need to be clear. Each of the organizations on the institutional framework must explain and take responsibility for what it does at whatever level.

ASSESSMENT OF THE INSTITUTIONAL IMPLICATIONS OF SCENARIOS

This section analyses the institutional implications of each scenario on the basis of the criteria and indicators identified above. The results presented are organised in terms of the strong features (advantages) and main challenges (weaknesses) of each end state (table 1). These results are based on a more comprehensive analysis detailed in Euromarket (2005), and which has included illustrative case studies.

Assessment of scenarios

In institutional and organisational terms, one can see that the major issue in the **delegation contracts and strong regulation scenario** pertains to the sustainability of the system. This can be explained for two main reasons:

1) The potential incoherencies arising from the overlap of responsibilities between the public authority and the regulatory entity;
2) The high transaction costs.

In order to ensure the permanency of the system, roles and responsibilities should be clearly allocated between the different actors in the legal framework, and the corresponding level of resources should be made available to properly execute the allocated functions (e.g., human and financial resources to the independent regulator). It is also very important to minimise transaction costs because the high number of actors involved in the framework may raise them to costly levels to society.

Table 1: Institutional and organisational Implications of scenarios: advantages, weaknesses and major issues

Scenarios	Advantages	Weaknesses	Major issue
Scenario 1 Delegation contracts and strong regulation	– Balance of power: well balanced between the operator, regulator and responsible entity – Open system: transparency of information guaranteed by the independent regulator – Incentives to perform: regular tendering process and regulation	– Coherence: in the relationship between regulator and responsible entity – High transaction costs – Low participation: the consumer-provider relationship is governed by the supply contract	SUSTAINABILITY OF THE SYSTEM
Scenario 2 Delegation contracts and extreme competition	– Coherence: separation of operating and responsible entity – Low likelihood of conflicts – Low transaction costs – High incentives to perform	– Possible confusion between regulatory and political functions – Unclear definition of capacity investment responsibility – Low participation as the consumer-provider relationship is governed by the supply contract	FULL RELIANCE ON CONTRACTS
Scenario 3 Outsourcing	– High flexibility of the supplier market arrangements – High incentives to perform – High accountability in the supplier market	– Low public participation as the decision to outsource and the type of services that are externalised are typically an operational decision	HOLLOWING OUT OF THE MANAGEMENT ENTITY'S TECHNICAL EXPERTISE
Scenario 4 Regulated Monopoly	– Performance incentives through benchmarking – Transparency of the system through benchmarking	– Low accountability of the regulator and the operator – Likelihood of conflicts between operators and the regulator/benchmarking authority	ACCOUNTABILITY OF REGULATORS AND PRIVATE OPERATORS
Scenario 5 Direct Public Management	– Low transaction costs – High accountability	– Weak conflict resolution mechanisms – Low system adaptability – Weak performance incentives	ORGANISATIONAL FLEXIBILITY
Scenario 6 Community Management	– High level of legitimacy	– High transaction costs	DECISION MAKING

The complete dependence of the system on short-term contracts in the **delegation contracts and extreme competition scenario** has many consequences in institutional and organisational terms. The major actors involved are the operator and the responsible entity and their relationships are governed by contracts negotiated every five years. The major issue of this scenario therefore pertains to the full reliance on the contract. This has three major consequences that need to be addressed by policy makers.

1) Public participation is likely to be low as the relationship between the responsible entity and the management one is governed through the contract;

2) Unclear definition of capacity investment responsibility (in order to ensure asset service ability over time and the development of the infrastructures);
3) The possible interference of political interest in the regulation of the management entity.

In order to avoid these organisational problems, different stakeholders and, in particular, consumers should be involved in the planning of long-term investment. This would resolve the major challenges of the scenario in that active participation would ensure less political interference and counterweight shortsighted approaches the system.

The drive towards **outsourcing** can be explained by the advantage of benefiting from highly qualified technical expertise. This drive, if uncontrolled and expanded to all the major services, may lead to the loss of internal competencies of the operator and its subsequent dependency on a limited number of outsourced firms. The major consequence would be the capture of the operator. In such a context, both the flexibility as well as the incentives to perform would decrease. In organisation terms, the management team must therefore ensure to keep a reasonable number of qualified personal in order to avoid these fallbacks.

The fact that the **regulated monopoly scenario** can encompass multiple configurations (e.g., diversity of types of operators) limits the assessment of the scenario in terms of stability, efficiency and legitimacy because some of these criteria depend on the local context. We have nonetheless tried to identify the major advantages and weaknesses of such a model as described in the table above.

The main preoccupation in organisational and institutional terms pertains to the accountability of the operator and the regulator. The fact that accountability goes through binding benchmarks organised by public bodies, not necessarily fully independent, constitute a clear limit in the legitimacy of the system. In this scenario, it is therefore important that accountability rules must be specified in the law (clear objectives, indicators of performance of public companies, liability vis-à-vis the public authority) as public control through shareholding is not sufficient.

The **direct public management** model has proven its advantages as the main institutional configuration for many years in the European water sector. The legitimacy and the coherence of the model are usually taken for granted, as well its supposed inefficiency. Our analysis showed that the performance of this model with regards to the different indicator is more complex.

In organisation and institutional terms, it is not efficiency *per se* but rather the lack of organisational flexibility that characterise the main challenge of this scenario. In this regard, it could be useful to develop incentives for the suppliers and their managers, not necessarily focusing on financial incentives, but including more qualitative preoccupations in terms of environmental protection, social equity, and quality of the service.

The fact that the **community management scenario** can encompasses multiple forms of the social enterprise (e.g., cooperative, mutual) limits the evaluation of the scenario especially in terms of stability and efficiency. The main challenges of this scenario concerns decision-making. In this model of management, compromise rather than majority rule drives the possible decisions that may be taken for the WSS system. Possible deadlock may be reached and some members of the community may reject the rules of the majority. Solid conflict resolution mechanisms should be developed in order to prevent a decision void.

Comparison

The scenarios on the evolution of the water sector in Europe presented in the paper can overlap. This is especially the case of *Outsourcing* and *Administrative Regulation* scenarios, whose competition processes may coincide with different forms of organising the sector. This has rendered difficult the assessment of their institutional consequences because they ultimately depend on the chosen organisation of the sector.

In terms of the storylines[3], there are strong similarities between *Outsourcing* and *Administrative Regulation* (which focus on links outside the operator circle) and between *Tendered Markets* and *Direct Public Management* (mainly built within the operator circle). We have therefore decided to compare these scenarios together.

On the one hand, it is interesting to first look at the *Outsourcing* and *Administrative Regulation* together because these two scenarios can occur with a mixture of operator types. The strong features of these scenarios pertain to the high levels of incentives to perform, created by the competition level in supplier market in *Outsourcing* and benchmarking in *Administrative Regulation*. Their main challenges pertain: (1) in *Outsourcing* to the possibility that competition in the supplier market is constrained by a tendency of firms to concentrate, and to the loss of internal competencies of the operator; and (2) in *Administrative Regulation* to the fact that accountability depends on benchmarks organised by public bodies not necessarily fully independent, and to the low level of public participation.

On the other hand, it is interesting to look at the *Tendered Market* (both scenario 1. and 2.) and *Direct Public Management* scenarios together because they are contrasting especially in the nature of the operators that predominate.

[3] For more details on the storyline, please see chapter five.

These contrasting scenarios have necessarily different institutional implications. Firstly, the strong features of *Tendered Market* scenarios pertain to the efficiency and stability-related indicators, while in *Direct Public Management* they are related to strong legitimacy. Secondly, the main challenges of *Tendered Market* pertain to legitimacy, especially in what concerns public participation, while in *Direct Public Management* they pertain to the potential lack of incentives to perform.

Finally, *Community Management* is possibly the most complex scenario in that it embraces both changes to the nature of the operator (from public/private enterprise to social enterprise) and associated adjustments in both the operator-supplier and operator-regulator relationship. The strong feature of this scenario is that it is characterised by high levels of legitimacy. The main challenge relate to the fact that transaction costs are generally high.

The main purpose was to analyse the institutional implications of the different scenarios. The aim was not to make judgements on the scenarios neither to recommend any one in particular, but it was to assess them across a set of criteria including various indicators. The identified scenarios are theoretical constructions on the possible evolution of the water sector in Europe, which means that the organisation of the water sector and its consequences in 15 years' time will as well depend on the specific characteristics of local systems. We mean by this that any of the conclusions presented in this report should not be extrapolated to all the situations. These conclusions pretend to improve the knowledge set of policy-making actors, namely calling their attention to important issues that are related to particular features of the scenarios.

CONCLUSION

Throughout the 20th century, the development and maintenance of urban water networks were a priority, making stability an essential feature of the system. Furthermore, the public nature of the majority of these systems granted an implicit legitimacy. More recently, civil society called for a different type of legitimacy that would provide transparency and accountability. In the same line, efficiency, and more particularly the incentives to perform, is becoming a central topic leading to the emergence of alternative modes of management (i.e.: compared to the 'traditional' direct public management model).

It is clear that there is no ideal model as each scenario examined represents a trade-off between stability, efficiency and legitimacy. Today, there is a push for policy makers to privilege immediate accountability and efficiency. The main danger for water supply systems is that their long-term sustainability is threatened by these short-term prerogatives.

REFERENCES

Baldwin, R. & M. Cave (1999), *Understanding Regulation: Theory, Strategy and Practice*, New York: Oxford University Press.

Crozier, M. (1963), *Le Phénomène Bureaucratique*, Paris : Editions du Seuil.

Etzioni, A (1964), *Modern Organizations*, Englewood Cliffs, N. J.: Prentice Hall.

Euromarket (2005), *Analysis of the Institutional, Legal and Organisational Implications of the Scenarios.* Deliverable 9: http://www.epfl.ch/mir/euromarket.

European Commission (2001), *European Governance. A White Paper,* COM (2001) 428final.

Finger, M (2005), "The New Water Paradigm: The Privatization of Governance and the Instrumentalization of the State", *in* D. Levy & P. Newell (eds.), *The Business of Global Environmental Governance*, Cambridge MA: MIT Press.

Selznick, P. (1985), "Focusing organizational research on regulation", In: Noll (ed.), *Regulatory Policy and the Social Sciences*, Berkeley and Los Angeles: The University of California Press.

Scharpf, F. (1997), *Games real actors play: Actor-Centered Institutionalism in Policy Research.*Boulder: Westview Press.

Williamson, O. (2005), *The Economics of Governance.* Berkeley: University of California.

Williamson, O. (2000), The New Institutional Economics: Taking Stock, Looking ahead. *Journal of Economic Literature*; 38 (3): 595-613.

Williamson, O. (1991), Comparative Economic Organization: the Analysis of Discrete Structural Alternatives. *Administrative Science Quarterly*, 36(2): 269-290.

INDEX

IWA Publishing's authorised EU representative for General Product Safety Regulations is Diane D'Arras, 15 rue Duret, 75116 Paris, France, e-mail: safety@iwap.co.uk.

Printed and bound by CPI Group (UK) Ltd, Croydon, CR0 4YY
12/05/2026
02108886-0017